Power Systems

Electrical power has been the technological foundation of industrial societies for many years. Although the systems designed to provide and apply electrical energy have reached a high degree of maturity, unforeseen problems are constantly encountered, necessitating the design of more efficient and reliable systems based on novel technologies. The book series Power Systems is aimed at providing detailed, accurate and sound technical information about these new developments in electrical power engineering. It includes topics on power generation, storage and transmission as well as electrical machines. The monographs and advanced textbooks in this series address researchers, lecturers, industrial engineers and senior students in electrical engineering.

Power Systems is indexed in Scopus

More information about this series at https://link.springer.com/bookseries/4622

Yao Sun • Xiaochao Hou • Jinghang Lu
Zhangjie Liu • Mei Su • Joseph M. Guerrero

Series-Parallel Converter-Based Microgrids

System-Level Control and Stability

Yao Sun
School of Automation
Central South University
Changsha, China

Xiaochao Hou
Department of Electrical Engineering
Tsinghua University
Beijing, China

Jinghang Lu
School of Mechanical Engineering and Automation
Harbin Institute of Technology
Shenzhen, China

Zhangjie Liu
School of Automation
Central South University
Changsha, China

Mei Su
School of Automation
Central South University
Changsha, China

Joseph M. Guerrero
AAU Energy, Aalborg University
The Villum Center for Research on Microgrids (CROM)
Aalborg, Denmark

ISSN 1612-1287 ISSN 1860-4676 (electronic)
Power Systems
ISBN 978-3-030-91513-1 ISBN 978-3-030-91511-7 (eBook)
https://doi.org/10.1007/978-3-030-91511-7

This Springer imprint is published by the registered company Springer Nature Switzerland AG
The registered company address is: Gewerbestrasse 11, 6330 Cham, Switzerland

Preface

Microgrid is a key concept for future energy distribution system that enables renewable energy integration and has elicited considerable research interest. It integrates dispersed DG units, energy storage systems (ESSs), demand-side management, and various loads in order to establish distribution networks which are largely connected to the upstream power grid. Additionally, in the event of grid downtimes or other external disturbances, these networks can be isolated from the utility grid, increasing the reliability and sustainability of the power supply. Despite its potential benefits, the development of microgrid has been fraught with challenges to achieve a stable and secure operation. The operation control and stability issues with microgrid are urgently in need for research.

Series-converter and parallel-converter are two basic power conversion systems. These two topologies have been a promising solution to the high-voltage and high-rating power conversion system due to their modularity, transformer-less feature, and high power quality. They have been widely studied in photovoltaic micro-converters, reactive power compensation and the applications of electrical traction, vehicle-to-grid, and solid-state transformer. When these converters are applied into the microgrid, the microgrid can be classified into series-type microgrid and parallel-type microgrid. The cooperative control and power regulation of these microgrids need to be further explored.

This book gives a comprehensive and in-depth introduction into the development of series-parallel converter applications in the microgrid system. For each topic, a theoretical introduction and overview are backed by very concrete programming examples that enable the reader to not only understand the topic but also develop simulation model.

With this book, we intend to enable the reader to: (1) get the latest knowledge on the operation and control of the series-parallel converter-based microgrid systems from the fundamental to the whole picture and (2) develop the microgrid system in the simulation and have access to the "hand-on" simulation examples in the book.

The main research results of this book have been originally taken from the authors who carried out the related research together for almost 8 years, which is a comprehensive summary of the authors' latest research results. This book is likely to be of interest to university researchers, R&D engineers, and graduate students in electrical engineering who wish to learn the core principles, control methods, and applications of microgrid system.

Changsha, China Yao Sun

Beijing, China Xiaochao Hou

Shenzhen, China Jinghang Lu

Changsha, China Zhangjie Liu

Changsha, China Mei Su

Aalborg, Denmark Josep M. Guerrero

Acknowledgments

This book is supported by the Nature Science Foundation of China (NSFC) under Grant 61933011, 61903383, and 62125308, the Hunan Provincial Key Laboratory of Power Electronics Equipment and Grid under Grant 2018TP1001, the Project of Innovation-driven Plan in Central South University under Grant 2019CX003, and the Major Projectcof Changzhutan Self-dependent Innovation Demonstration Area under Grant 2018XK2002.

The authors would like to thank postgraduate students Mr. Shimiao Chen, Ms. Siqi Fu, and Ms. Junlan Ou for their contributions and proofreading. Special thanks for the academic support and contributions go to Prof. Hua Han and Dr. Yajuan Guan for Chap. 1, Mr. Yao liu for Chap. 4, Dr. Lang Li for Chaps. 8–9, Mr. Guangze Shi for Chaps. 8, 10, Mr. Chao Luo for Chap. 8, Mr. Wenbin Yuan for Chap. 14, Mr. Xiaohai Ge for Chap. 15, Mr. Yuanhao Zhu for Chap. 16, and Mr. Qingping Xia for Chap. 17. Finally, the authors would like to thank the long-term support and encouragement from their families.

Contents

About the Authors

Yao Sun is currently a Professor at the School of Automation, Central South University, China. He received BS, MS, and PhD degrees from the School of Information Science and Engineering, Central South University, in 2004, 2007, and 2010, respectively. Professor Sun has worked on microgrids, coordinated control, and power electronic converters for more than ten years, and he has published his research results in more than 100 journals and conferences. He was a recipient of the first prize of the 2020 Hunan Natural Science Award of China. His research interests include matrix converter, microgrid, and wind energy conversion systems.

Xiaochao Hou received BS, MS, and PhD degrees in control science and engineering from the School of Automation, Central South University, China, in 2014, 2017, and 2020, respectively. From September 2018 to September 2019, he was a joint PhD student at the School of Electrical and Electronic Engineering of Nanyang Technological University, Singapore. He is currently working as Research Fellow at Tsinghua University, China. Dr. Hou served as a microgrid special session chair for the 12th IEEE International Conference on Electrical and Electromechanical Energy Conversion-ECCE Asia 2021 and a track chair for the 16th IEEE Conference on Industrial Electronics and Applications-ICIEA 2021. He was a recipient of the first prize of the 2020 Hunan Natural Science Award of China and the 2017 Hunan Outstanding Graduation Thesis. His research interests include the control and stability of distributed microgrid and more electronic power networks.

Jinghang Lu is currently an Assistant Professor at Harbin Institute of Technology (Shenzhen), China. He received a BSc degree in electrical engineering from Harbin Institute of Technology, China, in 2009, two MSc degrees in electrical engineering from Harbin Institute of Technology, in 2011, and University of Alberta, Canada, in 2014, respectively, and his PhD in Power Electronics from Aalborg University, Denmark, in 2018. From 2018 to 2019, Dr. Lu was a Research Fellow at Nanyang

Technological University, Singapore. His research interests include uninterruptible power supply, distributed control of microgrids, and control of power converters.

Zhangjie Liu is currently a Distinguished Associate Professor and Master Tutor at the School of Automation, Central South University, China. He received a bachelor's degree in detection guidance and control technology and a doctorate in control science and engineering from Central South University in 2013 and 2018, respectively. From May 2019 to May 2020, he was engaged in postdoctoral research at Nanyang Technological University in Singapore. He won First Prize in the 2020 Hunan Province Natural Science Awards; he has served as a review expert for the Hunan Science and Technology Award and a Database Expert at the Chinese Society of Automation. In recent years, he has published several papers in international journals, including IEEE Transactions on Automatic Control, IEEE Transactions on Smart Grid., IEEE Transactions on Power Systems, IEEE Transactions on Sustainable Energy, and Automatica. His research interests include power system stability, power electronic power network, DC microgrid, and distributed control.

Mei Su received BS, MS, and PhD degrees from the School of Information Science and Engineering, Central South University, Changsha, China, in 1989, 1992, and 2005, respectively. Since 2006, she has been a Professor with the School of Automation, Central South University. Her research interests include matrix converter, adjustable speed drives, and wind energy conversion system.

Josep M. Guerrero has been a Full Professor in the Department of Energy Technology, Aalborg University, Denmark, since 2011 where he is responsible for the Microgrid Research Program. He received a BS degree in telecommunications engineering, an MS degree in electronics engineering, and his PhD degree in power electronics from the Technical University of Catalonia, Barcelona, in 1997, 2000, and 2003, respectively. His research interests are oriented to different microgrid aspects, including power electronics, distributed energy storage systems, hierarchical and cooperative control, and energy management systems.

List of Symbols

S_i	The converter output apparent power of DG-i
P_i	The active power of DG-i
$P_{\max}$	The maximum values of the allowable active power
$P_{\min}$	The minimum values of the allowable active power
Q_i	The reactive power of DG-i
$Q_{\max}$	The maximum values of the allowable reactive power
$Q_{\min}$	The minimum values of the allowable reactive power
V_{pcc}	The amplitude of common ac bus voltage
θ_{pcc}	The common ac bus voltage angle
X_{line}	The line impedance
R_{line}	The line resistance
Z_{line}	The impedance between converter and the common bus
Z_{load}	The common bus load
V_g	The amplitude of grid voltage
δ_g	The grid voltage angle
δ_i	The angle of output voltage of DG-i
V_i	The amplitude of output voltage of DG-i
ω_i	The angular frequency of DG-i
$\omega_{\max}$	The maximum values of the allowable angular frequency
$\omega_{\min}$	The minimum values of the allowable angular frequency
m_i	The active droop coefficient
n_i	The reactive droop coefficient
ω^*	The nominal angular frequency value
V^*	The nominal voltage value
T_s	The sampling period
$C_i(P_i)$	The cost function
P_L	The total load
$P_{L,\max}$	The upper bound of load power
f_i	The DG-i actual frequency
f^*	The DG-i reference frequency
ζ	The damping factor

sgn ()	The sign function
$\hat{\cdot}$	A small perturbation around the steady state points
L_f	The filter inductance
R_f	The filter resistance
C_f	The filter capacitance
i	The grid current
i^*	The reference grid current
ω_c	The cut-off frequency

Chapter 1
Overview of Microgrid

1.1 Microgrid Concept and Challenges

1.1.1 Microgrid Concept

Power generation methods using nonconventional energy resources such as solar photovoltaic (PV) energy, wind energy, fuel cells, hydropower, combined heat and power systems (CHP), biogas, etc. are referred to as distributed generation (DG) [1–3]. The digital transformation of distributed systems leads to active distribution networks with bidirectional power flow transmission, distributed control, and bidirectional decision-making, providing a bidirectional network design for the transmission of electricity. As compared to the conventional synchronous generators, DG units are more controllable and easier to be operated.

In recent years, researchers have shown considerable interest in microgrid (MG), a new concept for future energy distribution systems that allows for renewable energy integration [4–6]. This technique integrates DG units with energy storage systems (ESSs), demand side management, and various loads into a central, organized control plan, thereby establishing distribution networks connected primarily to the upstream power system. Figure 1.1 shows the structure of an MG. Moreover, these networks can be disconnected from the utility grid and operated independently as isolated networks in case of grid failures or other external disturbances. In turn, this makes the power supply more reliable and sustainable. To provide the necessary flexibility within the operation of MGs and to ensure the specified power quality and power output, they have power electronic interfaces (PEIs) and appropriate control systems [7–9].

MGs are environmentally friendly because they utilize renewable energy sources (RESs). They can also provide benefits to the main grid and to the customers [10, 11]. MGs can be viewed as either a controlled entity within a power system or an integrated load from a central grid perspective. The controllability and flexibility of MGs enable to easily conform to grid regulations without negatively affecting utility

Y. Sun et al., *Series-Parallel Converter-Based Microgrids*, Power Systems,
https://doi.org/10.1007/978-3-030-91511-7_1

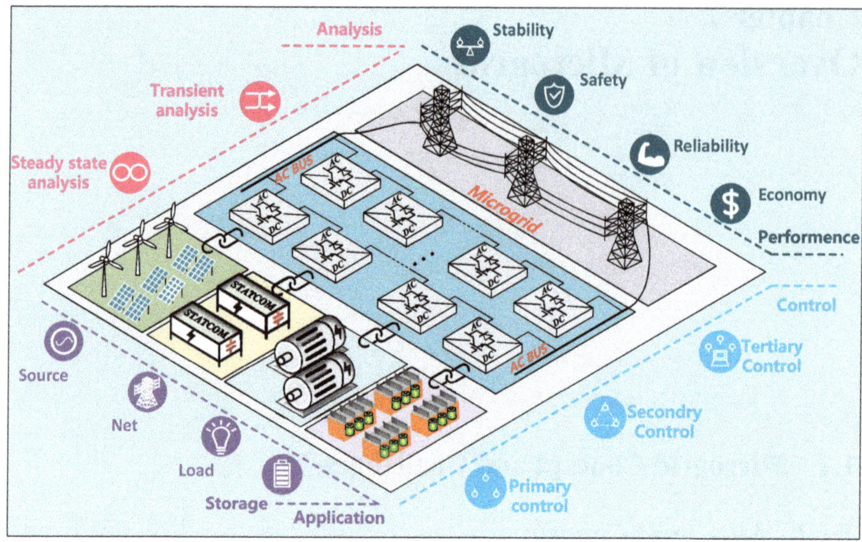

Fig. 1.1 The structure of an MG

grid stability, which in turn improves the reliability of power supplies. To meet consumer demands for electrical power and heat, MGs also provide uninterrupted power. Additionally, they can improve the local electrical reliability, reduce feeder losses, and support local voltage support. Microgrids will gradually be used to support the main grid and could even be a future trend for the power systems.

1.1.2 Challenges for Microgrid

Although the development of MGs has significant potential benefits, there have been several challenges to achieving a stable and secure operation. A number of technical and regulatory issues need to be resolved [12–15].

Low Inertia Issues [16] The replacement of synchronous machines has more general consequences. Power electronics converters introduce faster dynamics than conventional synchronous machines for both active and reactive power support. This may bring unexpected couplings, and the control approaches based on timescale separations may become more brittle and increasingly less valid.

Operation Control and Stability Issues with Multi-resource MGs [17] An MG typically includes different sources of energy with different inertia, capacities, and transient responses. As an example, hydropower produces large rotational inertia due to the presence of rotating devices, whereas power electronics interfaced with converter-based PV generation systems has very little physical inertia. To overcome the interactive effects of multi-resource MGs, comprehensive consideration should be given to the operation control strategy.

Seamless Transition Between Different Operation Modes It should be possible to operate MGs in both grid-connected and island-based models. Seamless transition between these two modes is required to prevent MGs from being intentionally/unintentionally damaged while protecting the integrity of the utility grid and the power quality of the local power supply.

Power Sharing Control To eliminate operation failures caused by overcurrent incidents, unintentional outages of DG units, and reductions in power transmission losses, power sharing control methods should take into account differences in nominal capacities, transient characteristics, and the distances between loads and DG units.

Power Quality Issues In MGs involving nonlinear or unbalanced loads, power quality issues like harmonics and unbalances will arise. The purpose of a PEI is to achieve various functions such as harmonic attenuation, voltage balance compensation, and interruption rejection, to improve power supply quality.

Protection for Active Networks DG units are becoming more prevalent, resulting in a few issues that should be considered on MG protection. RESs suffer from a variety of problems such as bidirectional power flow, fault currents, and climate dependence, causing variable infeed currents. Additionally, interactive influences are also present in high penetration distributed networks that result from extensive joints.

Optimum Sizing and Placement of Multiple Sources As power and energy losses, voltage profiles, and redundant capacities are decreased, the economic cost will be smaller by the effective placement of RES, ESS, and CHP in the distribution network.

Regulations for Operating Grid-Connected MGs In grid-connected MGs, an appropriate control strategy must be implemented to ensure synchronized behavior, to comply with grid codes, and to ensure precise compliance with regulatory requirements.

Communication Integration MGs require the development of a specific communication infrastructure and protocol. The IEC 61850 standard was published in the communication for media gateways and active distribution networks. In rural areas, there remains a lack of adequate communication infrastructure.

1.2 Converters Classification in Microgrid

Converters can be classified into two broad categories based on how they interact with the distribution grid: grid-forming converters and grid -following converters. Besides establishing and regulating the bus voltage of the distribution grid for the load, grid-forming converters are also equipped for power sharing among parallel converters in multiple source environments. As grid-following converters deliver energized power to a grid, they operate as expected.

1.2.1 Grid-Following Converter

Grid-following converters behave as current sources with a high parallel output impedance. Generally, phase-locked loop (PLL)-based grid-following converters are suitable to operate in parallel with these converters. Power converters in DG units typically operate in the grid-following mode, like in PV or wind power units. These converters can participate in the control of the microgrid AC voltage amplitude and frequency by adjusting, at a higher level control layer, the references of active and reactive powers, P^* and Q^*. Figure 1.2 shows the control structure of the grid-following converter and the equivalent circuit.

In the absence of a grid-forming converter, or an AC microgrid with a local synchronous generator generating the voltage amplitude and frequency, grid-following converters cannot operate in island mode. Also, a weak grid will affect the stability of grid-following converters connected to it.

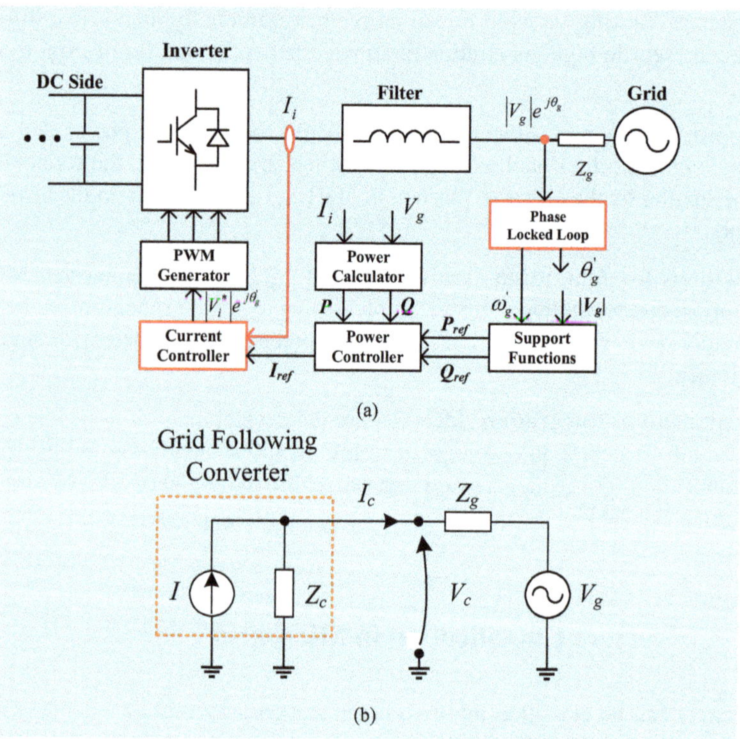

Fig. 1.2 Grid-following converter. (**a**) Block diagram. (**b**) The equivalent circuit

1.2.2 Grid-Forming Converter

A grid-forming converter is controlled in closed loop so that it can work as an ideal AC voltage source with a specified amplitude and frequency. Since they produce low voltages, parallel operation with other grid-forming converters requires extremely accurate synchronization systems. In parallel grid-forming converters, power sharing depends on the output impedances of the converters. In the stand-alone mode, grid-forming converters can track loads as voltage and frequency stay constant. Different control mechanisms have been proposed to parallelize the operation of multiple grid-forming converters, including droop control, virtual oscillator control [18], virtual synchronous machines, etc. Grid-forming converters can be used in standby UPS as a practical example. Within certain operating parameters, the system remains off of the main grid. Power converters of UPSs provide grid voltage during a grid failure. The grid-forming converter will provide a reference for the grid-following converters connected to it, thus making the DC voltage generated by the converter the basis for how the rest of the grid will be generated. Figure 1.3 shows the control structure of the grid-forming converter and the equivalent circuit.

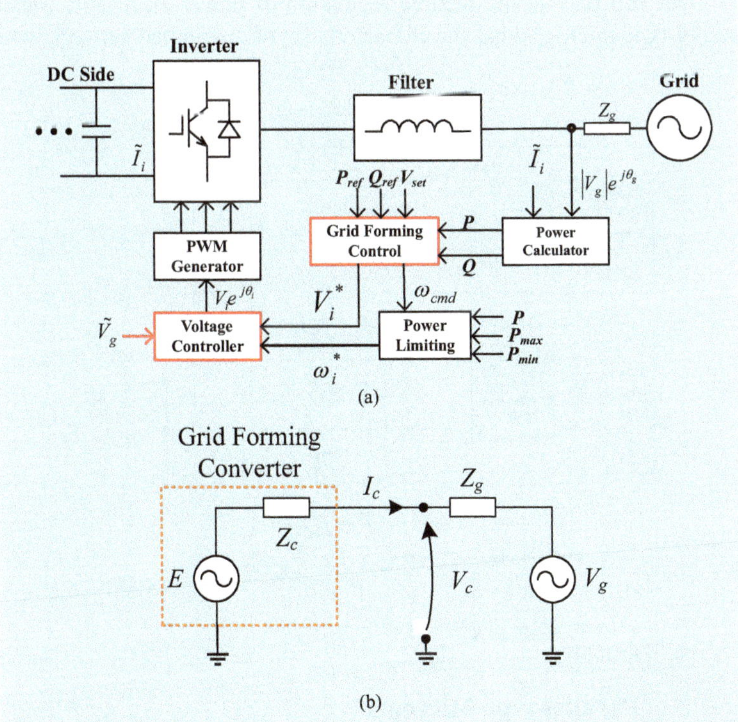

Fig. 1.3 Grid-forming converter. (**a**) Block diagram. (**b**) The equivalent circuit

There are certain limitations because grid-forming converters operate more like voltage sources than current sources. Control schemes that rely on simple grid formation are loosely regulated regarding converter current. Consequently, grid-forming converters may struggle to handle overcurrent during changes in load or abnormal conditions, such as faults or grid transients. It may be necessary to use advanced control schemes to achieve current limiting, such as virtual impedance current limiting, etc. In most cases, however, such schemes will affect the stability of the system as a whole. The converter may be overwhelmed by large current flows caused by simultaneous events such as a grid voltage sag or swell.

1.3 Architecture of Microgrid

1.3.1 Parallel-Type Microgrid

Parallel-type microgrid is the most common microgrid in current power system architecture. As is shown in Fig. 1.4, each DG unit is connected to the common bus in parallel through the converter. In this parallel-type microgrid, each unit can be controlled independently, and the energy can be distributed effectively, which greatly gives full play to the flexible regulation of power electronic. Meanwhile, the parallel-type microgrid has the characteristics of distributed network, which has

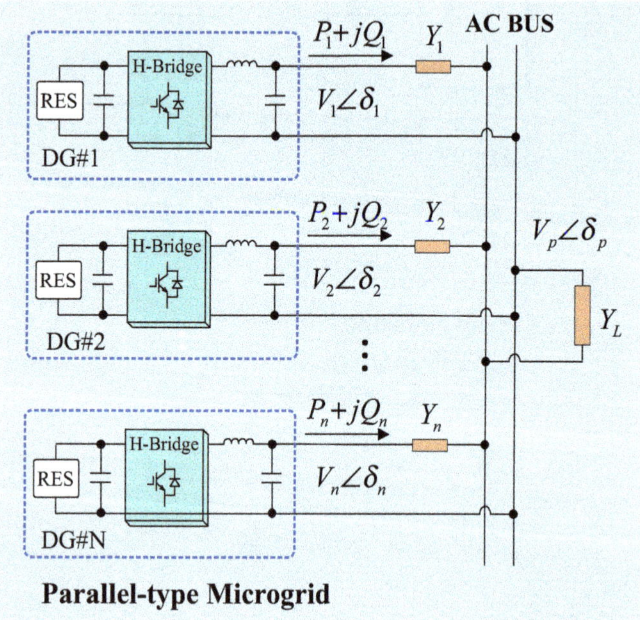

Fig. 1.4 Parallel-type microgrid

the ability for fault tolerance, extensively and the plug-and-play. However, its wide applications in medium and high voltage fields are limited due to the relatively low voltage grade of the DG unit.

1.3.2 Series-Type Microgrid

Series-type microgrid is a new type of microgrid system, and it is the vertical development of microgrid from the traditional single node in parallel to multi-nodes in series. As is shown in Fig. 1.5, each DG unit directly forms a microgrid system with a higher voltage level through the converter in series. This way of boosting voltage is not only direct and simple, requiring no complex boost circuits and bulky and expensive transformers, but also easy to control the overall microgrid system. In the series-type microgrid, each DG unit can still be controlled independently, and the power electronic equipment can be flexibly regulated to improve the power control level and the optimization management ability. Because the failure of a single unit in the series-type microgrid will lead to the operation failure of the whole system, more complex hardware circuits or control algorithms are needed to ensure the robustness of the series power electronic system.

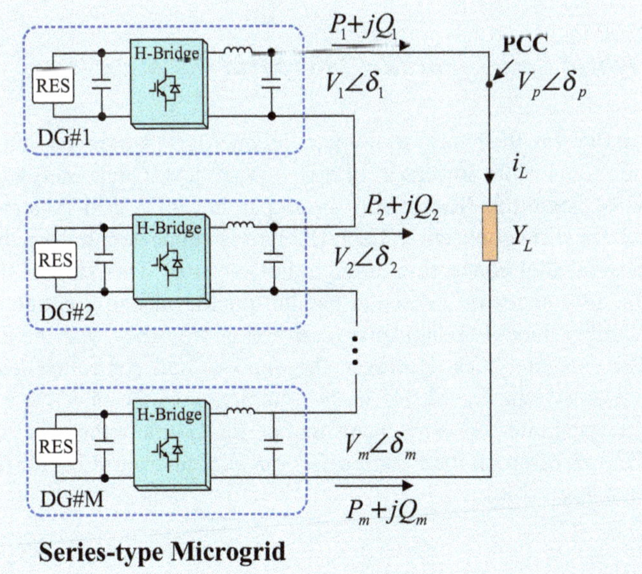

Fig. 1.5 Series-type microgrid

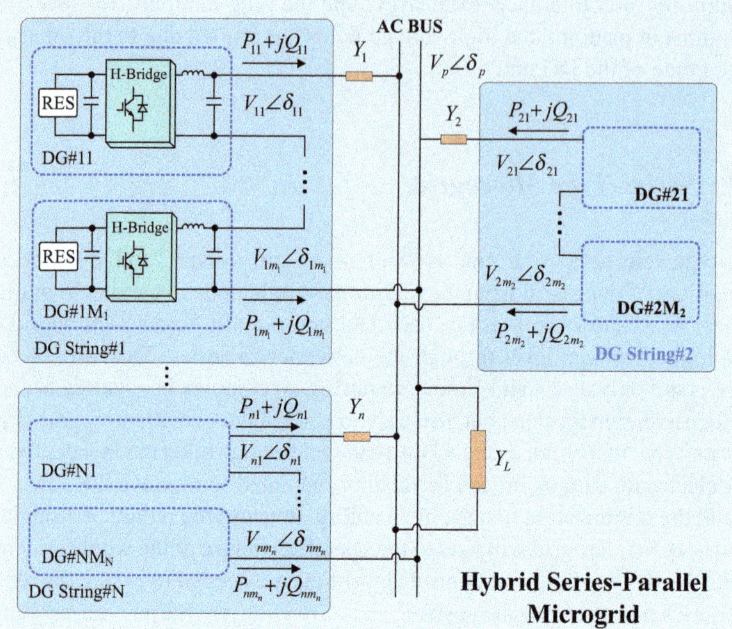

Fig. 1.6 Hybrid series–parallel microgrid

1.3.3 *Hybrid Series–Parallel Microgrid*

As shown in Fig. 1.6, the hybrid series–parallel microgrid system has attracted much attention due to its triple attributes for high voltage level, high energy density, and strong control flexibility. The hybrid series–parallel microgrid system adopts the series–parallel hybrid connection. Each DG unit is connected to the public bus in a hybrid series–parallel connection through the converter. This kind of the structure takes all the advantages of series and parallel microgrid into account, such as the control flexibility, module redundancy, conversion efficiency, and the medium and high voltage characteristics. However, the analysis and controller design of this structure are greatly increased due to the complexity of the physical transmission characteristics and interaction mechanism of each DG unit in the series and parallel structure. The research on the hybrid series–parallel microgrid is still in the initial exploration stage.

1.4 Hierarchical Control Theory—General Introduction and Motivation

A typical management and control system is expected in an MG setting to provide diverse benefits over a wide range of timescales, taking full consideration into account. As a result, MGs adopt and implement hierarchical control structures, which have been proven effective in power systems [17–21].

The MG hierarchical control system comprises three levels, as shown in Fig. 1.7. The primary control is responsible for DG unit internal control. By maintaining voltage, frequency, and active/reactive power output sharing among paralleled DG units, it can achieve voltage and frequency stability. Using the secondary control, voltage and frequency deviations that the primary control has caused can be corrected. As part of the secondary control, the tertiary control manages the flow of power between the MG and the utility grid. Additionally, economic and efficient optimization is taken into account.

Furthermore, MG hierarchical control also offers several special features: distributed intelligent management, compensation for voltage unbalance, self-healing networks, smart homes with cost-effective energy networks, and generation scheduling. Microgrids can be regarded as complex, multi-agent, and intelligent systems because of their hierarchical structure.

1.4.1 Primary Control

The primary control, or field control, is the first level [17]. Voltage and frequency control and stability are the focus at this level. It is possible to share the active and reactive power in this level without communicating through the droop control system or a similar system.

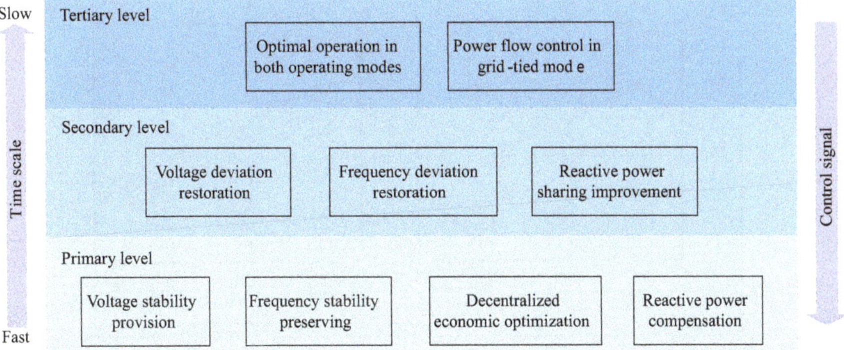

Fig. 1.7 Hierarchical control

1.4.1.1 Conventional Droop Control

The droop control method has been widely applied to converters control for decentralized control in many applications, such as parallel redundant uninterruptible power supplies (UPSs), distributed power systems and MGs that facilitate independent, autonomous, and wireless control through the elimination of intercommunication [8–10, 22–31]. Typical droop control expressions are

$$\begin{cases} \omega_i = \omega^* - m_P \cdot (P_i - P^*) \\ V_i = V^* - n_Q \cdot (Q_i - Q^*) \end{cases}$$

where ω^* and V^* are the normal angular frequency and output voltage amplitude, respectively, and m_P and n_Q are the droop coefficients.

It is the objective of the well-known control technique to control active and reactive power proportionally by locally adjusting the frequency and voltage amplitude of each DG to emulate the behavior of a synchronous generator, that is, to reduce frequency when active power increases and lower voltage when reactive power increases. The droop characteristic is shown in Fig. 1.8. The control algorithm with conventional droop control is illustrated in Fig. 1.9. LC filters and coupling line inductors make up the component of the power stage. The controller consists of three control loops: (i) a power sharing controller is used to generate the magnitude and frequency of the fundamental output voltage of the converter according to the droop characteristic, (ii) a voltage controller is used to synthesize the reference filter inductor current vector, and (iii) a current controller is adopted to generate the command voltage by a pulse width modulation (PWM) module.

It is worthy to notice that the droop coefficients for regulating voltage amplitude and frequency are typically proportional terms. In order to improve system dynamics, derivative terms are included to increase the range of coefficient values [32–35] and this control method can also be referred to as a virtual synchronous generator (VSG). When derivative terms are added to droop controllers, they yield two-

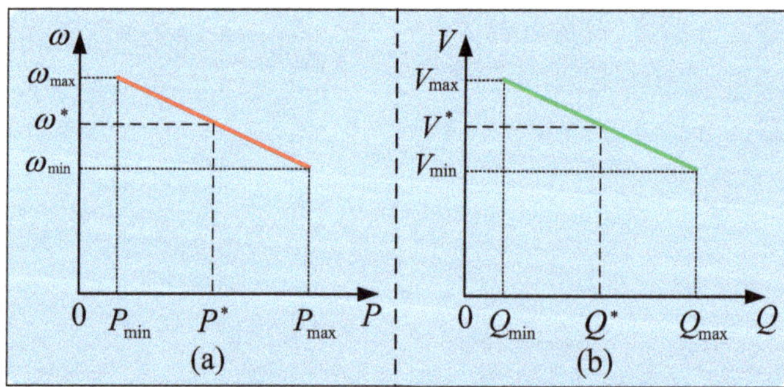

Fig. 1.8 Conventional droop characteristics. (**a**) $P--\omega$ droop control. (**b**) $Q--V$ droop control

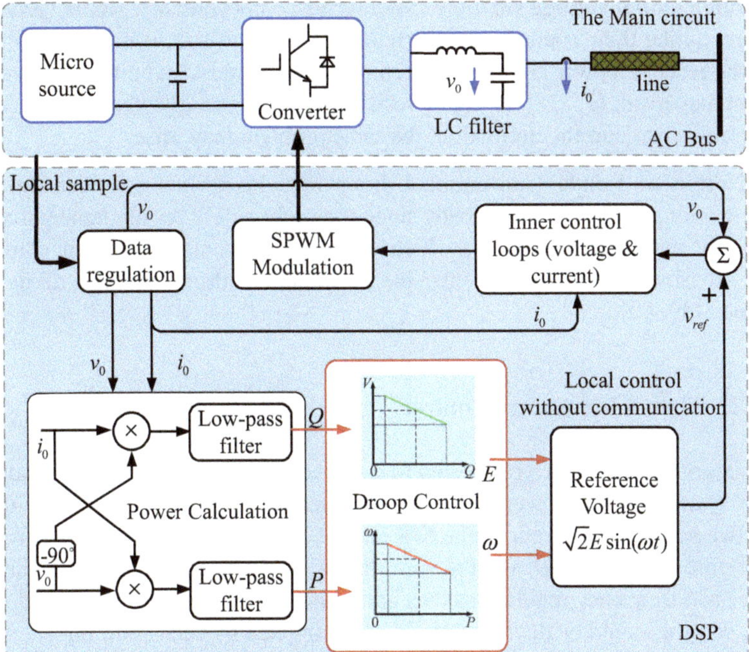

Fig. 1.9 Conventional droop control

dimensional (DOF) tunable control. Traditional droop methods, by contrast, can only control the movement in 1-DOF. Therefore, the dynamic behavior of the system can be improved by dampening the oscillations in the power sharing controllers while retaining the same static gain.

As discussed above, the conventional droop control method can be implemented without communication between modules. However, it has some drawbacks as listed below [17]:

Multiple Control Objectives Multiple control objectives are not possible since there is only one control variable for each droop characteristic. Among the voltage f/V regulations and load P/Q sharing, for example, there needs to be a design tradeoff taken into account [36].

Mixed Resistive and Inductive Line Impedance The conventional droop method assumes that the VSC (voltage source converter) and AC bus have highly inductive equivalent impedances. Due to the primarily resistive nature of low-voltage distribution lines in microgrid applications, this assumption is thrown into question. The active and reactive power will also be strongly coupled if the line impedance is mixed resistive and inductive [35, 36]. Microgrids that operate at medium voltage (MV) can have X/R ratios of almost one.

Without a global voltage variable Microgrids do not contain a global variable for voltage, unlike their counterparts for frequency. Due to this, it may be difficult to split the reactive power control between parallel inverters, and circulating reactive current may result [37, 38]. In highly resistive lines, especially when voltage is used to control active current circulation, the same problem may arise.

The Nonlinear Loads Conventional droop methods do not consider harmonics when working with nonlinear loads since they rely solely on fundamental values. This only measures Q and P, which are usually averaged over an entire line cycle. To share harmonic currents, the conventional droop method needs to be modified [39, 40].

1.4.1.2 Virtual Impedance Control

The majority of DERs are connected to distribution lines, as opposed to traditional power systems where power sources are connected to transmission lines—usually inductive with high X/R ratios. The X/R ratio of line impedances between generators is therefore relatively low, and the considerable resistance inevitably combines the voltage and frequency regulations over active and reactive power. $P-f$ and $Q-V$ droop controls could be degraded or even destabilized by such coupling.

When the system is not sufficiently inductive, a virtual impedance can be introduced between the converter and the external grid in order to facilitate the control of the $P-f$ and $Q-V$ droop, as shown in Fig. 1.10.

It is dependent upon the structure and mode of operation of the converters as to how a virtual impedance loop is implemented. As well as providing the decoupling necessary for droop control, a virtual impedance may also reduce the sub-synchronous oscillations resulting from phase-locked loops (PLLs) and AC current and voltage control loop harmonic instability. It is possible, generally, to decouple active and reactive power flows and help improve the sharing of reactive power in parallel converters by introducing a virtual negative resistance or virtual impedance. A virtual positive resistance could be introduced into power distribution

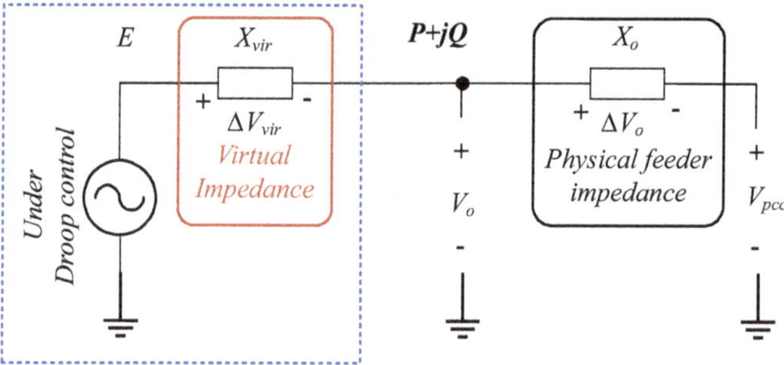

Fig. 1.10 Virtual impedance control

systems to dampen low-order harmonics and to improve the power sharing between converters. Furthermore, a virtual negative-sequence conductance can be introduced to mitigate grid voltage imbalances and improve power sharing for unbalanced loads in addition to the virtual impedance on the positive sequence.

1.4.2 Secondary Control

The primary control is responsible for power quality and reliable operation of the MG, and the secondary control was designed to compensate for frequency and voltage amplitude deviations resulting from primary control. Secondary control may be implemented in a centralized [41–43] or a decentralized [44–50] format as appropriate to the application.

1.4.2.1 Centralized Control

Centralized controller relies on centralized communication infrastructure. Through low-bandwidth communication links [41–43], the measured variables from each essential component of MGs are collected and transferred to the central controller. Control and energy management algorithms can be executed at high levels based on the data collected. As a result of the obtained compensation signals, each DG will be controlled appropriately and efficiently through the primary control. It is possible to monitor the system in real-time through centralized control. However, since the point-to-point communications are necessary for centralized control, a single point of failure (SPOF) issue, which will invalidate the whole high-level control system, is inescapable. MG controllers are therefore less reliable and expandable. Considering that, small or localized MGs are better suited for centralized control schemes.

1.4.2.2 Distributed Control and the Consensus Algorithm

MG's reliability and flexibility have been improved by introducing distributed controls in order to avoid SPOFs [44–50]. Instead of communicating directly with the MG, essential information will be exchanged directly among DG units using communication technologies such as wired and wireless technologies [46], as well as by using distributed information exchange algorithms such as gossip and consensus [47–50]. In addition to reducing communication complexity, distributed control provides more scalable and flexible operations. As a result of the effective means of sharing information among DGs and facilitating the distributed coordination control, consensus algorithms have been used in MGs [51–56]. Communication links are only needed between neighboring DGs when consensus algorithms are used, which enables plug-and-play performance and lowers communication costs. The state of each DG unit is only communicated to adjacent DGs in this technique.

The DGs update their states based on a linear equation of both their own and their neighbor's states. The desired average value is achieved when the states of all the DGs converge. MGs can be eliminated from dependence on a single DG by using a consensus algorithm to share information and coordinate among units of DGs.

1.4.3 Tertiary Control

Among the highest and lowest levels of control, a tertiary control is responsible for controlling the flow of power between MGs and the utility grid, while considering factors such as cost, benefit, and efficiency, among others [18, 20, 56, 57]. This control level is part of the host grid rather than the management grid itself. By taking both technical and economic factors into account, the tertiary control ensures that all MGs run efficiently. The distribution system operator exchanges information with tertiary control to make a feasible decision and optimize MG operation.

By regulating the voltage amplitude and frequency of the point of common coupling (PCC), the power flow between MGs and utility grids can be managed. A measurement of MGs' active and reactive power outputs is made first, and then these results are compared with the desired values. In primary control, the differences are sent to two independent PI controllers that function on the voltage, current, and power outputs measured by active and reactive power outputs, respectively. An MG can export or import active and reactive power independently, depending on the power references.

To achieve power sharing performance in addition to power flow management between MGs and the main grid, a tertiary control would be developed on the basis of the aforementioned novel synchronous-reference-frame (SRF) VR-based primary control.

1.5 Microgrid System Stability

1.5.1 Classification of Microgrid System Stability

Stability in microgrids can be classified by the physical cause of the instability, the size of the disturbance, the physical components that are involved, the duration of the instability, as well as the methodology used to understand or predict it [58].

Figure 1.11. illustrates the classification of microgrid system stability. Stability for microgrid systems is characterized by two distinct phenomena: phenomena associated with the equipment control systems and phenomenon associated with the power supply and balance [58]. Also, microgrid instability falls into either category and can be either a short-term or a long-term issue. The short-term stability of the system can be impacted within a few seconds, but other issues beyond this time frame affect its long-term stability [58].

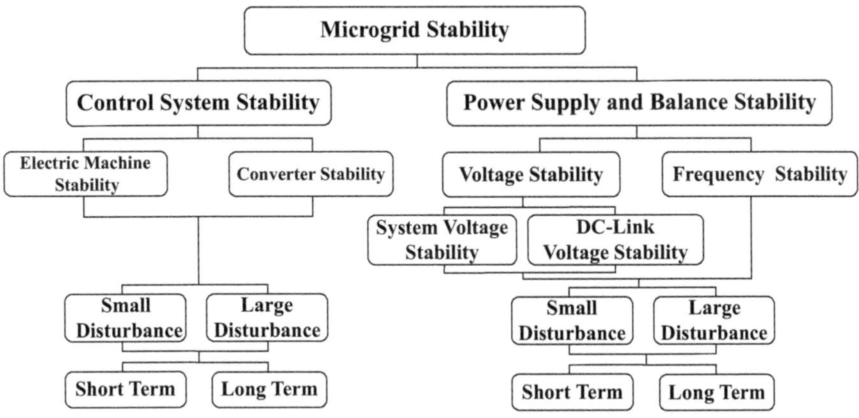

Fig. 1.11 Structure of microgrid system stability

1.5.1.1 Power Supply and Balance Stability

System balance and power supply stability consist of the ability of the system to maintain power balance among DERs and to effectively share demand power such that the system meets operational requirements. This type of stability issue occurs when a generation unit fails, DERs are exceeded, power sharing between multiple DERs is incorrectly selected, multiple DERs are overloaded, and slack(s) resources are not appropriately allocated [57]. The load tripping may be voluntary or involuntary. Moreover, certain load types, such as constant power loads or induction motors, can cause certain types of system instability, such as voltage fluctuations. Stability in frequency and voltage terms are subgroups of this class of stability [58].

(a) Frequency Stability

A major concern in islanded microgrids is frequency regulation. Microgrids are also vulnerable to large disruptions when generators go out due to their low number of generation units. Accordingly, for such disturbances, the system frequency may experience large excursions at a fast rate, potentially compromising system frequency stability [59, 60]. The frequency control technologies and techniques conventionally used may not be sufficient to deal with the rapid shift in frequency present in the system, even with sufficient generation reserve [61]. Throughout the world, such incidents have been reported [58].

(b) Voltage Stability

Voltage instability is a result of the limits of DERs and the sensitivity of load power consumption to supplied voltage in microgrids. There may be voltage instabilities in these systems due to their low voltages in steady state and in dynamic states. An unexpected change in the output or demand of renewable energy resources, or a sudden generator failure, may cause voltage instability based on the response of the system and the load characteristics. For systems that are close to their loading limits or highly unbalanced, small disturbances, such as incremental changes in demand, can also lead to voltage instabilities [62–64].

Voltage instability is a potentially short-term or long-term phenomenon, depending on the time frame. An active and/or reactive power mismatch can result in short-term voltage instability due to poor coordination of control actions. Long-term voltage instabilities, on the other hand, occur when DER output limits are gradually reached by a steady increase in demand, such as in the case of thermoelectric loads.

1.5.1.2 Control System Stability

A lack of control schemes (e.g., harmonic resonance of parallel DERs) and/or poor tuning of individual equipment controllers may lead to control system stability issues. When the system becomes unstable, the root of the problem is a poorly tuned controller. This controller must be retuned, or the entire system must be taken apart. Electrical machines and converters control loops, LCL filters, and PLLs all benefit from this type of stability [58].

(a) Electric Machine Stability

The typical aim of these types of stability studies is to examine the recovery of synchronous machines from angular acceleration during a fault, so they can regain their original synchronization. Microgrids, however, have not been observed to exhibit this phenomenon. Synchronic machines, for example, can decelerate as a result of short circuits if they are faced with faults at the end of the feeder due to the resistive nature of microgrids; this is demonstrated in the experiment described in [65].

The rotor angle of conventional synchronous generators can be observed oscillating without apprehension when there is a small perturbation in the system [66]. While the first forms because of insufficient synchronizing torque, the second results from insufficient damping torque. Microgrids, however, do not experience synchronizing and damping torque problems with generators equipped with well-tuned voltage regulators and governors. Microgrid stability is almost always attributed to inadequately tuned exciters and governors of synchronous machines.

(b) Converter Stability

There are two types of converter instabilities in microgrids: small perturbations and large perturbations. A major concern in small perturbation stability of a system is the inner voltage and current control loops, whose tuning is a challenging issue. Blackouts caused by DERs, especially low-frequency and undervoltage protection schemes, are also seriously concerning.

Harmonic oscillations can be caused by exchanges between inner current and voltage control loops which differ from low-frequency oscillations caused by outer power control loops [67, 68] and are referred to as harmonic instability. Harmonic instability, or harmonic oscillations, is a term in the technical literature that refers to a range of phenomena that lead to high harmonic frequencies or resonances. Multi-resonance peaks occur when a large number of converters are positioned close together [69]. Harmonic instability can also be caused by high-frequency switching, resulting in parallel and series resonances that are introduced by LCL power filters or parasitic feeder capacitors [67, 70]. It is also possible to trigger the resonance of an LCL converter filter by controlling it directly or by co-operating with nearby controllers [41]. The so-called active damping strategies can be used to prevent and/or mitigate harmonic instability [71].

1.5.2 Stability Analysis and Performance Assessment

1.5.2.1 Time-Scale Separation and Model Reduction

Power systems are typically analyzed and controlled using reduced-order models of various degrees of fidelity that exploit the contrast between the dynamics of synchronous machines and the transmission network to facilitate tractable design of multi-machine systems [16, 72]. In conventional multi-machine power system models, the singular perturbation theory and integral manifolds are used to approximate or neglect fast dynamics, like transmission line dynamics, to reduce the model order.

Despite their usefulness for stability analysis, these simplified ad hoc models have been contested when it comes to microgrids. A swing equation model crucially relies on synchronous machine properties, which makes it questionable for microgrids with low inertia and converters dominated. The dynamics and control loops of converters are fast, and a more thorough analysis is required to ensure that reduced-order conclusions apply to the full dynamics of the system [73].

Comparing the dynamics of conventional power systems with low-inertia and converter-dominated power systems, the timescales of the different dynamic phenomena overlap and cannot be analyzed separately. When fast acting grid-forming power converters dominate power systems, for example, it becomes impossible to ignore the dynamics of transmission lines [74]. The overlapping timescales can also create adverse interactions in grid-following controls, control loops inside converters, the DC bus of converters, and power system stabilizers [75].

An approach that simplifies the analysis even further is to assume that voltage dynamics and frequency dynamics are decoupled and can thus be studied separately. For practical purposes, this approach is often justified when the X/R ratio is high (for example, when a large amount of transmission is conducted) and the nominal operating range is narrow (for example, when the voltage magnitude and phase angle deviations are small). Although not many analytical studies have examined the general case, analyses of converter-dominated systems reveal a link between voltage dynamics and frequency dynamics that influences stability boundaries [76].

1.5.2.2 Stability of a Single Converter Connected to an Infinite Bus

Grids are often modeled as constant voltage sources in power electronics modeling (colloquially called infinite buses), owing to the voltage frequency and magnitude of grids at the converter terminals were relatively stiff signals, especially on the fast timescales (fast with respect to power grid timescales) of interest for power converter design and analysis.

The large-signal nonlinear stability analysis methods for grid-connected converters contain three steps.

1. Models using nonlinear stability methods generally apply Lyapunov/LaSalle functions (broader notions of energy) to rigid grids in a rotating frame, but with the goal of developing robust and scalable analysis methods that go beyond these simplified models [77]. The ability to analyze nonlinear stability using a single rotating reference frame usually associated with the angle in converter control architectures (e.g., the angle used for grid-forming control or estimated by a PLL) is crucial for nonlinear stability analysis.
2. This grid can also be modeled as a variable voltage source or equivalent synchronous machine, and the analysis methods are extended to this case as well. In many cases, the results obtained with these methods are optimistic, since they lead to overestimating the integrity boundary of a system [78].
3. To lift the analysis methods in (2) to a mixed multi-converter/multi-machine model.

Various studies have been conducted on the stability of converter-based resources that are subject to current limits and faults. Grid-following converters with an infinite bus have been well studied in terms of the current limiting problem. When grid-forming control is combined with current limiting, synchronism and stability are usually compromised [79]. In an ad hoc engineering approach to this problem, the current limiting of a single converter has been examined in the infinite bus setting and employs simulated increasing impedances for converters as their currents approach their limits but has not been widely studied beyond this strict setting.

1.5.2.3 Stability of Multi-Converter Systems

A multi-converter system with 100% grid-forming converters—each of which is equipped with frequency/active power and voltage/reactive power droop controllers—warrants further consideration for stability analysis [80–82].

It had been assumed that initial literature would focus on ideal models, in which each closed-loop converter would behave in accordance with its ideal reference behavior, i.e., a droop in output behavior would accompany any input-to-output converter. Moreover, this assumption can be justified by the timescale separation: the inner control loops work flawlessly beyond the dynamics of the network to prevent the development of unwanted interactions. The early presentations followed standard decoupling assumptions and introduced novel stability analysis approaches; for example, methods developed for synchronizing coupled oscillators and networked control proved beneficial for analyzing the transition for voltage and voltage variations of a multi-converter system. In subsequent works, Lyapunov/LaSalle approaches were combined with passivity arguments to generalize these approaches to the coupled active/reactive power and voltage frequency/amplitude dynamics. In a nutshell, although these approaches generalize classic energy functions developed for power system transient stability, many of their assumptions, for example, transmission lines being in steady state and transmission lines without conductance (i.e., lossless), require sufficiently simplified models. In short, these methods are effective for simple models, but they encounter limitations when there are transmission line losses, similar to classical energy function-based methods.

Though the above methods appear to have different methodological approaches, different models, and a variety of controllers, all of them reach qualitatively similar conclusions. Essentially, these classic insights from power system stability apply to converter stability when each converter is reduced to its reference dynamics, i.e., there should not be any transmission bottlenecks when reducing each converter to its reference dynamics. By moving to more detailed models by means of singular perturbations, these conclusions hold up provided that the transmission network dynamics are separated sufficiently from the converter's current and voltage control loops, which is in line with usual engineering practice [83].

Furthermore, multiple analytic studies have been conducted on the stability of small signals in multi-converter power systems, with representative results for grid forming converters and grid-following converters. The large-signal stability results are qualitatively similar to the converter and line dynamics needing to be separated sufficiently over time.

Standard stability analysis methods for multi-converter systems assume the DC bus to be connected to an infinite source, and hence they cannot be used to analyze stability of the DC bus (or a DC network in the case of high- and medium-voltage DC components) [84]. It is possible to overcome this limitation by using the so-called machine matching control, and it exhibits promising performance when the DC bus is not tightly controlled, but there are no quantitative results available on analytical stability so far.

1.5.2.4 Stability of Multi-Converter Multi-Machine Systems

Using the model-reduction methods, the (linearized) swing equation can approximate the majority of grid-forming controls. The frequency dynamics of multi-converter multi-machine systems can be analyzed by using standard energy function-based stability analysis [85, 86]. Although this baseline approach can be considered a reasonable place to start and grid-following resources could fit into this framework, it has several significant disadvantages. For instance, when mapping a wide range of converter-based resource dynamics to a unified swing equation model, fast dynamics and control loops with overlapping timescales (e.g., phase-locked loops) are neglected. This framework also cannot account for instabilities caused by the limitations of power converters, e.g., DC and AC currents. There is no general framework that can consider diverse converter, machine, and network dynamics, including volatility and frequency dynamics. Recent results enable us to research small-signal frequency stability for a wider range of machine dynamics and converter dynamics.

Since analytical approaches have limitations, current stability analysis on converter-based resources and synchronous machines emphasizes numerical methods. There are two broad types of works here: theoretical approaches for investigating transient stability (i.e., large contingencies) and small-signal methods for analyzing performance based on linearized models and various stability and performance metrics (such as eigenvalues and eigenvectors). By leveraging the fast response capabilities of converter-based resources, these studies demonstrate increased performance and resilience of low-inertia and converter-dominated systems [87].

In large systems, however, with a mixture of machines and converters, many considerations are no longer valid, and typical assumptions made for systems dominated by synchronous machines may no longer apply. It has been noted that the threat of decreasing inertia has triggered a great deal of attention, but too much virtual inertia at the wrong place can dilute the performance and robustness metrics of small-signal systems. It may not be straightforward to characterize the stability of a multi-machine multi-converter system with various types and levels of converters.

In addition, numerical studies that explore the transition from conventional multi-machine systems to converter-dominated power systems also highlight the potential for small-signal instabilities that arise from interactions between phase-locked loops, inner converter controls, network dynamics, and grid-forming controls of power systems. Grid-forming converter controls can cause adverse interactions with standard current limiting schemes, as well as slow machine dynamics [88].

1.6 Organization of the Book

According to the preliminary study [89–104], this book would bring latest research result on the system-level control and stability of the series–parallel converter-based microgrids. From the topology point of view, this book can be divided into three parts. The first part is focused on the controls for the parallel-type microgrid system, which is from Chaps. 2–7. The second part presents the controls and applications for the series-type microgrid system, which are included from Chaps. 8–13. And the third part introduces the controls for synchronous control, SOC balance, and economic optimization of the hybrid series–parallel microgrid, which is from Chaps. 14–17. The overall structure of the book is shown in Fig. 1.12.

Our researches for system-level control and stability of parallel-type microgrid system are introduced in Part I. Chapter 2 compares the similarities and differences among three different concepts, virtual impedance method, angle droop control,

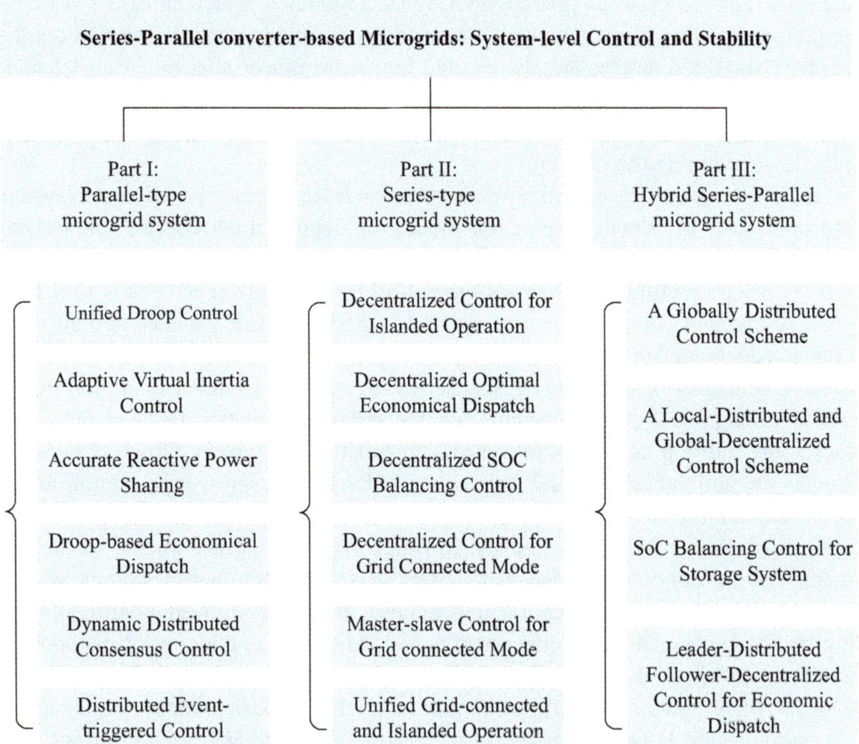

Fig. 1.12 The overall structure of the book

and frequency droop control. The modified droop control is introduced to unify these three independently developed droop control methods into a generalized theoretical framework. Chapter 3 introduces a virtual synchronous generator (VSG) control based on adaptive virtual inertia to improve dynamic frequency regulation of microgrid. The flexible inertia property combines the merits of large inertia and small inertia, which contributes to the improvement of dynamic frequency response. Chapter 4 introduces an improved droop control method to improve the reactive power sharing accuracy, enhance the power quality of the microgrid, and also have good dynamic performance. The needed communication in this method is very simple, and the plug-and-play is reserved. Chapter 5 studies the economical dispatch problems (EDPs) in islanded microgrid and presents a criterion to determine whether the global optimal dispatch (GOD) can be realized via a decentralized manner. Besides, an optimal/suboptimal decentralized economical-sharing scheme is presented when the criterion is met/not met. Chapter 6 introduces a dynamic consensus algorithm (DCA)-based virtual resistance control strategy to share the active power and harmonic power. In contrast with the previous method, this approach does not require a central controller, and the communication links are only required between the neighboring UPS modules, which enhances system's reliability and flexibility. Chapter 7 presents an event-triggered distributed control strategy for the reactive, unbalance, and harmonic power sharing in an islanded microgrid. The introduced control strategy could realize the accurate power sharing while highly reducing communication data exchange and achieving the plug-and-play feature among the DG units.

In Part II, several decentralized control methods for series-type Microgrid system are presented. In islanded operation mode, the control methods can suit for any types of loads. Moreover, economical dispatch operation, SOC balancing, series-type H-bridge rectifiers, and series-type H-bridge inverter-based STATCOM have been researched. In addition, the methods for grid-connected and unified grid-connected/islanded operation are studied. Chapter 8 introduces several decentralized control strategies of the single-phase series inverters in islanded operation mode, which are capable to achieve synchronization and power balance. Besides, the power factor angle droop control is introduced to suit for any types of loads. Chapter 9 studies the optimal economical operation problem of the series-type microgrids. A communication-free control scheme is introduced, which could achieve the global-optimal economical operation easily. In Chap. 10, a decentralized SOC balancing method is introduced to balance the SOC of series-type energy storage system (CESS). Without communication dependence, the decentralized control obtains higher reliability and lower construction cost. Meanwhile, the control method has good extensibility, and its application is not limited by the number of ESUs. Chapter 11 presents decentralized control strategies for series-connected inverters, series-connected H-bridge rectifiers, and series-connected H-bridge inverter-based static compensators (STATCOM) in medium-/high-voltage (MV/HV) power network without any communication, and each module makes decisions based on its own local information. In Chap. 12, a novel hybrid voltage/current control scheme with low-communication burden is introduced for series-type inverters in a decentralized manner. The control method can be realized with very low communication burden, which is more cost-effective and has higher reliability in terms of communication faults. In Chap. 13, a unified decentralized control strategy is presented for both grid-connected and islanded operation of the series-type microgrid (CMG). It can realize the smooth mode switching without the need of changing controllers. In terms of the scheme, the system always holds a unique equilibrium point regardless of the grid-connected or islanded operation.

Part III introduces the synchronization control methods and further applications for hybrid series–parallel microgrid. In Chap. 14, we briefly compare and analyze the decentralized power control strategy of parallel microgrid and series microgrid and present a globally distributed control strategy to implement power sharing control in hybrid series–parallel microgrid under both resistive-inductive and resistive-capacitive load, where a sign function is introduced to automatically match load characteristic. Chapter 15 presents a local distributed control for hybrid series–parallel microgrid in islanded mode. A number of local converters in series type are controlled as a consistent string by distributed cooperative control without a central controller, and power demand among parallel strings is shared in a decentralized manner. Hence, only an in-string sparse low-bandwidth communication network is sufficient for limited information exchange among local converters. Chapter 16 introduces a local-distributed and global-decentralized SOC balancing control for hybrid series–parallel energy storage. Under the proposed scheme, the SOC of the unit in each local string converges to the same value driven in a distributed manner, and the SOC balance among different strings and power sharing of each unit under both the charging and discharging modes are realized in a decentralized manner. In Chap. 17, we introduce a leader-distributed follower-decentralized control strategy for series–parallel microgrids, under which the economic dispatch is achieved with voltage quality guaranteed and frequency synchronization. In this control strategy, a distributed control method is presented for the leader of distributed generators and an improved droop control method is designed for other DGs (followers) to optimize the generation cost and frequency control of the system.

References

1. I.S. Bae, J.O. Kim, J.C. Kim, C. Singh, Optimal operating strategy for distributed generation considering hourly reliability worth. IEEE Trans. Power Syst. **19**(1), 287–292 (2004)
2. A. Yadav, L. Srivastava, Optimal placement of distributed generation: an overview and key issues, in *2014 International Conference on Power Signals Control and Computations (EPSCICON)* (2014), pp. 1–6
3. R. Lawrence, S. Middlekauff, Distributed generation: the new guy on the block, in *IEEE Industry Applications Society 50th Annual Petroleum and Chemical Industry Conference, 2003*. Record of Conference Papers (2003), pp. 223–228
4. R.H. Lasseter, P. Paigi, Microgrid: a conceptual solution, in *Proc. IEEE PESC, Aachen* (2004), pp. 4285–4290
5. R.H. Lasseter, Smart distribution: coupled microgrids. Proc. IEEE **99**(6), 1074–1082 (2011)
6. N. Hatziargyriou, H. Asano, R. Iravani, C. Marnay, Microgrids. IEEE Power Energy Mag. **5**(4), 78–94 (2007)
7. P. Piagi, R.H. Lasseter, Autonomous control of microgrids, in *Power Engineering Society General Meeting* (IEEE, Piscataway, 2006), pp. 2006
8. Y. Li, D.M. Vilathgamuwa, P.C. Loh, Design, analysis, and real-time testing of a controller for multibus microgrid system. IEEE Trans. Power Electron. **19**(5), 1195–1204 (2004)
9. J.M. Guerrero, L. Garcia De Vicuna, J. Matas, M. Castilla, J. Miret, Output impedance design of parallel-connected UPS inverters with wireless load-sharing control. IEEE Trans. Ind. Electron. **52**(4), 1126–1135 (2005)
10. S. Barsali, M. Ceraolo, P. Pelacchi, D. Poli, Control techniques of dispersed generators to improve the continuity of electricity supply, in *2002 IEEE Power Engineering Society Winter Meeting*. Conference Proceedings (Cat. No.02CH37309), vol. 2 (2002), pp. 789–794
11. P. Asmus, Why microgrids are moving into the mainstream: improving the efficiency of the larger power grid. IEEE Electrif. Mag. **2**(1), 12–19 (2014)
12. N. Hadjsaid, J.-F. Canard, F. Dumas, Dispersed generation impact on distribution networks. IEEE Comput. Appl. Power **12**(2), 22–28 (1999)
13. S. Chowdhury, S.P. Chowdhury, P. Crossley, *Microgrids and Active Distribution Networks* (The Institution of Engineering and Technology, London, 2009)
14. N. Hatziargyriou, *Microgrids-Architectures and Control* (Wiley, New York, 2014)
15. N.-C. Sintamarean, F. Blaabjerg, H. Wang, F. Iannuzzo, P. de Place Rimmen, Reliability oriented design tool for the new generation of grid connected PV-inverters. IEEE Trans. Power Electron. **30**(5), 2635–2644 (2015)

16. F. Milano, F. Dorfler, G. Hug, D. Hill, G. Verbic, Foundations and challenges of low-inertia systems, in *Power Systems Computation Conference* (2018)
17. Y. Guan et al., Novel control strategies for parallel-connected inverters in AC microgrids.
18. A. Bidram, A. Davoudi, Hierarchical structure of microgrids control system. IEEE Trans. Smart Grid **3**, 1963–1976 (2012)
19. J.M. Guerrero, J.C. Vasquez, J. Matas, L.G. De Vicuna, M. Castilla, Hierarchical control of droop-controlled AC and DC microgrids — a general approach toward standardization. IEEE Trans. Ind. Electron. **58**, 158–172 (2011)
20. J.M. Guerrero, M. Chandorkar, T.-L. Lee, P.C. Loh, Advanced control architectures for intelligent microgrids—part I: decentralized and hierarchical control. IEEE Trans. Ind. Electron. **60**(4), 1254–1262 (2013)
21. M. Guerrero Josep, J.C. Vásquez, M. Savaghebi, J. de la Hoz, H. Martín, Hierarchical control of power plants with microgrid operation, in *IECON 2010 - 36th Annual Conference on IEEE Industrial Electronics Society, Glendale, AZ*, Nov 2010, pp. 3006–3011
22. M.C. Chandorkar, D.M. Divan, R. Adapa, Control of parallel connected inverters in standalone AC supply systems. IEEE Trans. Ind. Appl. **29**(1), 136–143 (1993)
23. J. Guerrero, L. de Vicuna, J. Matas, M. Castilla, J. Miret, A wireless controller to enhance dynamic performance of parallel inverters in distributed generation system. IEEE Trans. Power Electron. **19**(5), 1205–1213 (2004)
24. J.C. Vasquez, J.M. Guerrero, J. Miret, M. Castilla, L.G. de Vicuna, Hierarchical control of intelligent microgrids. IEEE Ind. Electron. Mag. **4**(4), 23–29 (2010)
25. J.A. Peas Lopes, C.L. Moreira, A.G. Madureira, Defining control strategies for microgrids islanded operation. IEEE Trans. Power Syst. **21**, 916–924 (2006)
26. M. Marwali, J.-W. Jung, A. Keyhani, Control of distributed generation systems—Part II: load sharing control. IEEE Trans. Power Electron. **19**(6), 1551–1561 (2004)
27. J.C. Vasquez, J.M. Guerrero, A. Luna, P. Rodriguez, Adaptive droop control applied to voltage-source inverters operating in grid-connected and islanded modes. IEEE Trans. Ind. Electron. **56**(10), 4088–4096 (2009)
28. F. Katiraei, M.R. Iravani, P.W. Lehn, Micro-grid autonomous operation during and subsequent to islanding process. IEEE Trans. Power Del. **20**, 248–257 (2005)
29. J. Kim, J.M. Guerrero, P. Rodriguez, R. Teodorescu, K. Nam, Mode adaptive droop control with virtual output impedances for an inverter-based flexible AC microgrid. IEEE Trans. Power Electron. **26**(3), 689–701 (2011)
30. F. Katiraei, M.R. Iravani, Power management strategies for a microgrid with multiple distributed generation units. IEEE Trans. Power Syst. **21**(4), 1821–1831 (2006)
31. F. Blaabjerg, R. Teodorescu, M. Liserre, A.V. Timbus, Overview of control and grid synchronization for distributed power generation systems. IEEE Trans. Ind. Electron. **53**, 1398–1409 (2006)
32. Y.A.-R.I. Mohamed, E.F. El-Saadany, Adaptive decentralized droop controller to preserve power sharing stability of paralleled inverters in distributed generation microgrids. IEEE Trans. Power Electron. **23**(6), 2806–2816 (2008)
33. C. Sao, P. Lehn, Autonomous load sharing of voltage source converters. IEEE Trans. Power Del. **20**(2), 1009–1016 (2005)
34. S.M. Ashabani, Y.A.-R.I. Mohamed, New family of microgrid control and management strategies in smart distribution grids–analysis, comparison and testing. IEEE Trans. Power Syst. **29**(5), 2257–2269 (2014)
35. Y. Guan, W. Wu, X. Guo, H. Gu, An improved droop controller for grid-connected voltage source inverter in microgrid, in *2nd International Symposium on Power Electronics for Distributed Generation Systems, PEDG 2010* (2010), pp. 823–828
36. J. He, Y.W. Li, An enhanced microgrid load demand sharing strategy. IEEE Trans. Power Electron. **27**(9), 3984–3995 (2012)
37. J.C. Vasquez, J.M. Guerrero, A. Luna, Adaptive droop control applied to voltage-source inverters operating in grid-connected and islanded modes. IEEE Trans. Ind. Electron. **56**(10), 4088–4096 (2009)

38. Y.W. Li, C.N. Kao, An accurate power control strategy for power-Electronic-interfaced distributed generation units operating in a low-voltage multibus microgrid. IEEE Trans. Power Electron. **24**(12), 2977–2988 (2009)
39. A. Tuladhar, H. Jin, T. Unger, Parallel operation of single phase inverter modules with no control interconnections, in *Proc. of Twelfth Annual IEEE Applied Power Electron. and Exposition* (1997), pp. 94–100
40. A. Tuladhar, H. Jin, T. Unger, Control of parallel inverters in distributed AC power systems with consideration of line impedance effect. IEEE Trans. Ind. Appl. **36**(1), 131–138 (2000)
41. J. Vasquez, J. Guerrero, J. Miret, M. Castilla, L. de Vicuãsa, Hierarchical control of intelligent microgrids. IEEE Ind. Electron. Mag. **4**, 23–29 (2010)
42. A. Micallef, M. Apap, C. Spiteri-Staines, J.M. Guerrero, Secondary control for reactive power sharing in droop-controlled islanded microgrids, in *IEEE Intern. Symp. Ind. Electron. (ISIE)*, May 2012, pp. 1627–1633
43. A. Mehrizi-Sani, R. Iravani, Potential-function based control of a microgrid in islanded and grid-connected modes. IEEE Trans. Power Syst. **25**, 1883–1891 (2010)
44. Q. Shafiee, C. Stefanović, T. Dragičević, P. Popovski, J.C. Vasquez, J.M. Guerrero, Robust networked control scheme for distributed secondary control of islanded microgrids. IEEE Trans. Ind. Electron. **61**(10), 5363–5374 (2014)
45. H. Liang, B.J. Choi, W. Zhuang, X. Shen, A.S.A. Awad, A. Abdr, Multiagent coordination in microgrids via wireless networks. IEEE Wirel. Commun. **19**, 14–22 (2012)
46. D. Niyato, L. Xiao, P. Wang, Machine-to-machine communications for home energy management system in smart grid. IEEE Commun. Mag. **49**, 53–59 (2011)
47. V.C. Gungor, D. Sahin, T. Kocak, S. Ergut, C. Buccella, C. Cecati, G.P. Hancke, Smart grid technologies: communication technologies and standards. IEEE Trans. Ind. Inf. **7**, 529–539 (2011)
48. S.D.J. McArthur, E.M. Davidson, V.M. Catterson, A.L. Dimeas, N.D. Hatziargyriou, F. Ponci, T. Funabashi, Multi-agent systems for power engineering applications—part II: technologies, standards, and tools for building multi-agent systems. IEEE Trans. Power Syst. **22**(4), 1753–1759 (2007)
49. R.C. Qiu, Z. Hu, Z. Chen, N. Guo, R. Ranganathan, S. Hou, G. Zheng, Cognitive radio network for the smart grid: experimental system architecture, control algorithms, security, and microgrid testbed. IEEE Trans. Smart Grid **2**, 724–740 (2011)
50. R. Olfati-Saber, R.M. Murray, Consensus problems in networks of agents with switching topology and time-delays. IEEE Trans. Autom. Control **49**, 1520–1533 (2004)
51. R. Olfati-Saber, J.A. Fax, R.M. Murray, Consensus and cooperation in networked multi-agent systems. Proc. IEEE **95**(1), 215–233 (2007)
52. M. Kriegleder, R. Oung, R. D'Andrea, Asynchronous implementation of a distributed average consensus algorithm, in *2013 IEEE/RSJ International Conference on Intelligent Robots and Systems (IROS)*, 3–7 Nov 2013, pp. 1836–1841
53. D. Spanos, R. Olfati-Saber, R. Murray, Dynamic consensus for mobile networks, in *IFAC World Congress* (2005)
54. L. Xiao, S. Boyd, Fast linear iterations for distributed averaging, in *Proc. 42nd IEEE Int. Conf. Decis. Control, Maui, HI*, vol. 5 (2003), pp. 4997–5002
55. A. Kaveh, *Optimal Analysis of Structures by Concepts of Symmetry and Regularity* (Springer, Wien, 2013)
56. L. Meng, T. Dragicevic, J.M. Guerrero, J.C. Vásquez, Tertiary and secondary control levels for efficiency optimization and system damping in droop controlled DC-DC converters. IEEE Trans. Smart Grid **6**(6), 2615–2626 (2015)
57. L. Meng, F. Tang, M. Savaghebi, J.C. Vasquez, J.M. Guerrero, Tertiary control of voltage unbalance compensation for optimal power quality in islanded microgrids. IEEE Trans. Energy Convers. **29**(4), 802–815 (2014)
58. M Farrokhabadi et al., Microgrid stability definitions, analysis, and examples. IEEE Trans. Power Syst. **PP.99**, 1 (2019)

59. A. Bernstein, J.L. Boudec, L. Reyes-Chamorro, M. Paolone, Realtime control of microgrids with explicit power setpoints: unintentional islanding, in *Proc. IEEE PowerTech, Eindhoven*, July 2015, pp. 1–6
60. G. Dellile, B. Francois, G. Malarange, Dynamic frequency control support by energy storage to reduce the impact of wind and solar generation on isolated power system's inertia. IEEE Trans. Sustain. Energy **3**(4), 931–939 (2012)
61. A. Borghetti, C.A. Nucci, M. Paolone, G. Ciappi, A. Solari, Synchronized phasors monitoring during the islanded maneuver of an active distribution network. IEEE Trans. Smart Grid **2**(1), 82–91 (2011)
62. NERC Transmission Issues Subcommittee and System Protection and Control Subcommittee, A technical reference paper fault-induced delayed voltage recovery. North American Electric Reliability Corporation, Princeton, NJ, Tech. Rep. 1.2, June 2009
63. J. Zhao, D. Shi, R. Sharma, C. Wang, Microgrid reactive power management during and subsequent to islanded process, in *Proc. of IEEE PES T & D Conf. Expo., Chicago, IL*, April 2014, pp. 1–5
64. J. Elizondo, P. Huang, J.L. Kirtley, M.S. Elmoursi, Enhanced critical clearing time estimation and fault recovery strategy for an inverter-based microgrid with IM load, in *Proc. of IEEE PES General Meeting, Boston, MA*, July 2016, pp. 1–5
65. R. Belkacemi, S. Zarrabian, A. Babalola, R. Craven, Experimental transient stability analysis of microgrid systems: lessons learned. IEEE Trans. Power Del. **28**(4), 2428–2436 (2013)
66. P. Kundur, *Power System Stability and Control* (McGraw-Hill Professional, New York, 1994)
67. X. Wang, F. Blaabjerg, W. Wu, Modeling and analysis of harmonic stability in an AC power-electronic-based power system. IEEE Trans. Power Electron. **29**(12), 6421–6432 (2014)
68. X. Wang, F. Blaabjerg, Z. Chen, Autonomous control of inverter-interfaced distributed generation units for harmonic current filtering and resonance damping in an islanded microgrid. IEEE Trans. Ind. Appl. **50**(1), 452–461 (2014)
69. J. He, Y.W. Li, D. Bosnjak, B. Harris, Investigation and active damping of multiple resonances in a parallel-inverter-based microgrid. IEEE Trans. Power Electron. **28**(1), 234–246 (2013)
70. X. Wang, F. Blaabjerg, M. Liserre, Z. Chen, J. He, Y. Li, An active damper for stabilizing power-electronic-based AC systems. IEEE Trans. Power Electron. **29**(7), 3318–3329 (2014)
71. M. Liserre, F. Blaabjerg, S. Hansen, Design and control of an LCL-filter based three-phase active rectifier. IEEE Trans. Ind. Appl. **41**(5), 1281–1291 (2005)
72. G. Peponides, P. Kokotovic, J. Chow, Singular perturbations and time scales in nonlinear models of power systems. IEEE Trans. Circuits Syst. **29**(11), 758–767 (1982)
73. O. Ajala, A.D. Dominguez-Garcia, P.W. Sauer, *A Hierarchy of Models for Inverter-Based Microgrids* (Springer, Berlin, 2018), pp. 307
74. J.G. Rueda-Escobedo, J.A. Moreno, J. Schiffer, Finite-time estimation of time-varying frequency signals in low-inertia power systems, in *2019 18th European Control Conference (ECC)* (IEEE, Piscataway, 2019)
75. J. Schiffer, D. Zonetti, R. Ortega, A.M. Stankovic, T. Sezi, J. Raisch, A survey on modeling of microgrids—from fundamental physics to phasors and voltage sources. Automatica **74**, 135–150 (2016)
76. A. Tayyebi, A. Anta, F. Dörfler, Hybrid angle control and almost global stability of grid-forming power converters. Preprint, arXiv:2008.07661 (2020)
77. V. Mariani, F. Vasca, J.C. Vasquez, J.M. Guerrero, Model order reductions for stability analysis of islanded microgrids with droop control. IEEE Trans. Ind. Electron. **62**(7), 4344–4354 (2015)
78. P. Vorobev, P.-H. Huang, M. Al Hosani, J.L. Kirtley, K. Turitsyn, High-fidelity model order reduction for microgrids stability assessment. IEEE Trans. Power Syst. **33**(1), 874–887 (2017)
79. P. Vorobev, P. Huang, M. Al Hosani, J.L. Kirtley, K. Turitsyn, A framework for development of universal rules for microgrids stability and control, in *IEEE Conference on Decision and Control* (2017)
80. L. Huang, H. Xin, F. Dörfler, H-control of grid-connected converters: design, objectives and decentralized stability certificates. IEEE Trans. Smart Grid **11**(5), 3805–3816 (2020)

81. J. Schiffer, R. Ortega, A. Astolfi, J. Raisch, T. Sezi, Conditions for stability of droop-controlled inverter-based microgrids. Automatica **50**(10), 2457–2469 (2014)
82. J.W. Simpson-Porco, F. Dörfler, F. Bullo, Synchronization and power sharing for droop-controlled inverters in islanded microgrids. Automatica **49**(9), 2603–2611 (2013)
83. M. Sinha, F. Dörfler, B.B. Johnson, S.V. Dhople, Uncovering droop control laws embedded within the nonlinear dynamics of Van der Pol oscillators. IEEE Trans. Control Netw. Syst. **4**(2), 347–358 (2015)
84. J.H. Chow, Slow coherency and aggregation, in *Power System Coherency and Model Reduction* (Springer, New York, 2013), pp. 39–72
85. B.K. Poolla, D. Groß, F. Dörfler, Placement and implementation of grid- forming and grid-following virtual inertia and fast frequency response. IEEE Trans. Power Syst. **34**(4), 3035–3046 (2019)
86. C. De Persis, N. Monshizadeh, Bregman storage functions for microgrid control. IEEE Trans. Autom. Control **63**(1), 53–68 (2017)
87. V. Purba, B.B. Johnson, M. Rodriguez, S. Jafarpour, F. Bullo, S.V. Dhople, Reduced-order aggregate model for parallel-connected single-phase inverters. IEEE Trans. Energy Convers. **34**(2), 824–837 (2018)
88. J. Schiffer, D. Goldin, J. Raisch, T. Sezi, Synchronization of droop-controlled microgrids with distributed rotational and electronic generation, in *IEEE Conference on Decision and Control* (2013)
89. Y.W. Li, C.N. Kao, An accurate power control strategy for power-electronics-interfaced distributed generation units operating in a low-voltage multibus microgrid. IEEE Trans. Power Electron. **24**(12), 2977–2988 (2009)
90. A. Engler, N. Soultanis, Droop control in LV-grids, in *2005 International Conference on Future Power Systems*, 18 Nov 2005, p. 6
91. S.J. Chiang, C.Y. Yen, K.T. Chang, A multimodule parallelable series-connected PWM voltage regulator. IEEE Trans. Ind. Electron. **48**(3), 506–516 (2001)
92. T.L. Vandoorn, B. Meersman, J.D.M. De Kooning, L. Vandevelde, Directly-coupled synchronous generators with converter behavior in islanded microgrids. IEEE Trans. Power Syst. **27**(3), 1395–1406 (2012)
93. K. De Brabandere, B. Bolsens, J. Van den Keybus, et al., A voltage and frequency droop control method for parallel inverters. IEEE Trans. Power Electron. **22**(4), 1107–1115 (2007)
94. J.M. Guerrero, J. Matas, L.G. de Vicuna, M. Castilla, J. Miret, Decentralized control for parallel operation of distributed generation inverters using resistive output impedance. IEEE Trans. Ind. Electron. **54**(2), 994–1004 (2007)
95. W. Yao, M. Chen, J.M. Guerrero, Z.-M. Qian, Design and analysis of the droop control method for parallel inverters considering the impact of the complex impedance on the power sharing. IEEE Trans. Ind. Electron. **58**(2), 576–588 (2011)
96. H. Laaksonen, P. Saari, R. Komulainen, Voltage and frequency control of inverter based weak LV network microgrid, in *Proc. 2005 Int. Conf. Future Power Systems, Amsterdam*, 18 Nov 2005
97. T.L. Vandoorn, B. Meersman, L. Degroote, B. Renders, L. Vandevelde, A control strategy for islanded microgrids with DC-Link voltage control. IEEE Trans. Power Delivery **26**(2), 703–713 (2011)
98. J. Hu, J. Zhu, D.G. Dorrell, J.M. Guerrero, Virtual flux droop method—a new control strategy of inverters in microgrids. IEEE Trans. Power Electron. **29**(9), 4704–4711 (2014)
99. A.H. Hajimiragha, M.R. Dadash Zadeh, S. Moazeni, Microgrids frequency control considerations within the framework of the optimal generation scheduling problem. IEEE Trans. Smart Grid **6**(2), 534–547 (2015)
100. K. Christakou, J.Y. LeBoudec, M. Paolone, D.C. Tomozei, Efficient computation of sensitivity coefficients of node voltages and line currents in unbalanced radial electrical distribution networks. IEEE Trans. Smart Grid **4**(2), 741–750 (2013)
101. M. Farrokhabadi, C.A. Cañizares, K. Bhattacharya, Frequency control in isolated/islanded microgrids through voltage regulation. IEEE Trans. Smart Grid **8**(3), 1185–1194 (2015)

102. M. Diaz-Aguilo, J. Sandraz, R. Macwan, F. de Leon, D. Czarkowski, C. Comack, D. Wang, Field-validated load model for the analysis of CVR in distribution secondary networks: energy conservation. IEEE Trans. Power Del. **28**(4), 2428–2436 (2013)
103. G. Delille, L. Capely, D. Souque, C. Ferrouillat, Experimental validation of a novel approach to stabilize power system frequency by taking advantage of load voltage sensitivity, in *Proc. of IEEE PowerTech., Eindhoven*, June 2015
104. N. Pogaku, M. Prodanovic, T.C. Green, Modeling, analysis and testing of autonomous operation of an inverter-based microgrid. IEEE Trans. Power Electron. **22**(2), 613–625 (2007)

Part I
Parallel-Type Microgrid System

Chapter 2
Unified Droop Control Under Different Impedance Types

2.1 Different Droop Control Under Different Impedance Types

Microgrid is a future trend of integrating renewable generation units in a distribution energy system, which consists of various inverter-based distributed generators (DGs). In microgrid, the voltage/frequency stability and accurate load sharing are two important tasks. The droop control is the basic scheme to achieve the active and reactive power sharing without using communication channels. Based on the type of system impedance, Table 2.1 establishes the linear relationship for each impedance type based on the approximate expressions of P and Q. Hence, the appropriate droop relation for a known system impedance can be derived from this table.

The system impedance is comprised of the line impedance and the inverter output impedance. The low-voltage rural distribution networks have a high R/X ratio, whereas the inverter output impedance is mainly inductive due to the output LC or LCL filters. The system impedance is therefore subject to change based on the microgrid control and operation.

Virtual impedance method is early introduced to shape desired output impedances in uninterruptible power systems [1]. Then, it is widely utilized to decouple P–Q and eliminate reactive power differences in microgrid due to the line impedance mismatch [2, 3].

The angle droop control is developed to ensure proper load sharing in a rural distribution network with highly resistive lines [4]. As it directly regulates the converter output voltage angle, a significant steady-state frequency drop is avoided.

The conventional P–ω frequency droop control is to achieve power sharing in parallel inverters without communication [5]. The basic idea of this control manner is to mimic the behavior of synchronous generators [6]. In addition, a larger value of droop gains improves power sharing accuracy, but it increases the deviation of frequency/voltage from their normal values, resulting in a tradeoff [7].

Y. Sun et al., *Series-Parallel Converter-Based Microgrids*, Power Systems,
https://doi.org/10.1007/978-3-030-91511-7_2

Table 2.1 Droop relations for various impedance types

Impedance types	P–Q relations	Droop laws
L, $\theta_{line} \approx 90°$	$\begin{cases} P \simeq V_g V_i / X \\ Q \simeq V_g \left(V_i - V_g\right) / X \end{cases}$	$\begin{cases} \omega_i = \omega^* - mP \\ V_i = V^* - nQ \end{cases}$
RL, $\theta_{line} \approx 45°$	$\begin{cases} P - Q \simeq \sqrt{2} V_g V_i / Z \\ P + Q \simeq \sqrt{2} V_g \left(V_i - V_g\right) / Z \end{cases}$	$\begin{cases} \omega_i = \omega^* - m\left(P - Q\right) \\ V_i = V^* - n\left(P + Q\right) \end{cases}$
R, $\theta_{line} \approx 0°$	$\begin{cases} P \simeq V_g \left(V_i - V_g\right) / R \\ Q \simeq V_g V_i / R \end{cases}$	$\begin{cases} V_i = V^* - nP \\ \omega_i = \omega^* + mQ \end{cases}$
RC, $\theta_{line} \approx -45°$	$\begin{cases} P + Q \simeq -\sqrt{2} V_g V_i / Z \\ P - Q \simeq \sqrt{2} V_g \left(V_i - V_g\right) / Z \end{cases}$	$\begin{cases} \omega_i = \omega^* + m\left(P + Q\right) \\ V_i = V^* - n\left(P - Q\right) \end{cases}$
C, $\theta_{line} \approx -90°$	$\begin{cases} P \simeq -V_g V_i / X \\ Q \simeq -V_g \left(V_i - V_g\right) / X \end{cases}$	$\begin{cases} \omega_i = \omega^* + mP \\ V_i = V^* + nQ \end{cases}$

2.2 Basic Droop Control

With different purposes in microgrid, basic droop control can be divided into three categories: virtual impedance method, angle droop control, and frequency droop control. Sometimes they produce similar effects: (1) both virtual impedance and angle droop control are practicable to the highly resistive lines of microgrid and (2) the reactive power sharing can be ameliorated by regulating virtual impedance and Q–V droop gain, respectively. To explain these phenomena, the analogous relationships among them are discussed.

2.2.1 *Fundamental Concept of Frequency Droop*

The conventional frequency droop control is expressed as follows in the inductive wires of AC microgrid [5].

$$\omega_r = \omega^* - mP\,,\, m \leq \frac{\omega_{\max} - \omega_{\min}}{P^*} \tag{2.1}$$

$$V_r = V^* - nQ\,,\, n \leq \frac{V_{\max} - V_{\min}}{Q^*} \tag{2.2}$$

where ω_r and V_r are the angular frequency and voltage amplitude references of a voltage source inverter (VSI), respectively. ω^* and V^* represent values of ω and V at no load, and m and n are droop gains of P–ω and Q–V, respectively. P^* and Q^* are the rated active and reactive power. ω_{max} and ω_{min} are maximum and minimum values of the allowable angular frequency, and V_{max} and V_{min} are maximum and minimum values of the allowable voltage amplitude, respectively. For a system with

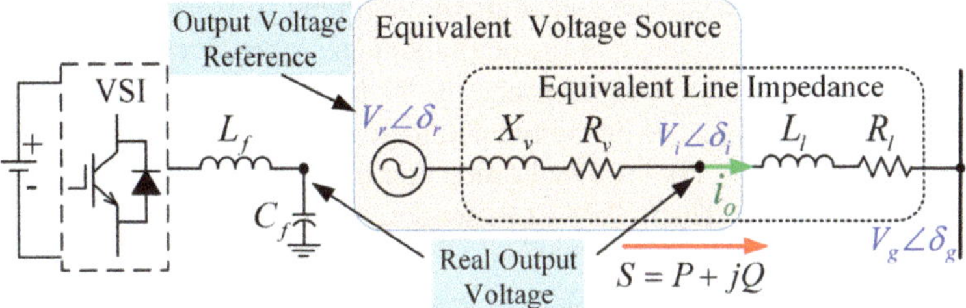

Fig. 2.1 Equivalent output voltage source considering virtual impedance

the same steady-state operation frequency, the P–ω droop coefficients of multiple DGs should satisfy the relationship ($m_1 P_1^* = m_2 P_2^* = \cdots = m_n P_n^*$) [5]. Thus, it can be ensured that load demands are taken up among DGs in proportion to their power ratings.

2.2.2 *Equivalence of Virtual Impedance and Angle Droop*

The virtual impedance method is used to shape the output impedance of VSI, as shown in Fig. 2.1. It drops the output voltage reference proportionally to the output current

$$v_i = v_r - Z_v i \tag{2.3}$$

where $Z_v = R_v + jX_v$ is the virtual impedance. $v_i = V_i\angle\delta_i$ and i are the output voltage and current, respectively. $v_r = V_r\angle\delta_r$ is the voltage reference of voltage–current dual closed loop.

According to Fig. 2.1, we have

$$V_i\angle\delta_i\left(\frac{V_r\angle\delta_r - V_i\angle\delta_i}{R_v + jX_v}\right)^* = P + jQ \tag{2.4}$$

By substituting output power for output current in (2.3), power flowing through virtual impedance yields the associated voltage drop ΔV and phase angle difference δ_v. Simplifying (2.4) yields the following equations:

$$\Delta V = V_r - V_i \cong \frac{R_v P + X_v Q}{V_i} \tag{2.5}$$

$$\delta_v = \delta_r - \delta_i \cong \frac{X_v P - R_v Q}{V_i V_r} \tag{2.6}$$

where V_r and δ_r are magnitude and angle of the reference voltage, respectively. V_i and δ_i are magnitude and angle of the output voltage, respectively.

For simplicity, V_r and V_i are replaced by V^* because their voltage magnitudes lie in the acceptable range of the nominal voltage deviation. Moreover, to meet the application condition of conventional droop control, virtual impedance is designed as pure inductance [8]. Then, (2.7)–(2.8) are derived from (2.5)–(2.6)

$$\delta_i = \delta_r - m_d P \tag{2.7}$$

$$V_i = V_r - n_d Q \tag{2.8}$$

where

$$m_d = \frac{X_v}{V*^2};\, n_d = \frac{X_v}{V*} \tag{2.9}$$

From (2.7)–(2.8), virtual inductance is regarded as a P–δ and Q–V feedback control. Especially, (2.7) is equivalent to angle droop in [4], and (2.8) is the conventional Q–V droop control. Reference [4] has proved that larger coefficients m_d and n_d can greatly improve the power sharing. Actually, it means that a larger virtual inductance is adopted to ameliorate line impedance mismatch. This equivalence provides a physical-based insight to tune the parameters of angle droop control.

2.2.3 *Analogy Between Angle Droop and Frequency Droop*

By taking the derivative from the both sides of (2.7), the equivalent character of virtual inductance is given by

$$\omega_i = \omega_r - m_d \frac{dP}{dt} \tag{2.10}$$

where ω_r is the angular frequency reference. Usually, a derivative term of active power is replaced by a high-pass filter to suppress interference. Thus, (2.10) takes the form

$$\omega_i = \omega_r - \frac{m_d s}{s + \omega_c} P \tag{2.11}$$

where ω_c is the cutoff frequency of the high-pass filter.

From (2.11), virtual inductance method can be viewed as a special P–ω frequency droop control, whose droop gain is a washout high-pass filter [9]. In contrast to the static feedback of (2.1), the washout filter-based active power sharing does not cause the frequency deviation. In addition, it should be noted that the washout filter-based reactive power sharing in [9] cannot improve the reactive power sharing.

2.3 Unified Droop Control Under Different Impedance Types

2.3.1 *Unified Droop Control*

Usually, virtual inductance and frequency droop control are simultaneously adopted. Therefore, a modified droop control is presented as follows by substituting (2.1)–(2.2) into (2.8)–(2.10):

$$\omega_i = \omega^* - mP - m_d \frac{dP}{dt} \tag{2.12}$$

$$V_i = V^* - (n + n_d)Q \tag{2.13}$$

Clearly, the P–ω droop is changed to a PD type frequency droop control in (2.12). According to (2.13), an equivalent Q-V droop gain n_d resulting from virtual inductance is added to improve the reactive power sharing.

For the power angle dynamics, the PD type control of frequency droop (2.12) is basically equivalent to the PI type control of phase droop (2.14), which is more practical to implement without introducing external disturbance.

$$\delta_i = \delta^* - m \int_{-\infty}^{t} P d\tau - m_d P \tag{2.14}$$

2.3.2 *Small-Signal Analysis*

Small-signal analysis of (2.12) is an effective tool to reflect the power angle response. According to Fig. 2.1, the output instantaneous active power p of VSI is expressed as [1]

$$p = \frac{V_i V_g}{Z_{line}} \cos(\theta_{line} - \delta_l) - \frac{V_g^2}{Z_{line}} \cos\theta_{line} \tag{2.15}$$

where Z_{line} and θ_{line} are the magnitude and phase of the output line impedance. $V_g \angle \delta_g$ is the common bus voltage. θ_l is the power angle, expressed as

$$\delta_l = \delta_i - \delta_g \tag{2.16}$$

Using the linearized model (2.15)–(2.16), the corresponding transient model around the steady state is formed.

$$\Delta p = k_{P\delta}\Delta\delta_l \tag{2.17}$$

$$\Delta\delta_l = \Delta\delta_i - \Delta\delta_g = \frac{1}{s}(\Delta\omega_i - \Delta\omega_g) \tag{2.18}$$

where $k_{P\delta} = \partial p/\partial\delta_l$ is a differential coefficient.

In the small-signal modeling, the low-pass power filter is usually added to obtain the average active power and avoid the external disturbance from the power derivation [10–12]. The output characteristic of modified droop in (2.12) is given by

$$\Delta\omega_i = \Delta\omega^* - \frac{m + m_d s}{\tau s + 1}\Delta p \tag{2.19}$$

Substituting (2.19) in (2.17)–(2.18) yields

$$\Delta p = \frac{(\tau s + 1)k_{P\delta}}{\tau s^2 + (1 + m_d k_{P\delta})s + mk_{P\delta}}(\Delta\omega^* - \Delta\omega_g) \tag{2.20}$$

For a typical second-order model of characteristic equation in (2.20), the damping ratio ζ is obtained in (2.21) and usually chosen between 0.4 and 0.8 [12].

$$0.4 \le \zeta = \frac{1 + m_d k_{P\delta}}{2\sqrt{mk_{P\delta}\tau}} \le 0.8 \tag{2.21}$$

By tuning parameters, τ and m_d, the transient response can be regulated appropriately without compromising steady state. The function of a derivative feedback is to enhance dynamic stability.

To design the droop gains (m, n, m_d, n_d) properly, the issues such as proportional power sharing, voltage quality, and stability should be considered.

1. When choosing the proportional P-ω droop coefficient m, the real power sharing is guaranteed among DGs according to their power ratings [5]. Meanwhile, the frequency deviation in (2.1) should be limited in a feasible range.
2. The coefficient m_d should be designed according to (2.21) for a better dynamic response.
3. The coefficients n and n_d have the same effect on improving reactive power sharing. But, there is a tradeoff between the reactive power accuracy and the voltage deviation [7]. The voltage deviation in (2.2) should be limited in the feasible range.
4. At last, the designed droop gains must guarantee stability. For more details, please refer to references [10, 13, 14].

Furthermore, as virtual inductance only provides one degree of freedom (DOF) in (2.9), m_d and n_d are dependent. Therefore, transient response and reactive power sharing cannot be separately regulated by virtual inductance. Alternatively, the modified droop control in (2.12)–(2.13) should be adopted.

2.4 Simulation Results

To verify the unified control law between the conventional droop control with virtual inductance and the modified droop control (2.12)–(2.13), the control scheme and simulation model with three parallel-connected DGs are built in Fig. 2.2. The reference voltage at no load is 220 V/50 Hz.

Firstly, the frequency droop control (2.1)–(2.2) with gains is tested. As shown in Fig. 2.3, the dynamic responses of the power have several cyclical oscillations and the system arrives to the steady state within a second. In addition, the accuracy of reactive power sharing is low due to the mismatch of the line impedance.

Secondly, to overcome the drawbacks of conventional frequency droop control, a virtual reactance of 0.9 Ω is added in Fig. 2.2. From simulation results in Fig. 2.4, the active power has a fast and good response during load change. And the accuracy of reactive power sharing is acceptable.

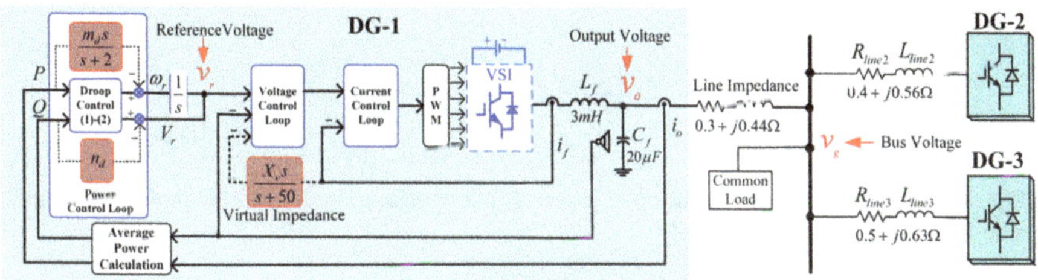

Fig. 2.2 Control schematic and test model of simulations in MATLAB/Simulink

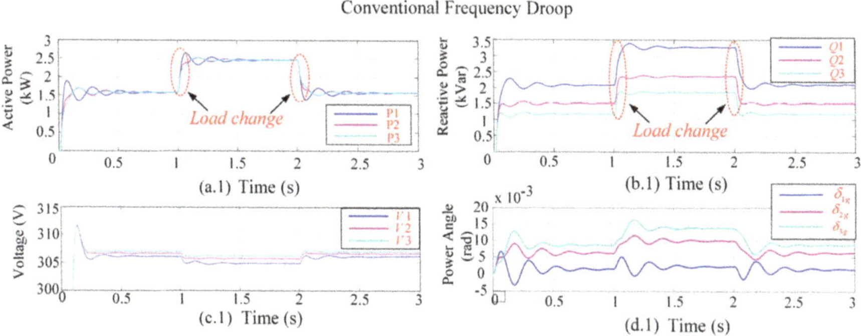

Fig. 2.3 (**a**) Active power, (**b**) reactive power, (**c**) voltage amplitude, and (**d**) power angle under conventional frequency droop control

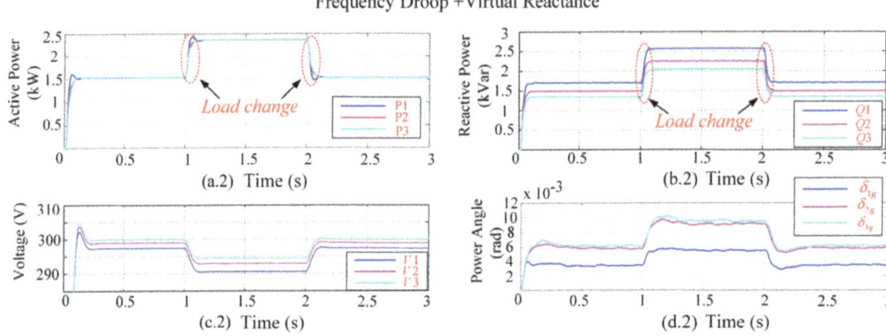

Fig. 2.4 (**a**) Active power, (**b**) reactive power, (**c**) voltage amplitude, and (**d**) power angle under frequency droop and virtual reactance control

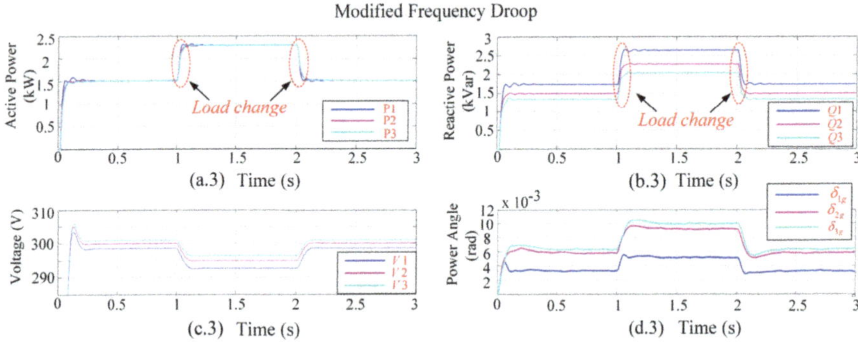

Fig. 2.5 (**a**) Active power, (**b**) reactive power, (**c**) voltage amplitude, and (**d**) power angle under modified frequency droop control

Finally, the modified droop control (2.12)–(2.13) is adopted to verify the inherent analogy with virtual inductance method. According to (2.9), $m_d = 1 \times 10^{-5}$ and $n_d = 3 \times 10^{-3}$. Figure 2.5 reveals that the modified droop (2.12)–(2.13) has the equivalent functions to the frequency droop plus virtual inductance. Compared with the second case of virtual inductance, m_d and n_d are dependent, and the system transient response of and reactive power sharing can be regulated, respectively.

2.5 Experimental Results

To verify the advantages of the modified method (2.12), a microgrid prototype is built based on two single-phase inverters. The experiment parameters are listed in Table 2.2. Similar to the simulations, the frequency droop control (2.1)–(2.2), the virtual inductance method (2.3), and the modified droop method (2.12)–(2.13) are

Table 2.2 The experiment parameters

Parameter	Symbol	Value
Nominal frequency	f^*	50 Hz
Nominal voltage	V^*	311 V
Line parameters	Z_1	$0.1 + j0.18\,\Omega$
Line parameters	Z_2	$0.1 + j0.47\,\Omega$
Virtual inductance	X_v	$1.5\,\omega$
P–ω droop coefficient	m	4×10^{-3}
Q–V droop coefficient	n	0.01
Modified P–ω droop coefficient	m_d	6×10^{-4}
Modified Q–V droop coefficient	n	0.03
Load 1	Z_{L1}	$20 + j3.14\,\Omega$
Load 2	Z_{L2}	$10 + j2.51\,\Omega$

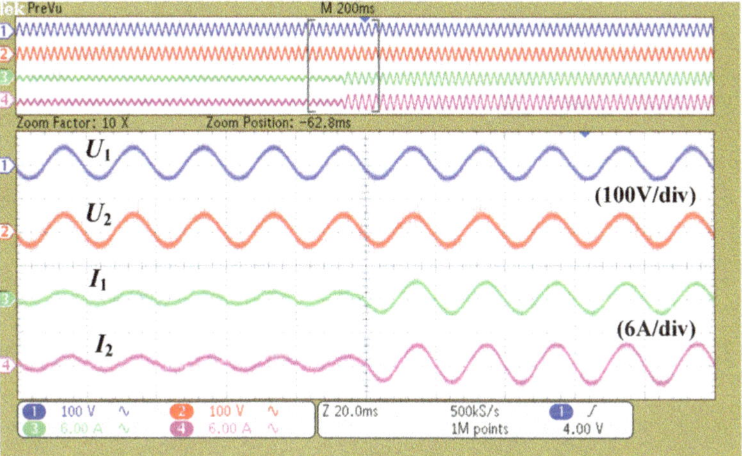

Fig. 2.6 Voltage/current waveforms in conventional frequency droop (2.1)–(2.2)

compared with experiments. During the initial stage, only load 1 is connected to the common bus. To show the transient response, load 2 is added after a while.

Figure 2.6 shows the measured voltage and current waveforms under the frequency droop control (2.1)–(2.2). The waveforms from top to bottom are the output voltage (U_1) of inverter 1, the output voltage (U_2) of inverter 2, the output current (I_1) of inverter 1, and the output current (I_2) of inverter 2, respectively. The calculated output active and reactive powers are illustrated in Fig. 2.7. As seen, an obvious power oscillation occurs during load change and there is a large reactive power sharing error.

To investigate the effects of virtual reactance, the frequency droop with virtual reactance is implemented. The experiment results are shown in Fig. 2.8. Compared with Fig. 2.7, Fig. 2.8 reveals that power oscillation has been well suppressed. And the power sharing accuracy has been improved significantly. To verify the similarity

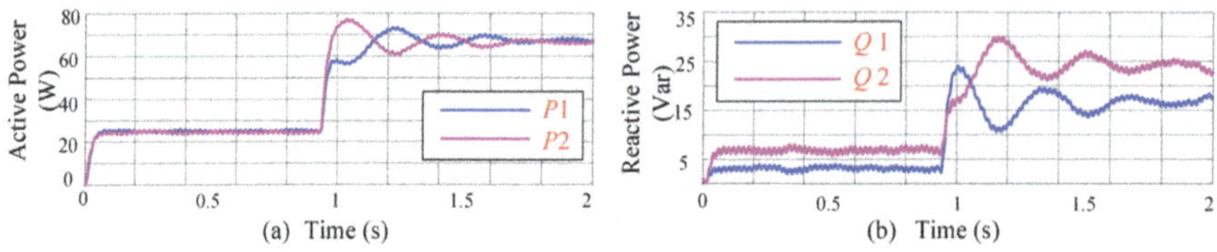

Fig. 2.7 Power response during load change in conventional frequency droop. (**a**) Active power and (**b**) reactive power

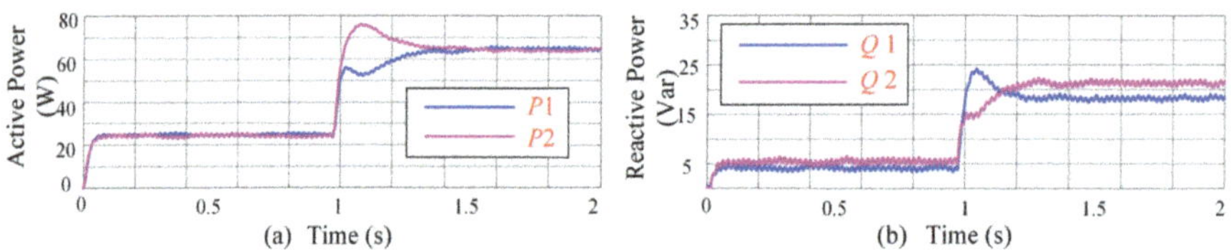

Fig. 2.8 Power response during load change in frequency droop plus virtual reactance. (**a**) Active power and (**b**) reactive power

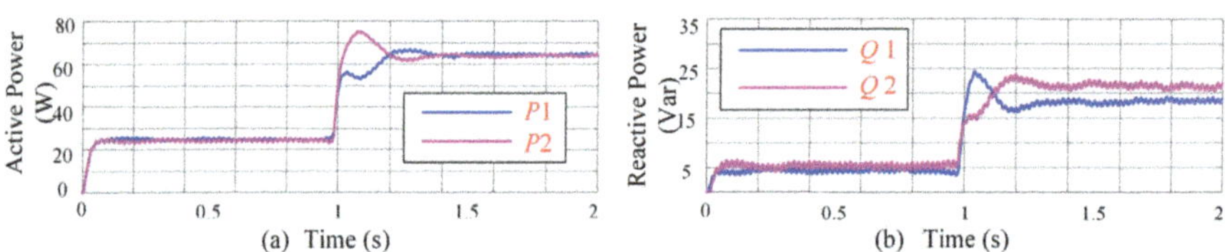

Fig. 2.9 Power response during load change in modified frequency droop. (**a**) Active power and (**b**) reactive power

between the virtual inductance method (2.3) and the modified droop control (2.12)–(2.13), md and nd are carried out according to the deduced equation (2.9). Figure 2.9a reveals that the power derivative feedback could effectively increase the power oscillation damping without reducing power sharing accuracy. Figure 2.9b shows that the added droop gain nd obtains the same reactive power sharing accuracy as the virtual reactance method of Fig. 2.8b.

2.6 Conclusion

This chapter compares the similarities and differences among three different concepts, virtual impedance method, angle droop control, and frequency droop control. Although each of them has been well researched, some new findings are established as follows: (1) the angle droop control is intrinsically a virtual inductance method, (2) virtual inductance method can also be regarded as a special frequency droop control with a power derivative feedback, and (3) the combination of virtual inductance method and frequency droop control is equivalent to the proportional–derivative (PD) type frequency droop, which is introduced to enhance the power oscillation damping. These relationships provide new insights into the design of the control methods for DGs in microgrid. Thus, the inherent relationships are established, and new insights into the controller design are provided. Finally, the modified droop control unifies these three independently developed droop control methods into a generalized theoretical framework.

References

1. J.M. Guerrero, L. GarciadeVicuna, J. Matas, Output impedance design of parallel-connected UPS inverters with wireless load-sharing control. IEEE Trans. Ind. Electron. **52**(4), 1126–1135 (2005)
2. J. He, Y. Li, Analysis, design, and implementation of virtual impedance for power electronics interfaced distributed generation. IEEE Trans. Ind. Appl. **47**(6), 2525–2538 (2011)
3. H. Mahmood, D. Michaelson, J. Jiang, Accurate reactive power sharing in an islanded microgrid using adaptive virtual impedances. IEEE Trans. Power Electron. **30**(3), 1605–1617 (2015)
4. R. Majumder, G. Ledwich, A. Ghosh, S. Chakrabarti, F. Zare, Droop control of converter-interfaced microsources in rural distributed generation. IEEE Trans. Power Del. **25**(4), 2768–2778 (2010)
5. M.C. Chandorkar, D.M. Divan, R. Adapa, Control of parallel connected inverters in standalone AC supply systems. IEEE Trans. Ind. Appl. **29**(1), 136–143 (1993)
6. S. D'Arco, J.A. Suul, Equivalence of virtual synchronous machines and frequency-droops for converter-based microgrids. IEEE Trans. Smart Grid **5**(1), 394–395 (2014)
7. J.M. Guerrero, J.C. Vasquez, J. Matas, L. G. de Vicuna, M. Castilla, Hierarchical control of droop-controlled AC and DC microgrids—a general approach toward standardization. IEEE Trans. Ind. Electron. **58**(1), 158–172 (2011)
8. M. Yazdanian, A. Mehrizi-Sani, Washout filter-based power sharing. IEEE Trans. Smart Grid **7**(2), 967–968 (2016)
9. C. Li, S.K. Chaudhary, M. Savaghebi, J.C. Vasquez, J.M. Guerrero, Power flow analysis for low-voltage AC and DC microgrids considering droop control and virtual impedance. IEEE Trans. Smart Grid **8**(6), 2754–2764 (2016)
10. E.A.A. Coelho, P.C. Cortizo, P.F.D. Garcia, Small signal stability for parallel-connected inverters in stand-alone AC supply systems. IEEE Trans. Ind. Appl. **38**(2), 533–542 (2002)
11. E.A.A. Coelho, D. Wu, J.M. Guerrero, J.C. Vasquez, T. Dragičević, Č. Stefanović, P. Popovski, Small-signal analysis of the microgrid secondary control considering a communication time delay. IEEE Trans. Ind. Electron. **63**(10), 6257–6269 (2016)
12. K. Ogata, *Modern Control Engineering [M]* (Prentice Hall PTR, Upper Saddle River, 2001), pp. 170–171
13. G. Diaz, C. Gonzalez-Moran, J. Gomez-Aleixandre, A. Diez, Scheduling of droop coefficients for frequency and voltage regulation in isolated microgrids. IEEE Trans. Power Syst. **25**(1), 489–496 (2010)
14. Y. Sun, X. Hou, J. Yang, H. Han, M. Su, J.M. Guerrero, New perspectives on droop control in AC microgrid. IEEE Trans. Ind. Electron. **64**(7), 5741–5745 (2017)

Chapter 3
Dynamic Frequency Regulation Via Adaptive Virtual Inertia

3.1 Analogy Between Droop Control and Virtual Synchronous Generator

To facilitate load sharing and improve system reliability, the conventional droop control methods are very popular in parallel inverter systems, as shown in Fig. 3.1 The frequency and magnitude of the output voltage reference depend on output active power and reactive power, respectively.

$$\omega = \omega^* - \frac{m}{\tau s + 1}(P - P^*) \tag{3.1}$$

$$V = V^* - \frac{n}{\tau s + 1}(Q - Q^*) \tag{3.2}$$

where ω^* and V^* indicate the reference values of ω and V at nominal condition, P^* and Q^* stand for the nominal power references, τ is the time constant of the low-pass filter (LPF) which filters out the averaged active and reactive powers, P and Q are output active and reactive powers, and m and n are the P–ω and Q–V droop coefficients, which are chosen as follows:

$$m = \frac{\omega_{\max} - \omega_{\min}}{P_{\max} - P_{\min}} \tag{3.3}$$

$$n = \frac{V_{\max} - V_{\min}}{Q_{\max} - Q_{\min}} \tag{3.4}$$

where $\omega_{\max}$ and $\omega_{\min}$ are the maximum and minimum values of the allowable angular frequency, $V_{\max}$ and $V_{\min}$ are the maximum and minimum values of the permissible voltage amplitude, $P_{\max}$ and $P_{\min}$ are the maximum and minimum

Y. Sun et al., *Series-Parallel Converter-Based Microgrids*, Power Systems,
https://doi.org/10.1007/978-3-030-91511-7_3

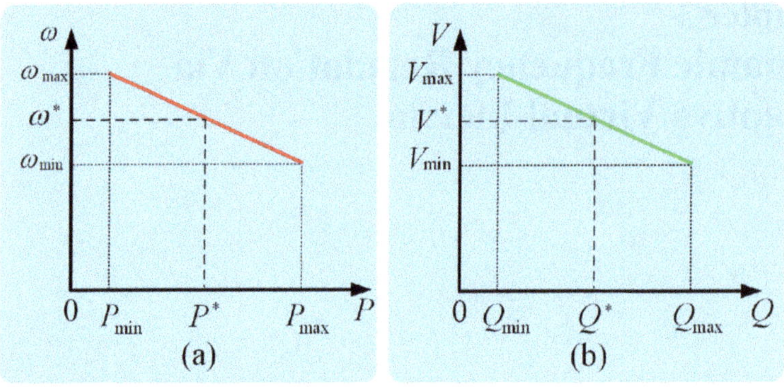

Fig. 3.1 Conventional droop characteristics for AC microgrid. (**a**) P–ω droop control. (**b**) Q–V droop control

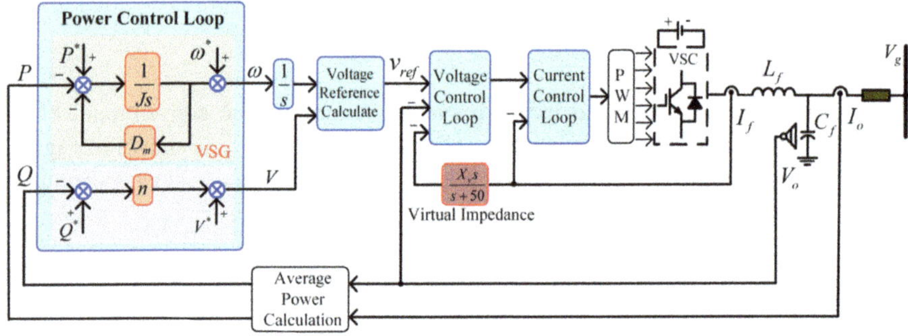

Fig. 3.2 Typical VSG control diagram of an inverter-based DG

capacities of the active power, and $Q_{\max}$ and $Q_{\min}$ are maximum and minimum capacities of the reactive power.

Rewrite (3.1) as follows:

$$\frac{\tau}{m}\frac{d(\omega-\omega^*)}{dt} = P^* - P - \frac{1}{m}(\omega-\omega^*) \tag{3.5}$$

By comparing (3.5) with the traditional 2-order swing equation of an SG, the inertia term and the damp term are equivalent to

$$J = \frac{\tau}{m}; D_m = \frac{1}{m} \tag{3.6}$$

From (3.5)–(3.6), the droop control is functionally equivalent to a VSG with a small inertia [1]. Meanwhile, inertia moment depends on the time constant of the LPF.

Figure 3.2 presents a typical VSG control scheme of inverter-based DG, which includes a power control loop and dual closed voltage–current loops. The outer power control loop includes an active power control of VSG and a reactive power droop control. A fixed virtual impedance is adopted to decouple *P/Q* and to reduce the impact of the line impedance mismatch [2]. It is implemented by using the high-pass filter instead of a pure derivative operation [2]. Moreover, virtual impedance also has an effect on the system stability, transient response, and power flow performance [3, 4].

3.2 Algorithm of Adaptive Virtual Inertia

3.2.1 Comparison Between SG and Droop-Based DG

Inertia is a measure of an object's reaction to changes. In conventional SG of power system, the rotor can slowly release rotational kinetic energy (around 10 s) when the disturbance occurs, such as, unbalanced supply–demand power. In other words, the SG has a large inertia, which implies a capability of overload and disturbance rejection. However, for a microgrid, the inverter-based DGs have a fast response speed (about 10 ms). If only conventional droop control is adopted in the inverter-based DGs, a small inertia would lead to sharp frequency variation, with load change and source uncertainty. To improve the dynamic frequency regulation, the control strategies of DGs should mimic not only the primary frequency control but also the virtual inertia control.

A relatively large inertia can decrease the frequency deviation in transient process, but the corresponding storage is required and power oscillations are triggered easily. Especially when the system operates with a large frequency deviation, it aggravates the process of frequency returning. On the other hand, system with a small inertia can react quickly to ensure the transient load sharing and ameliorate the frequency returning process, but it may lead to severe frequency deviation when load demand suddenly changes.

To fully integrate advantages of large inertia and small inertia, this chapter focuses on two issues: (1) how to design a proper value of inertia moment J according to real-time operation states? and (2) how to realize the control algorithm for DGs in a practical way to avoid a derivative action?

3.2.2 Adaptive Virtual Inertia

To address the first issue, an adaptive virtual inertia algorithm is presented in this section. Figure 3.3 shows the frequency curve deviating from the nominal steady-state value (50 Hz) and returning to nominal value under a small disturbance. The nominal steady state (50/60 Hz) is unchanged. As shown in Fig. 3.3 and Table 3.1, the system should have a slow response when the frequency deviates from the nominal reference, and thus a large inertia should be adopted. On the other hand, a

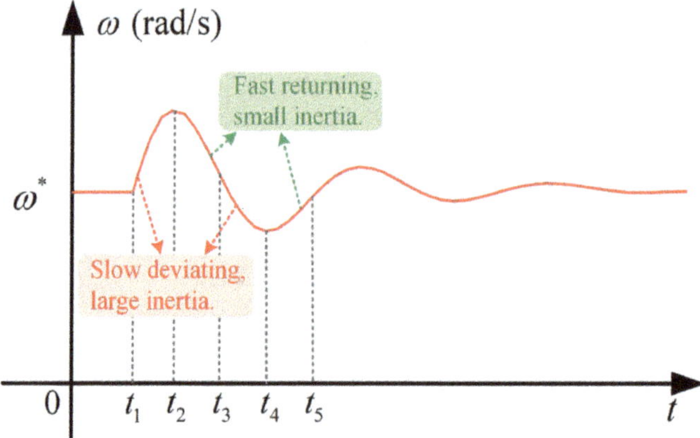

Fig. 3.3 Adaptive virtual inertia with a large inertia in frequency deviating and a small inertia in frequency returning

Table 3.1 Design principles of virtual inertia at different operation states

Segment	$\omega_s = \omega - \omega^*$	$(d\omega/dt)$	ω state	Inertia J
$t_1 - t_2$	>0	>0	Deviating	Large value
$t_2 - t_3$	>0	<0	Returning	Small value
$t_3 - t_4$	<0	<0	Deviating	Large value
$t_4 - t_5$	<0	>0	Returning	Small value

small inertia should be adopted to accelerate system dynamics when the frequency returns back the nominal frequency. To that end, a concise and unified mathematical equation of the adaptive virtual inertia is constructed as follows:

$$J = J_0 + k(\omega - \omega^*)\frac{d\omega}{dt} \tag{3.7}$$

From (3.7), the constructed inertia has two terms. The first term J_0 is the nominal constant inertia, and the second term $k(\omega - \omega^*)(d\omega/dt)$ is the adaptive compensation inertia. k is a positive inertia compensation coefficient, which can adjust the response speed of the frequency dynamic. Actually, the total moment of inertia is modified based on the relative angular velocity $(\omega - \omega^*)$ and its change rate $(d\omega/dt)$ in real time. Specially, in the nominal steady state $(\omega = \omega^*)$, the second term of adaptive compensation inertia would be 0, and the total inertia is equal to J_0.

3.2.3 Practical Control Scheme Without Derivative Action

In (3.7), it is worth noting that the adaptive inertia value would be inaccurate if we calculate it directly since the frequency derivative is sensitive to measurement noise [5]. Thus, we need to find a practical and effective method to address the second issue.

Substituting the constructed inertia (3.7) into the typical VSG control (3.5)–(3.6) yields

$$\left(J_0 + k\omega_s \frac{d\omega_s}{dt}\right) \frac{d\omega_s}{dt} = P^* - P - D_m \omega_s \tag{3.8}$$

where P^* is the nominal power reference, and $\omega_s = \omega - \omega^*$ represents the slip frequency.

Rewriting (3.8) yields

$$k\omega_s (\dot{\omega}_s)^2 + J_0 \dot{\omega}_s + D_m \omega_s = P_{rsrv} \tag{3.9}$$

$$P_{rsrv} = P^* - P \tag{3.10}$$

where P is the output active power. $P_r srv$ is the reserved power, which implies the difference between nominal power and actual output power.

Obviously, Eq. (3.9) is a quadratic equation in the variable $\dot{\omega}_s$. According to the Vieta theorem, two roots are solved

$$\dot{\omega}_s = \frac{-J_0 \pm \sqrt{J_0{}^2 - 4k\omega_s (D_m \omega_s - P_{rsrv})}}{2k\omega_s} \tag{3.11}$$

As both $\omega_s(\dot{\omega}_s) > 0$ and $\omega_s(\dot{\omega}_s) < 0$ may exist in (3.7), only one root of (3.11) is effective, derived as follows:

$$\dot{\omega}_s = f(\omega_s, P_{rsrv}) = \frac{-J_0 + \sqrt{J_0{}^2 - 4k\omega_s (D_m \omega_s - P_{rsrv})}}{2k\omega_s} \tag{3.12}$$

To avoid the singular point ($\omega_s = 0$), rewrite (3.12) by numerator rationalization

$$\dot{\omega}_s = f(\omega_s, P_{rsrv}) = \frac{-2(D_m \omega_s - P_{rsrv})}{\sqrt{J_0{}^2 - 4k\omega_s (D\omega_s - P_{rsrv})} + J_0} \tag{3.13}$$

Then, the improved active power-frequency ($P - \omega$) control based on adaptive virtual inertia algorithm is obtained by combining (3.9) and (3.13).

$$\omega = \omega^* + \omega_s = \omega^* + \int f(\omega_s, P_{rsrv}) dt \tag{3.14}$$

From (3.14), the angular frequency reference is a function of output active power. The detailed control scheme with adaptive virtual inertia is presented in Fig. 3.4. The control input is the real-time active power. The control output is the angular frequency reference. The control function (3.14) is derived from (3.7)–(3.8), and its design principle is shown in Table 3.1. Compared with the power loop of a typical VSG in Fig. 3.2, the control algorithm with adaptive virtual inertia is added.

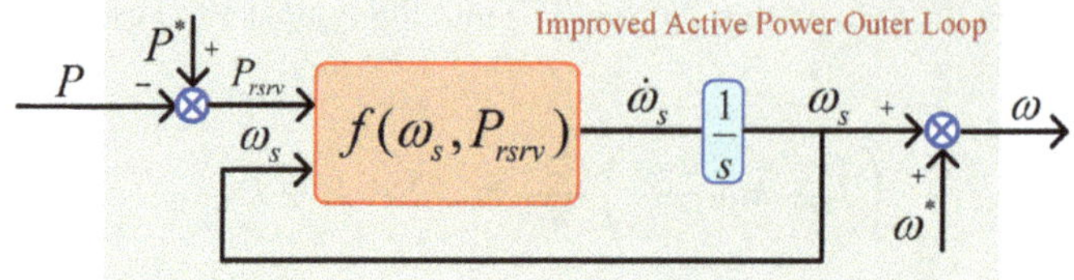

Fig. 3.4 Improved active power outer loop based on adaptive virtual inertia

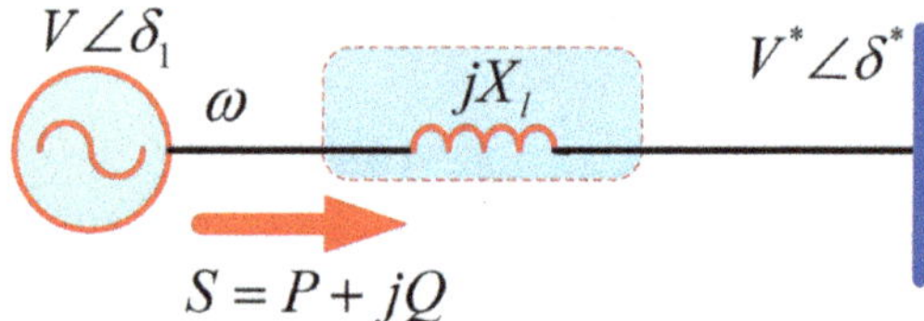

Fig. 3.5 Equivalent circuit of a DG unit connected to an infinite bus

It is worth noting that only output active power is feedback in Fig. 3.4, where the frequency derivative term is avoided. Thus, the presented control scheme is simple and practical.

3.3 Stability Proof

3.3.1 Single Inverter-Based DG in Grid-Connected Mode

The stability of the control algorithm will be investigated based on the Lyapunov stability theorem. The model of a single DG connected to an infinite bus is firstly built to study the steady and transient states [6] as shown in Fig. 3.5.

Assume that the line impedance is highly inductive [2], and the inverter-based DG is well designed with a salient timescale separation [6]. Then, the delivered power from a DG to the bus is given by (3.15).

$$P = \frac{VV^*}{X_l}\sin\delta \tag{3.15}$$

$$\delta = \delta_1 - \delta^* = \int(\omega - \omega^*)dt \tag{3.16}$$

where V, δ_1, and ω are the output voltage amplitude, angle, and angular frequency of an inverter-based DG, and V^*, δ^*, and ω^* are the voltage amplitude, angle, and angular frequency of the bus, respectively. X_l is the line reactance, and δ is the power angle.

Combining (3.9) and the power transmission characteristic (3.15) yields

$$k\omega_s(\dot{\omega}_s)^2 + J_0\dot{\omega}_s + D_m\omega_s = P^* - \frac{VV^*}{X_l}\sin\delta \tag{3.17}$$

Rewrite (3.17) in the form as

$$\begin{cases} \dot{\delta} = \omega - \omega^* = \omega_s \\ \dot{\omega}_s = \frac{1}{J_0}\left(P^* - \frac{VV^*}{X_l}\sin\delta - (D_m + k\dot{\omega}_s^2)\omega_s\right) \end{cases} \tag{3.18}$$

Note that the term of adaptive virtual inertia can also be regarded as a positive damping $k\dot{\omega}_s^2$ in (3.18). In other words, the system damping changes from original D_m to $(D_m + k\dot{\omega}_s^2)$ after using adaptive virtual inertia control.

Then, the state variables $[x_1\ x_2]^T = [(\delta - \delta_0)\ \omega_s]^T$ are chosen. Rewrite (3.18) as

$$\begin{cases} \dot{x}_1 = x_2 \\ \dot{x}_2 = a\sin\delta_0 - a\sin(x_1 + \delta_0) - b\left(D_m + k(\dot{x}_2)^2\right)x_2 \end{cases} \tag{3.19}$$

where

$$\begin{cases} \delta_0 = \arcsin\frac{P^*X_l}{VV^*} \\ a = \frac{VV^*}{J_0X_l} > 0;\ b = \frac{1}{J_0} > 0 \end{cases} \tag{3.20}$$

A candidate Lyapunov function is constructed as follows:

$$E(x) = \frac{1}{2}(x_2)^2 + a\int_0^{x_1}[\sin(x_1 + \delta_0) - \sin\delta_0]dx_1 \tag{3.21}$$

Equation (3.21) is positive definite under the condition that $-\pi \leq x_1 \leq \pi - 2\delta_0$. $\dot{E}(x)$ is obtained as

$$\dot{E}(x) = -b\left(D_m + k(\dot{x}_2)^2\right)(x_2)^2 \leq 0 \tag{3.22}$$

According to (3.22) and La Salle's Invariance Principle, we have proved that the control algorithm is convergent under the domain of attraction

$$-\pi + \delta_0 \leq \delta \leq \pi - \delta_0 \tag{3.23}$$

3.3.2 Synchronization of Multiple DGs in Islanded Mode

Multiple inverter-based DGs must synchronize with each other to guarantee the stable operation [7]. The case in islanded mode is different from that in grid-connected mode where an infinite bus is assumed. But, in islanded mode, there is an interaction among DGs, and the common bus is slack, which is determined by all DGs [3].

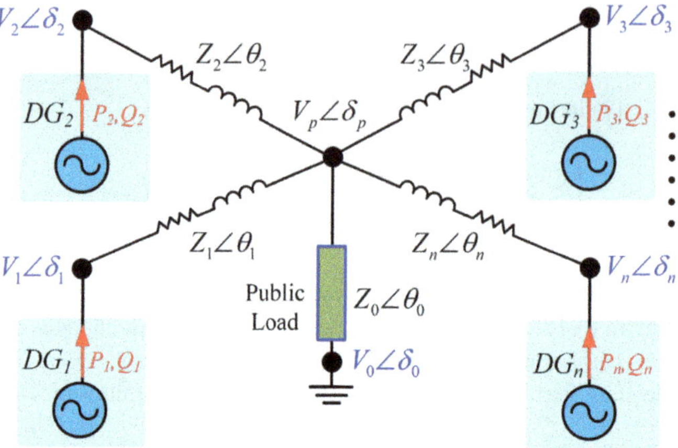

Fig. 3.6 Schematic of multiple parallel DGs with a public load

Herein, the model of multiple DGs using adaptive virtual inertia is analyzed to verify the frequency synchronization.

As shown in Fig. 3.6, V_i, δ_i, and ω_i are output voltage amplitude, angle, and angular frequency of ith DG, respectively. Z_0 and θ_0 are the load impedance amplitude and angle at the public point. Z_i and θ_i are the line impedance amplitude and angle between i-th DG and the public point. According to the power flow calculation, the output real power P_i of i-th DG can be obtained as follows:

$$P_i = \frac{V_i^2}{|Z_{ii}|}\cos\theta_{ii} - \sum_{j=1, j\neq i}^{n} \frac{V_i V_j}{|Z_{ij}|}\cos(\delta_i - \delta_j + \theta_{ij}) \tag{3.24}$$

where

$$\begin{cases} \vec{Z}_{ij} = \vec{Z}_i \vec{Z}_j \sum\limits_{k=0}^{n} (1/\vec{Z}_k);\ \vec{Z}_{ii} = 1\Big/\left(\sum\limits_{k=0, k\neq i}^{n} (1/\vec{Z}_{ik})\right) \\ \theta_{ij} = \arctan\dfrac{\mathrm{Im}(\vec{Z}_{ij})}{\mathrm{Re}(\vec{Z}_{ij})};\ \theta_{ii} = \arctan\dfrac{\mathrm{Im}(\vec{Z}_{ii})}{\mathrm{Re}(\vec{Z}_{ii})} \end{cases} \tag{3.25}$$

Usually, as the line impedance is mainly inductive ($\theta_i \approx \pi/2$) and the load impedance is far greater than the line impedance ($Z_0 \gg Z_i, i \in \{1, 2 \cdots n\}$), (3.26) can be derived from (3.25)

$$\theta_{ij} \cong \frac{\pi}{2} \tag{3.26}$$

Substituting (3.26) into (3.24) yields

$$P_i = k_{ii} + \sum_{j=1, j \neq i}^{n} k_{ij} \sin(\delta_i - \delta_j) \tag{3.27}$$

where k_{ii} and k_{ij} are positive coefficients.

$$k_{ii} = \frac{V_i{}^2}{|Z_{ii}|} \cos \theta_{ii};\ k_{ij} = \frac{V_i V_j}{|Z_{ij}|} \tag{3.28}$$

As the power angle $\delta_{ij} = \delta_i - \delta_j$ is always small [6], $\sin \delta_{ij} \cong \delta_{ij}$. Then, (3.27) can be simplified as

$$P_i = k_{ii} + \sum_{j=1, j \neq i}^{n} k_{ij} (\delta_i - \delta_j) \tag{3.29}$$

According to (3.9)–(3.10), the dynamic of the control algorithm can be accessed for i-th DG.

$$k\omega_{si} (\dot{\omega}_{si})^2 + J_0 \dot{\omega}_{si} + D_m \omega_{si} = P^* - P_i \tag{3.30}$$

where $\omega_{si} = \omega_i - \omega^*$. Set $\dot{\delta}_{si} = \omega_{si}$, and the dynamic of i-th DG is obtained by combining (3.29) and (3.30)

$$k\dot{\delta}_{si} (\ddot{\delta}_{si})^2 + J_0 \ddot{\delta}_{si} + D_m \dot{\delta}_{si} = P^* - k_{ii} - \sum_{j=1, j \neq i}^{n} k_{ij} (\delta_{si} - \delta_{sj}) \tag{3.31}$$

where

$$\delta_{si} - \delta_{sj} = \delta_i - \delta^* - (\delta_j - \delta^*) = \delta_i - \delta_j \tag{3.32}$$

For a system with n parallel DGs, the system dynamics are presented as follows:

$$\begin{cases} g(\dot{\delta}_{s1}, \ddot{\delta}_{s1}) = P^* - k_{11} - \sum\limits_{j=1, j \neq 1}^{n} k_{1j} (\delta_{s1} - \delta_{sj}) \\ g(\dot{\delta}_{s2}, \ddot{\delta}_{s2}) = P^* - k_{22} - \sum\limits_{j=1, j \neq 2}^{n} k_{2j} (\delta_{s2} - \delta_{sj}) \\ \vdots \\ g(\dot{\delta}_{sn}, \ddot{\delta}_{sn}) = P^* - k_{nn} - \sum\limits_{j=1, j \neq n}^{n} k_{nj} (\delta_{sn} - \delta_{sj}) \end{cases} \tag{3.33}$$

where

$$g(\dot{\delta}_{si}, \ddot{\delta}_{si}) = k\dot{\delta}_{si}(\ddot{\delta}_{si})^2 + J_0\ddot{\delta}_{si} + D_m\dot{\delta}_{si} \tag{3.34}$$

The form of (3.33)–(3.34) is subject to the two-way coupling configuration of Van der Pol oscillators [8, 9]. The convergence for a single DG has been proved by (3.17)–(3.23). For the coupling of multiple DGs, Eqs. (3.33)–(3.34) meet the commonly studied update rule of (3.35) in multi-agent system and nonlinear networked system [10, 11].

$$\dot{x}_i = -\sum_{j=1}^{n} a_{ij}(x_i - x_j) \tag{3.35}$$

As a result, the angles $\delta_{s1}, \delta_{s2}, \ldots, \delta_{sn}$ will converge and synchronize with each other, which means that $\omega_1 = \omega_2 = \cdots = \omega_n$ in steady state [8].

3.4 Design Guidelines for Key Control Parameters

In this section, the design guidelines for some key control parameters are given, including the droop damping coefficient D_m, inertia coefficient J_0, and inertia compensation coefficient k. Generally, the inertia moment implies a capability of the instant maximum power output. Thus, the inertia coefficient J_0 should be designed according to the power capacity of the individual inverter [12]. In addition, the coefficient D_m should be designed by the power sharing among the multiple inverters in the microgrid.

3.4.1 Design Guideline for Droop Damping Coefficient D_m

According to the droop characteristic, the system angular frequency should lie in the allowable range $[\omega_{\min}, \omega_{\max}]$. Thus, the P–ω droop coefficient m in (3.1)–(3.4) should meet

$$0 \le m \le \frac{\omega_{\max} - \omega_{\min}}{P_{\max} - P_{\min}} \tag{3.36}$$

From (3.6), that is,

$$D_m = \frac{1}{m} \ge \frac{P_{\max} - P_{\min}}{\omega_{\max} - \omega_{\min}} \tag{3.37}$$

Moreover, when choosing D_m, a general design guideline should be guaranteed to ensure the power sharing among multiple inverters according to $m_1 P_1^* = m_2 P_2^* = \cdots = m_i P_i^*$ [13].

$$D_{m1} : D_{m2} : \cdots : D_{mi} = P_1^* : P_2^* : \cdots : P_i^* \tag{3.38}$$

where P_i^* stands for the rated power capacity of DG-i.

3.4.2 Design Guideline for Inertia Coefficient J_0

Improper virtual inertia may lead to the power oscillation [1]. So it is necessary to investigate the frequency dynamic in consideration of inertia and damping function together. In the nominal steady state ($\omega = \omega^*$), the term of adaptive compensation inertia $k(\omega - \omega^*)(d\omega/dt)$ would be 0, and the total inertia J is equal to J_0 in (3.7). Neglecting the positive damping effect of adaptive compensation inertia ($k = 0$), the dynamic of the nominal steady state is obtained from (3.17)

$$J_0\ddot{\delta} + D_m\dot{\delta} + \frac{VV^*}{X_l}\sin\delta = P^* \tag{3.39}$$

Linearization of (3.39) at the steady-state point yields

$$J_o\Delta\ddot{\delta} + D_m\Delta\dot{\delta} + \frac{VV^*\cos\delta_0}{X_l}\Delta\delta = 0 \tag{3.40}$$

For a typical second-order model of (3.40), the natural frequency ω_n and damping ratio ζ are obtained as

$$\omega_n = \sqrt{\frac{VV^*\cos\delta_0}{J_0X_l}};\zeta = \frac{D_m}{2}\sqrt{\frac{X_l}{J_0VV^*\cos\delta_0}} \tag{3.41}$$

From (3.41), the damping ratio of the system depends on the operation points, the values of inertia term J_0 and damping term D_m. As $\zeta \in [0.1,\ 1.414]$ should be met to get a satisfactory transient response [5], the inertia coefficient J_0 should be chosen as follows:

$$\frac{0.125D_m^2X_l}{(V^*)^2} \le J_0 = \frac{D_m^2X_l}{4\zeta^2VV^*\cos\delta_0} \le \frac{25D_m^2X_l}{(V^*)^2} \tag{3.42}$$

3.4.3 Design Guideline for Inertia Compensation Coefficient k

In (3.13), the angular acceleration $\dot{\omega}_s$ must be a real number rather than an imaginary number to ensure the validity of the control. Hence, the following condition must hold identically:

$$J_0{}^2 - 4k\omega_s(D_m\omega_s - P_{rsrv}) \geq 0 \tag{3.43}$$

Especially, for two worst cases in Fig. 3.1,

$$\begin{cases} J_0{}^2 \geq 4k\omega_s(D_m\omega_s - P_{rsrv}); \text{ when } \omega_s = \omega^* - \omega_{\max}; \; P_{rsrv} = P^* - P_{\min} \\ J_0{}^2 \geq 4k\omega_s(D_m\omega_s - P_{rsrv}); \text{ when } \omega_s = \omega^* - \omega_{\min}; \; P_{rsrv} = P^* - P_{\max} \end{cases} \tag{3.44}$$

In the steady state as shown in Fig. 3.1, there exists

$$\begin{cases} P^* - P_{\min} = -D_m(\omega^* - \omega_{\max}) \\ P^* - P_{\max} = -D_m(\omega^* - \omega_{\min}) \end{cases} \tag{3.45}$$

Combining (3.44) and (3.45) yields

$$J_0{}^2 \geq \frac{8k(P^*_{\text{err}})^2}{D_m} \tag{3.46}$$

where P^*_{err} is the permissible maximum power error

$$P^*_{\text{err}} = \max\left\{P^* - P_{\min}, P_{\max} - P^*\right\} \tag{3.47}$$

From (3.46), the range of the inertia compensation coefficient k is given by

$$0 < k \leq \frac{D_m(J_0)^2}{8(P^*_{\text{err}})^2} \tag{3.48}$$

In (3.7), a relatively large value of compensation coefficient k is favorable to exhibit the effectiveness of adaptive inertia control. Thus, k should be chosen as an upper bound from (3.48).

3.4.4 Parameter Design to Limit Excessive RoCoF

Overfast returning of frequency may trigger the undesirable rate-of-change-of-frequency (RoCoF) protection relays of generator units [5]. Thus, the local control variable ($\dot{\omega}_s$) of RoCoF in (3.12) should be less than the permissible maximum RoCoF value $\dot{\omega}_s^{\max}$.

$$\dot{\omega}_s = \frac{-J_0 + \sqrt{J_0{}^2 - 4k\omega_s(D_m\omega_s - P_{rsrv})}}{2k\omega_s} < \dot{\omega}_s^{\max} \tag{3.49}$$

In (3.49), $\omega_s \in [\omega^* - \omega_{\max}, \omega^* - \omega_{\min}]$; $P_{rsrv} \in [P^* - P_{\max}, P^* - P_{\min}]$. Considering two worst cases where the load is switched from no load/full load to normal load, the DG frequency would have a fastest returning and the RoCoF would have a maximum value. That is, equation (3.49) should hold under the conditions: (1) $\omega_s = \omega^* - \omega_{\max}$; $P = P_{\min} \rightarrow P^*$ and (2) $\omega_s = \omega^* - \omega_{\min}$; $P = P_{\max} \rightarrow P^*$. Then, rewriting (3.49) yields

$$\bar{\omega}_s \left(D_m + k(\dot{\omega}_s^{\max})^2 \right) < J_0 \dot{\omega}_s^{\max} \tag{3.50}$$

where $\bar{\omega}_s$ is the permissible maximum frequency deviation.

$$\bar{\omega}_s = \max \left\{ \omega^* - \omega_{\min}, \omega_{\max} - \omega^* \right\} \tag{3.51}$$

According to (3.50), the coefficients D_m, J_0, and k should be synthetically designed to prevent excessive RoCoF levels.

3.4.5 *Adaptive Inertia Bound [$J_{\min}$, $J_{\max}$] to Avoid Long-Term Overcapacity of Converters*

Inertia provision is closely related to the available capacity of power sources and inverters [14]. Thus, a bound [$J_{\min}$, $J_{\max}$] of adaptive inertia value is necessary. Then, the parameter constraint is derived from (3.7)

$$J_{\min} \leq J = (J_0 + k\omega_s \dot{\omega}_s) \leq J_{\max} \tag{3.52}$$

where $J_{\max}$ is indicated by the available power capacity of converters [14]. $J_{\min}$ is indicated as the minimum value in (3.46) and (3.50) to ensure the effectiveness of control algorithm [5].

In order to guarantee the bound in (3.52), k should also meet (3.53) by combining (3.49)–(3.52).

$$0 \leq k \leq \min \left\{ \frac{J_0 - J_{\min}}{\bar{\omega}_s \dot{\omega}_s^{\max}}, \frac{J_{\max} - J_0}{\bar{\omega}_s \dot{\omega}_s^{\max}} \right\} \tag{3.53}$$

3.5 Hardware-In-Loop (HIL) Results

The adaptive virtual inertia control is verified by real-time HIL tests. The HIL system includes two sections: physical circuits and controller. The physical circuits are realized by the real-time simulator OP5600 whose time-step is 20 μs, which can accurately mimic the dynamics of the real power components. The controller is the real hardware dSPACE 1202 Microlab-Box, whose sampling frequency is 20 kHz.

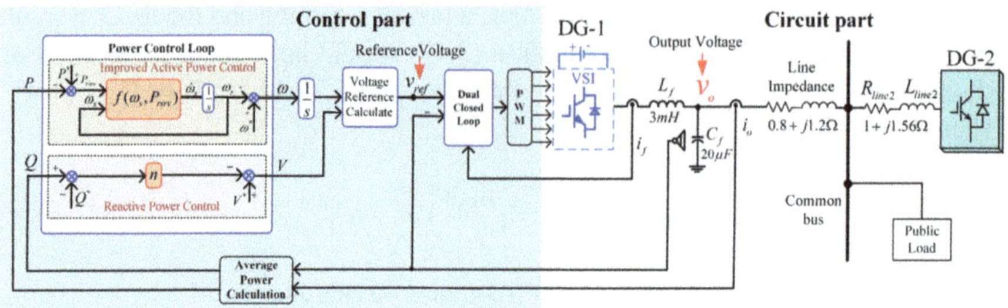

Fig. 3.7 Schematic diagram of improved power outer loop based on adaptive virtual inertia control algorithm

Table 3.2 HIL test parameters

Parameter	Symbol	Value
Nominal frequency	f^*	50 Hz
Nominal voltage	V^*	311 V
Rated active power	P^*	2 kW
Rated reactive power	Q^*	2 kvar
Virtual inductance	X_v	1.8 Ω
Power filter time constant	τ	1/60
P–ω droop coefficient	m	1/600
Q–V droop coefficient	n	0.01
Droop damp coefficient	D_m	600
Small inertia coefficient	J_0	10
Large inertia coefficient	J_0	100
Compensation coefficient	k	0.18

Figure 3.7 shows the model of two parallel DGs. The HIL parameters are listed in Table 3.2. All control parameters of two DGs are identical except different line impedances ($Z1 = 0.8 + j1.2\,\Omega$; $Z2 = 1 + j1.56\,\Omega$). The damp coefficient D_m is chosen according to (3.37)–(3.38). The small/large inertia coefficients J_0 are designed from (3.42). The inertia compensation coefficient k is calculated by (3.48). To avoid oscillation, the system is designed to be over-damped.

3.5.1 Case 1: Under Resistive Time-Varying Load

Figures 3.8 and 3.9 show the HIL results under resistive time-varying load. To verify the validity of the adaptive virtual inertia control, three groups of parameters are applied: (a) small constant inertia $J_{0-sml} = 10, k = 0$, (b) large constant inertia $J_{0-lrg} = 100, k = 0$, and (c) adaptive inertia $J_{0-adp} = 100, k = 0.18$. The first case with a small constant inertia represents the conventional droop control from (3.5)–(3.6). The second case with a large constant inertia implies the

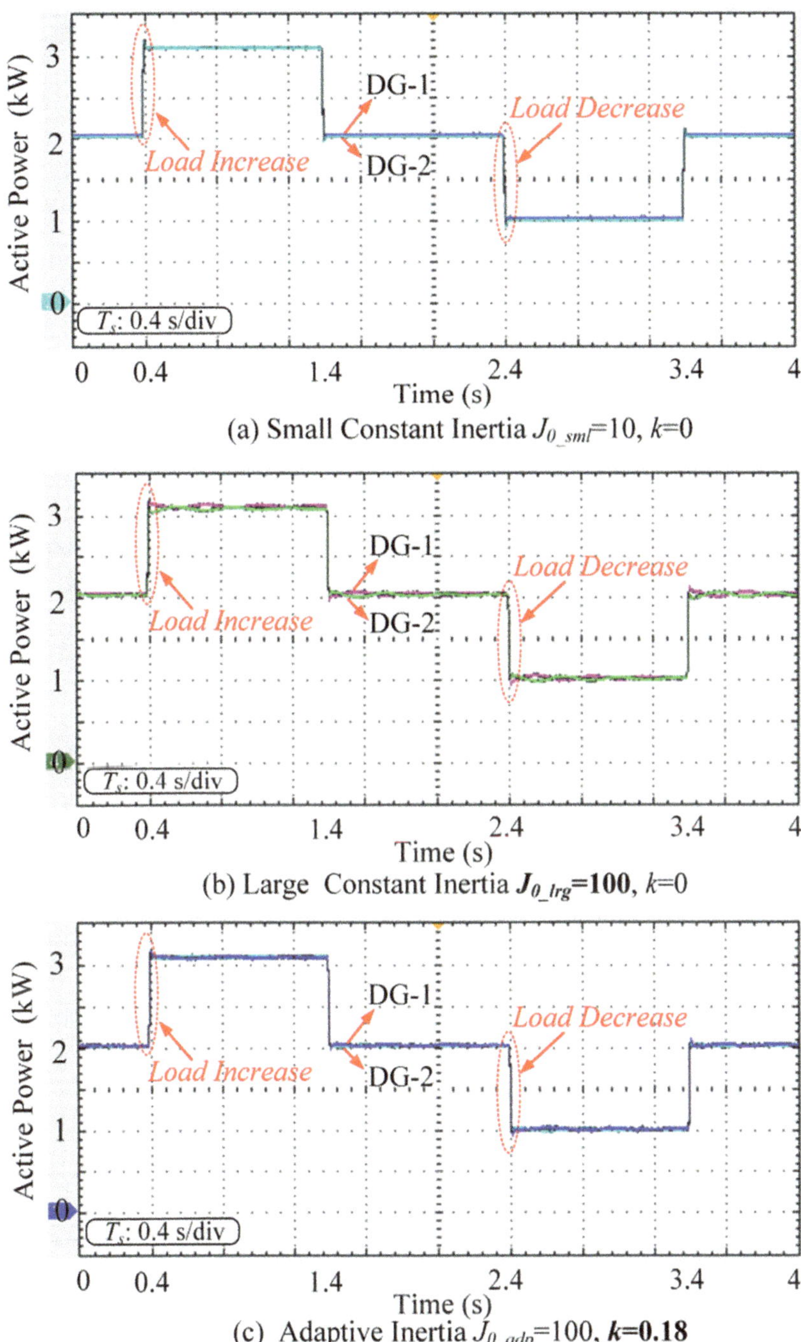

Fig. 3.8 Output active powers of two DGs under resistive time-varying load. (**a**) Small inertia ($J_{0-sml} = 10, k = 0$). (**b**) Large inertia ($J_{0-lrg} = 100, k = 0$). (**c**) Adaptive inertia ($J_{0-adp} = 100$, $k = 0.18$)

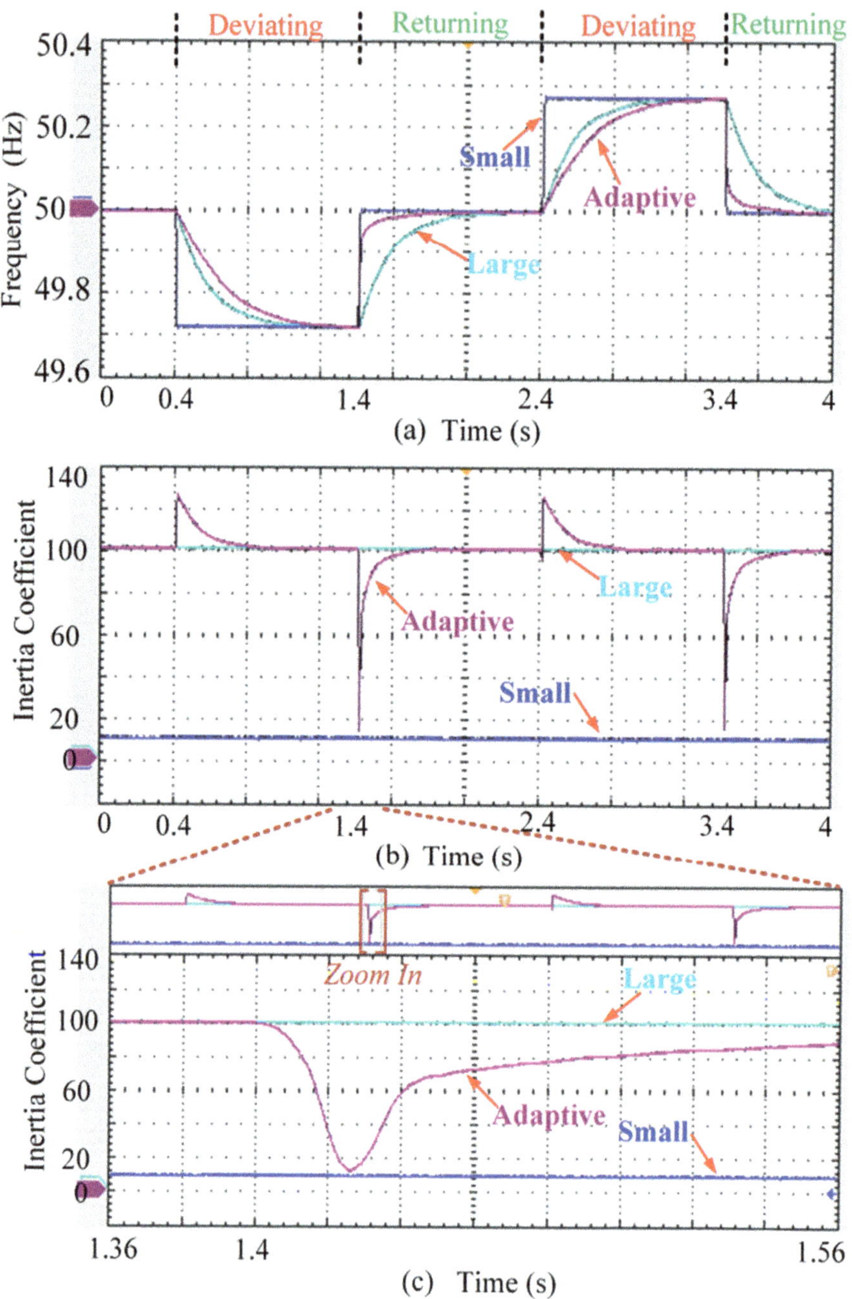

Fig. 3.9 Comparison of three tests under time-varying load. (**a**) System frequency. (**b**) Inertia value of DG-1. (**c**) Zoomed-in inertia at $t \in [1.36\,\text{s}, 1.56\,\text{s}]$

conventional VSG control. The third case with an adaptive inertia represents the control scheme.

Figure 3.8 presents the output active power of two DGs. The load demand changes every 1 s. It has an increase in 0.4 s and has a decrease in 2.4 s with respect to the normal load power 4 kW. As two DGs have the same capacity and same control parameters, the active power responses of three cases are similar in Fig. 3.8, and the accurate active power sharing is always achieved. Furthermore, it is worth noting that the active power under large constant inertia has a slight oscillation in Fig. 3.8b. However, after adopting the adaptive inertia control, the power oscillation is ameliorated in Fig. 3.8c, which reveals that adaptive virtual inertia also has a function of power oscillation damping as shown in (3.18).

Figure 3.9 illustrates the comparison of three HIL results under time-varying load. In the first case with a small constant inertia control, both the returning time and deviating time are about 0.05 s, which means that the system has a very fast response. In the second case with a large constant inertia, both the returning time and deviating time are about 0.6 s, which reveals that the system has a very slow response. When adopting the control, the system has a shorter returning time of 0.2 s and a longer deviating time of 0.75 s than that of the large inertia. That is, the system frequency can be deviated slowly with a relatively large inertia and returned quickly with a relatively small inertia in Fig. 3.9b. Moreover, it is noted that both DGs change their inertia simultaneously as they have same control parameters and power rating. Figure 3.9b just shows the inertia of DG-1 for comparisons. Since the frequency derivative term is not enabled in the control (3.13)–(3.14), the inertia regulation exhibits a smooth dynamic process from the thumbnail of Fig. 3.9c.

3.5.2 *Case 2: Under Frequent-Variation Load*

To further test the performances of the control, three comparative cases are carried out under frequent-variation load. Figure 3.10 shows the total power demand of the

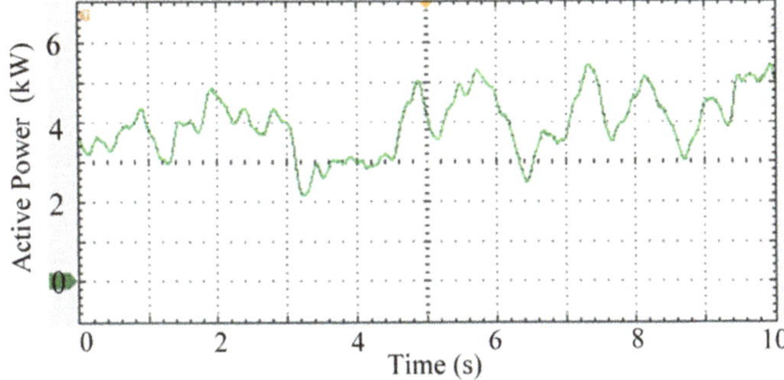

Fig. 3.10 Total load power demand under frequent-variation load

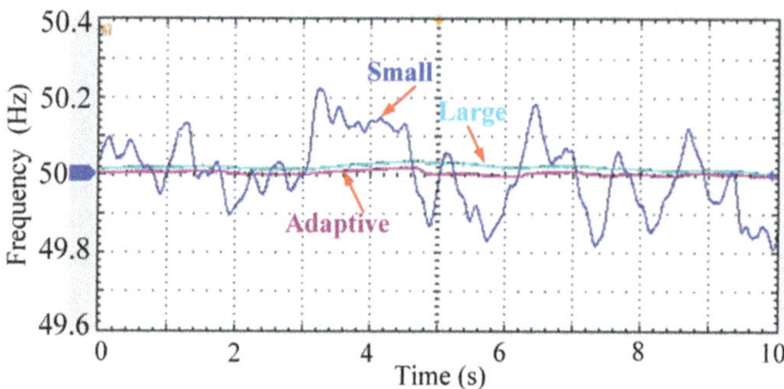

Fig. 3.11 Frequency comparison of three tests under frequent-variation load

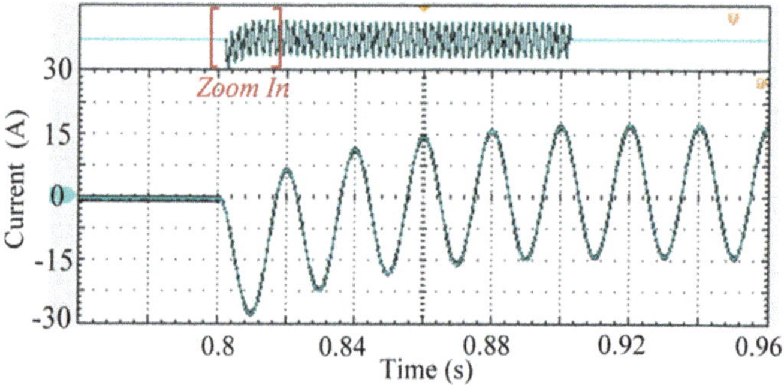

Fig. 3.12 Phase A stator current of IM at starting-up time

variable load. In the first case with a small constant inertia control, the maximum frequency deviation is about 0.22 Hz. In the second case with a large constant inertia control, the maximum frequency deviation is about 0.038 Hz. However, for the adaptive virtual inertia, the maximum frequency deviation is just about 0.016 Hz. As a result, the nominal operation frequency is guaranteed as much as possible under variable loads. Thus, compared with the conventional droop and VSG controls, the method improves the frequency nadir and dynamic response under load variation (Fig. 3.11).

3.5.3 Case 3: Under Induction Motor (IM)

In this subsection, three comparative tests are carried out under the load of squirrel-cage induction motor (IM). The inertia moment of IM is $0.089\,\mathrm{kg}\cdot\mathrm{m}^2$. The nominal power of IM is 2.2 kW. Figure 3.13 shows the total load power when IM is connected to the system at $t = 0.8$ s and is switched out at $t = 2.8$ s. Before 0.8 s, the system operates with a 4 kW resistive load. After starting-up the IM load at $t = 0.8$ s, a large inrush current is observed in Fig. 3.12, and the peak value of starting-up stator

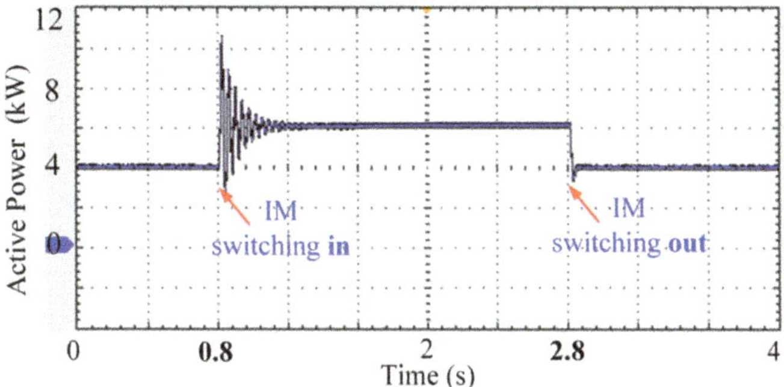

Fig. 3.13 Total load power demand under IM load

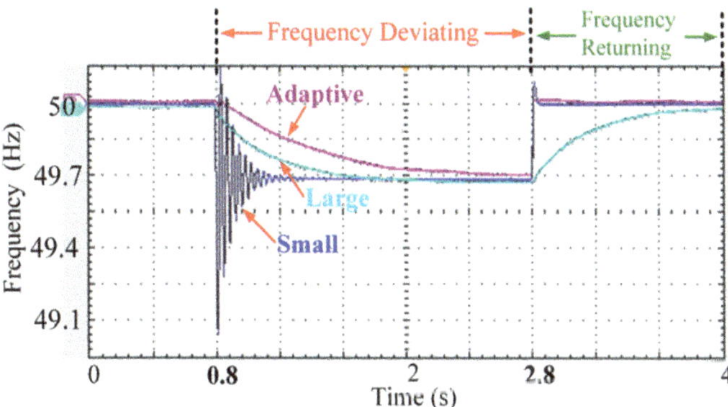

Fig. 3.14 System frequency comparison of three tests under IM load

current is almost twice as large as the nominal current. It is noted that a method of rotor series resistance is adopted to avoid an overlarge current [15] (Fig. 3.13).

Figure 3.14 shows the frequency performances of three HIL results under IM load. In the first case with a small constant inertia control, the system frequency has a very fast response and is sensitive to the load oscillation during IM starting-up. In the second case with a large constant inertia, the system frequency has a very slow response in both frequency deviating and returning process. When adopting the adaptive control, the system frequency can be deviated slowly and returned quickly in Fig. 3.14. As a result, the adaptive inertia control is still effective under IM load.

3.5.4 Case 4: Comparisons with Alternating Inertia Method

The purpose of this case is to verify the advantages of adaptive virtual inertia control strategy compared with the existing alternating inertia method. In [16],

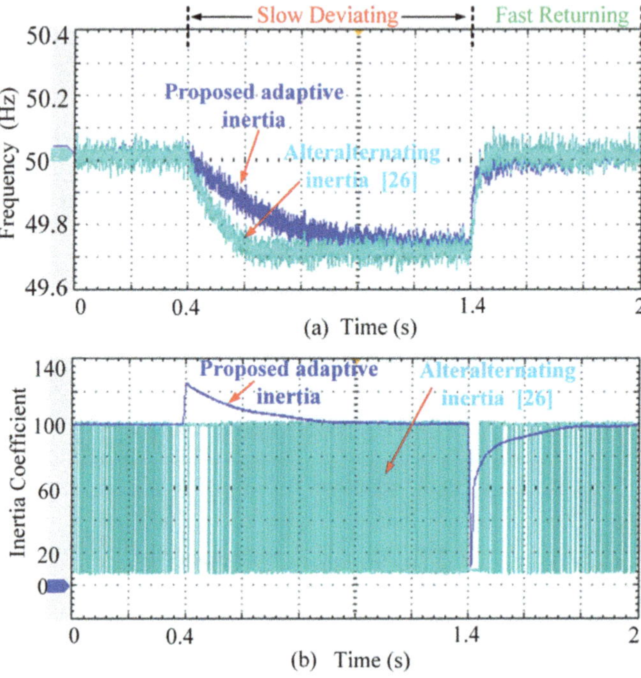

Fig. 3.15 Comparison with alternating inertia method. (**a**) System frequency with high-frequency noises. (**b**) Inertia value of DG-1

two large/small inertia values are indicated by judging states of the relative angular frequency difference ($\omega_s = \omega - \omega^*$) and its change rate ($\dot{\omega}_s$). Thus, the method in [16] has to acquire the frequency derivative (df/dt) to realize the alternating inertia. In the contrast test, the basic control parameters are set to the same with the former, such as a small constant inertia $J_{0-sml} = 10$ and a large constant inertia $J_{0-lrg} = 100$. To test the robustness performances of the adaptive virtual inertia control, high-frequency noises are imposed into the system operation frequency. The comparison results are shown in Fig. 3.15. Both the methods have a slow frequency deviating and a fast frequency returning under load changes in Fig. 3.15a. Meanwhile, the adaptive virtual inertia control method has a slightly better performance during the two dynamic processes. The main cause of this phenomenon is that the alternating inertia in Fig. 3.15b is sensitive to high-frequency noises. For example, during the frequency deviating process at t $\in$[0.4 s, 1.4 s], the alternating inertia is not always a large inertia value due to the inaccurate interference of $\dot{\omega}_s$. Similarly, during the frequency returning process at t $\in$[1.4 s, 2 s], the alternating inertia is not always a small inertia value, which has an adverse effect on dynamic response.

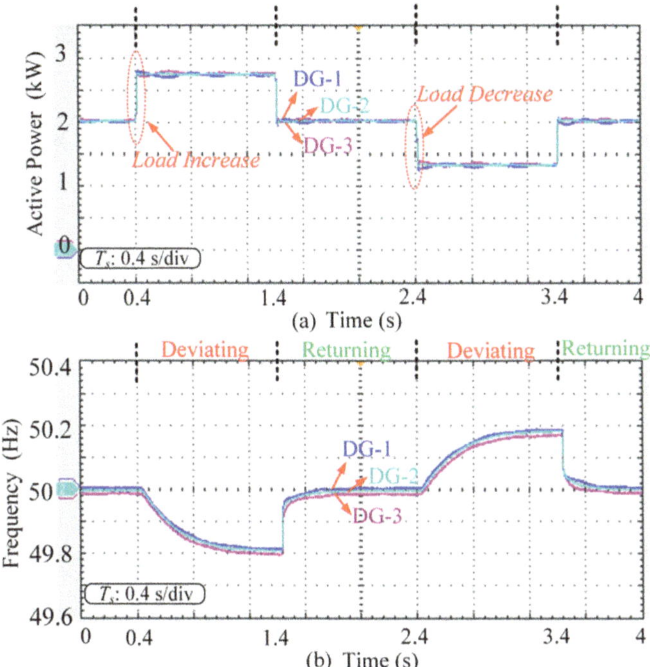

Fig. 3.16 Results of control with three DGs. (**a**) Output active powers. (**b**) Output frequencies

3.5.5 *Case 5: Adaptive Inertia Control with Three DGs*

Figure 3.16 shows the performances of adaptive inertia with three DGs. From Fig. 3.16a, the proper active power sharing is achieved among three DGs. Moreover, DGs can synchronize with each other in Fig. 3.16b, which verifies the synchronization analysis of multiple DGs in (3.30)–(3.35).

3.5.6 *Case 6: Adaptive Inertia Control with RoCoF Limitation*

To validate the practical RoCoF limitation, a permissible maximum RoCoF value of 2 Hz/s is set [5], and the coefficient k is changed from 0.18 to 0.08 in this case according to (3.50). The comparison results with/without RoCof limitation are shown in Fig. 3.17. From the frequency response in Fig. 3.17a, an overfast frequency returning is restrained after considering the RoCof limitation. Thus, an undesirable RoCoF protection tripping can be overcome in practical application. Moreover, Fig. 3.17b reveals that the inertia coefficients have a decreased value under frequency returning and an increased value under frequency deviating, which

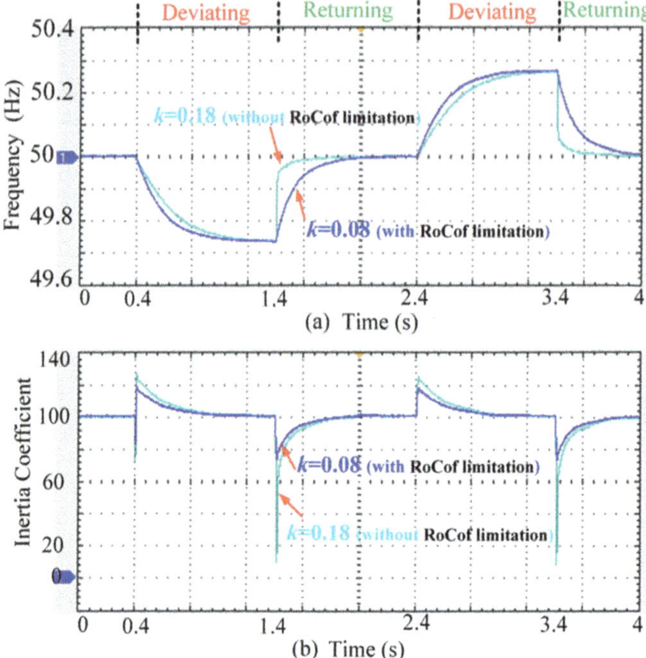

Fig. 3.17 Comparison results of control with/without RoCoF limitation. (**a**) System frequency. (**b**) Inertia value of DG-1

verifies the effectiveness of adaptive inertia control. Meanwhile, it is clear that the inertia variation range of $k = 0.08$ is smaller than that of $k = 0.18$. Particularly, for frequency returning at $t \in [1.4\,\text{s}, 2.4\,\text{s}]$ and $t \in [3.4\,\text{s}, 4\,\text{s}]$, an oversmall inertia value is prevented to guarantee a lower bound of inertia and limit excessive RoCoF [17, 18].

3.6 Conclusion

This chapter introduces an adaptive virtual inertia control algorithm to improve dynamic frequency regulation of VSG-based microgrids. The advantages of the control algorithm include (1) a concise and unified mathematical equation of the adaptive virtual inertia is constructed and (2) a practical control algorithm is presented to avoid the direct frequency derivative action. Under certain power disturbances, the control method has the advantages of both large inertia and small inertia. When the system frequency is deviating away from the nominal value, a large inertia is performed to slow the dynamic process and improve frequency nadir. When the frequency is returning to the nominal value, a small inertia is shaped to quickly accelerate system dynamics. Thus, the frequency regulation performance is greatly improved. The effectiveness of the control method is verified under three load types, including resistive time-varying load, frequent-variation load, and induction motor load. On the whole, the method supports frequency dynamics and promotes high penetrations of distributed generations.

References

1. J. Liu, Y. Miura, T. Ise, Comparison of dynamic characteristics between virtual synchronous generator and droop control in inverter-based distributed generators. IEEE Trans. Power Electron. **31**(5), 3600–3611 (2016)
2. Y. Sun, X. Hou, J. Yang, H. Han, M. Su, J.M. Guerrero, New perspectives on droop control in AC microgrid. IEEE Trans. Ind. Electron. **64**(7), 5741–5745 (2017)
3. J. He, Y.W. Li, Analysis, design, and implementation of virtual impedance for power electronics interfaced distributed generation. IEEE Trans. Ind. Appl. **47**(6), 2525–2538 (2011)
4. H. Zhang, S. Kim, Q. Sun, J. Zhou, Distributed adaptive virtual impedance control for accurate reactive power sharing based on consensus control in microgrids. IEEE Trans. Smart Grid **8**(4), 1749–1761 (2017)
5. J. Fang, H. Li, Y. Tang, F. Blaabjerg, On the inertia of future more-electronics power systems. IEEE J. Emerging Sel. Top. Power Electron. **7**(4), 2130–2146 (2019)
6. J. Machowski, J.W. Bialek, J.R. Bumby, *Power System Dynamics: Stability and Control*, 2nd edn. (Wiley, Chippenham, Wiltshire, 2008)
7. X. Hou, Y. Sun, X. Zhang, G. Zhang, J. Lu, F. Blaabjerg, A self-synchronized decentralized control for series-connected H-bridge rectifiers. IEEE Trans. Power Electron. **34**(8), 7136–7142 (2019)
8. W. Wang, J.J.E. Slotine, On partial contraction analysis for coupled nonlinear oscillators. Biol. Cybern. **92**(1),38–53 (2004)
9. M. Sinha, F. Dorfler, B.B. Johnson, S.V. Dhople, Uncovering droop control laws embedded within the nonlinear dynamics of Van der Pol Oscillators. IEEE Trans. Control Netw. Syst. **4**(2), 347–358 (2017)
10. J.W. Simpson-Porco, Q. Shafiee, F. Dorfler, J.C. Vasquez, J.M. Guerrero, F. Bullo, Secondary frequency and voltage control of islanded microgrids via distributed averaging. IEEE Trans. Ind. Electron. **61**(11), 7025–7038 (2015)
11. L. Lin-Yu, C. Chia-Chi, Consensus-based secondary frequency and voltage droop control of virtual synchronous generators for isolated AC micro-grids. IEEE J. Emerging Sel. Top. Circuits Syst. **5**(3), 443–455 (2015)
12. P.M. Anderson, A.A. Fouad, *Power System Control and Stability*, Chap. 2 (Wiley, New York, 2018), pp. 13–17
13. M.C. Chandorkar, D.M. Divan, R. Adapa, Control of parallel connected inverters in standalone AC supply systems. IEEE Trans. Ind. Appl. **29**(1), 136–143 (1993)
14. Y. Ma, W. Cao, L. Yang, F. Wang, L.M. Tolbert, Virtual synchronous generator control of full converter wind turbines with short-term energy storage. IEEE Trans. Ind. Electron. **64**(11), 8821–8831 (2017)
15. K. Ledoux, P.W. Visser, J.D. Hulin, H. Nguyen, Starting large synchronous motors in weak power systems. IEEE Trans. Ind. Appl. **51**(3), 2676–2682 (2015)
16. J. Alipoor, Y. Miura, T. Ise, Power system stabilization using virtual synchronous generator with alternating moment of inertia. IEEE J. Emerg. Sel. Topics Power Electron. **3**(2), 451–458 (2015)
17. Y.A.I. Mohamed, E.F. El-Saadany, Adaptive decentralized droop controller to preserve power sharing stability of paralleled inverters in distributed generation microgrids. IEEE Trans. Power Electron. **23**(6), 2806–2816 (2008)
18. X. Hou, Y. Sun, X. Zhang, J. Lu, P. Wang, J.M. Guerrero, Improvement of frequency regulation in VSG-based AC microgrid via adaptive virtual inertia. IEEE Trans. Power Electron. **35**(2), 1589–1602 (2020)

Chapter 4
Accurate Reactive Power Sharing

4.1 Analysis of Conventional Droop Control Method

4.1.1 Conventional Droop Control

A classic configuration of a microgrid that consists of multiple DG units and dispersed loads is shown in Fig. 4.1. The microgrid is connected to the utility through a static transfer switch at the point of common coupling (PCC). Each DG unit is connected to the microgrid through power electronic converter and its respective feeder.

Figure 4.2 shows the equivalent model of a DG unit, which is interfaced to the common bus of the AC microgrid through a power inverter with an output LCL filter. As shown in Fig. 4.2, $V_i = V_i \angle \delta_i$ is the voltage across the filter capacitor, and $V_{\text{pcc}} \angle 0°$ is the common AC bus voltage. Compared with the inductance of the LCL filter, the line resistance can be ignored. Then, the impedance between inverter and the common bus can be described as X_{line-i} ($X_{line-i} = \omega L_{line-i}$).

According to the equivalent circuit in Fig. 4.2, the inverter output apparent power is S_i, and it can be given by

$$S_i = P_i + jQ_i = \frac{V_i V_{\text{pcc}}}{X_{line-i}} \sin \delta_i + j\left[\frac{V_i V_{\text{pcc}} \cos \delta_i - V_{\text{pcc}}^2}{X_{line-i}}\right] \tag{4.1}$$

From (4.1), the output active and reactive powers of the DG units are shown as

$$\begin{cases} P_i = \frac{V_i V_{\text{pcc}}}{X_{line-i}} \sin \delta_i \\ Q_i = \frac{V_i V_{\text{pcc}} \cos \delta_i - V_{\text{pcc}}^2}{X_{line-i}} \end{cases} \tag{4.2}$$

Y. Sun et al., *Series-Parallel Converter-Based Microgrids*, Power Systems,
https://doi.org/10.1007/978-3-030-91511-7_4

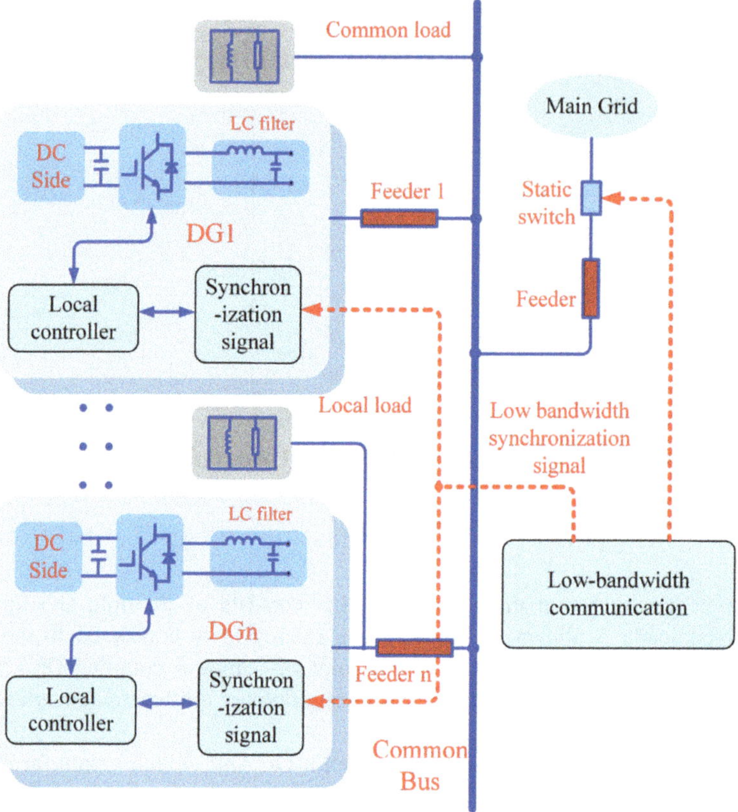

Fig. 4.1 Lustration of the AC microgrid configuration

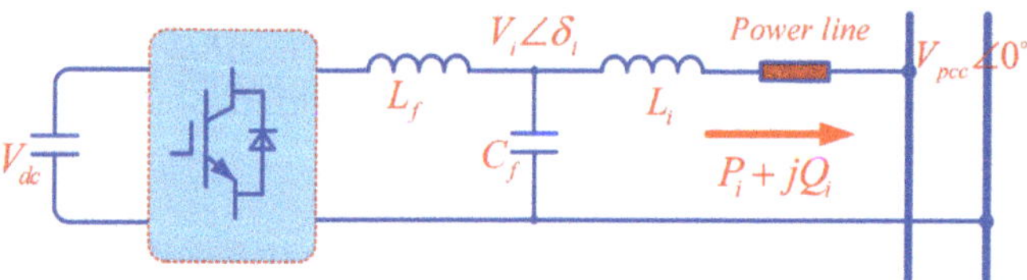

Fig. 4.2 Model of a DG unit

Usually, the phase shift angle δ_i is small. Therefore, the real power P_i and reactive power Q_i of each DG can be regulated by δ_i and the output voltage amplitude V_i, respectively [1]. Then, the conventional droop control is given by

$$\begin{cases} \omega_i = \omega^* - m_i \cdot P_i \\ V_i = V^* - n_i \cdot Q_i \end{cases} \tag{4.3}$$

where ω^* and V^* are the nominal values of DG angular frequency and DG output voltage amplitude, and m_i and n_i are the active and reactive droop slopes, respectively. P_i and Q_i are the measured averaged real and reactive power values through a low-pass filter, respectively.

4.1.2 *Reactive Power Sharing Errors Analysis*

For simplicity, a simplified microgrid with two DG units is considered in this section. According to (4.2) and (4.3), the reactive power of the ith DG unit is obtained

$$Q_i = \frac{V_{\text{pcc}}(V^* \cos\delta_i - V_{\text{pcc}})}{X_{line-i} + V_{\text{pcc}} n_i \cos\delta_i} \tag{4.4}$$

Assume the ith and jth DG unit are working in parallel with the same nominal capacity and droop slope. Note that shift angle is usually very small ($\sin\delta_i \approx \delta_i, \cos\delta_i \approx 1$), and then the reactive power sharing relative error with respect to Q_i can be expressed as follows:

$$\Delta Q_{\text{err}} = \frac{Q_i - Q_j}{Q_i} \approx \frac{X_{line-j} - X_{line-i}}{X_{line-j} + V_{\text{pcc}} n_j} \tag{4.5}$$

It shows that the reactive power sharing relative error is related to some factors, which include the impedance X_j, the impedance difference $(X_{line-j} - X_{line-i})$, the voltage amplitude V_{pcc} of the PCC, and the droop slope n_j. According to (4.2), there are two main approaches to improve the reactive power sharing accuracy: increasing impedance X_j and the droop gain n_j. Usually, increasing impedance is achieved by the virtual impedance [2–4], which requires a high-bandwidth control for inverters. Increasing the droop gain n_j is a simpler way to reduce the sharing error. However, it may degrade the quality of the microgrid bus voltage and even affects the stability of the microgrid system [5–7].

4.2 Reactive Power Sharing Error Compensation Method

4.2.1 *Droop Controller*

The introduced Q–V droop control method is given as follows:

$$V_i(t) = V^* - n_i Q_i(t) - \sum_{n=1}^{k-1} K_i Q_i^n + \sum_{n=1}^{k} G^n \Delta V \tag{4.6}$$

where k denotes the time of synchronization event until time t. According to (4.6), the control is a hybrid system with continuous and discrete traits. In the digital implementation of the method, the continuous variables $V_i(t)$ and $Q_i(t)$ are discretized with sampling period T_s, and T_s is greatly less than the time interval between two consecutive synchronization events. Therefore, the droop (4.6) at the kth synchronization interval could be expressed as

$$V_i^k = V^* - n_i Q_i^k - \sum_{n=1}^{k-1} K_i Q_i^n + \sum_{n=1}^{k} G^n \Delta V \tag{4.7}$$

where ω^* and V^* are the values of DG angular frequency and output voltage amplitude at no-load condition, m_i and n_i are the droop gains of frequency and voltage of DG-i unit, and G^n is the voltage recovery operation signal at the nth synchronization interval; G^n has two possible values: 1 or 0. If $G^n = 1$, it means the voltage recovery operation is performed. Q_i^n represents the output reactive power of DG-i unit at the nth synchronization interval. K_i is a compensation coefficient for the DG-i unit, and ΔV is a constant value for voltage recovery. For simplicity of description, the third term of (4.7) is referred to the sharing error reduction operation, and the last term is called the voltage recovery operation. For simplicity, the output voltage for the DG-i unit in (4.7) is written as follows in iterative method:

$$V_i^k = V_i^{k-1} - n_i(Q_i^k - Q_i^{k-1}) - K_i Q_i^{k-1} + G^k \Delta V \tag{4.8}$$

Therefore, for its implementation, only V_i^{k-1} and Q_i^{k-1} should be stored in DSP. To better understand the method, a specific example is given. If there are two DG units with the same capacity working in parallel, and the conventional droop is only used, there will exist some reactive power sharing error due to some factors. If the sharing error reduction operation for each DG unit is performed at the time, the resulting reactive power sharing error will decrease. The principle behind the sharing error reduction operation can be understood with the aid of Fig. 4.3. If the aforementioned operation is repeated with time, the reactive power sharing error will converge. However, the associated operations will result in a decrease in PCC voltage. To cope with the problem, the voltage recovery operation will be performed. That is to say if the output voltage of one DG unit is less than its allowed low limit, then the DG unit will trigger the voltage recovery operation until its output voltage is restored to rating value. The output voltage of all the DG units will be added an identical value ΔV to increase the PCC voltage. The idea for the voltage recovery operation can be comprehended with the aid of Fig. 4.4.

4.2.2 Communication Setup

A DG unit can communicate with other DG units by RS232 serial communication. Each DG unit has the opportunity to trigger a synchronization event on the condition

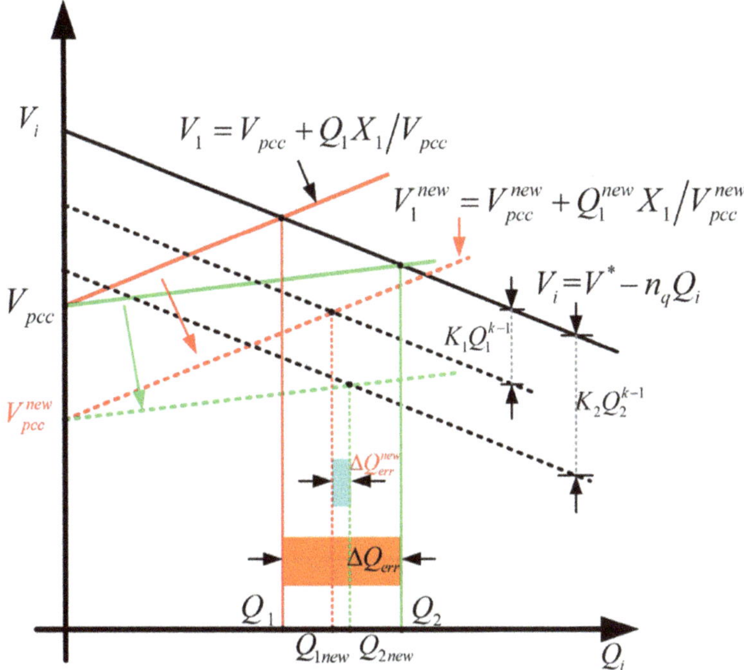

Fig. 4.3 Schematic diagram of the sharing error reduction operation

that the time interval between two consecutive synchronization events is greater than a permissible minimum value, and the output voltage of each DG unit is in the reasonable range. If the output voltage of one DG unit is less than its allowed low limit, it will ask for having the priority to trigger a synchronization event at once. Until the constraint that the interval between two consecutive synchronization events is greater than a permissible minimum value is satisfied, the DG unit with the priority will trigger a synchronization event, and in this event, the command for voltage recovery operation will be sent to other DG units. If the communication fails (the time interval between two consecutive synchronization events is greater than a permissible maximum value), all the error reduction operations and voltage recovery operations should be disabled and the control method reverts back to the conventional one.

According to the aforesaid analysis, such a microgrid system only needs a low-bandwidth communication. And it is robust to the delay of communication. To illustrate this point, the control timing diagram is shown in Fig. 4.5. The sharing error operation and the voltage recovery operation are performed in update interval. Sampling operation occurs in sampling interval. There is a time interval τ, which is long enough to guarantee the system being in steady state. It is obvious that method is robust to the time delay because all the necessary operations only need to be completed in an interval, not a critical point.

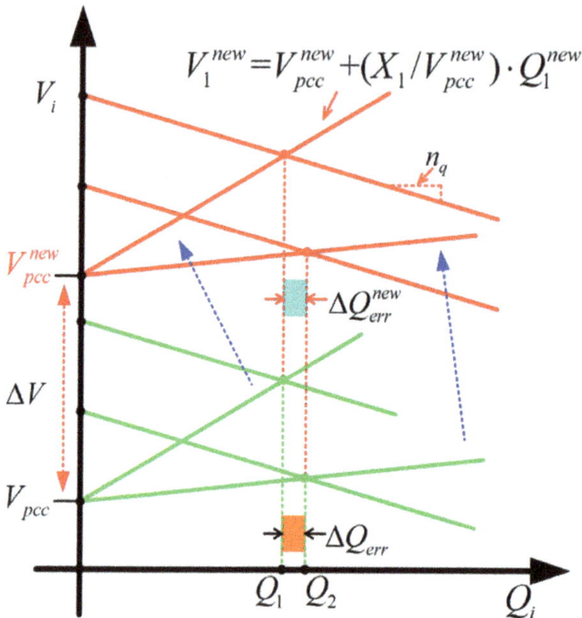

Fig. 4.4 Schematic diagram of voltage recovery mechanism

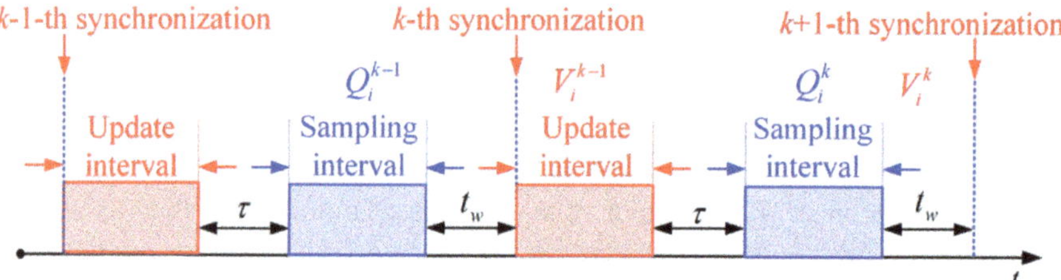

Fig. 4.5 Control timing diagram of one DG with the two consecutive synchronization events

4.2.3 Convergence Analysis

In this subsection, the convergence of the introduced method will be proved [11]. Without loss of generality, the sharing reactive power error between DG-i and DG-j with the same capacity will be analyzed. According to (4.7), the reactive droop equation for DG-j can be expressed as

$$V_j^k = V^* - n_j Q_j^k - \sum_{n=1}^{k-1} K_j Q_j^n + \sum_{n=1}^{k} G^n \Delta V \tag{4.9}$$

Subtracting (4.9) from (4.7), then

$$\Delta V_{ij}^{k} = -n\Delta Q_{ij}^{k} - \sum_{n=1}^{k-1} K\Delta Q_{ij}^{n} \tag{4.10}$$

where $n = n_j = n_i$, $K = K_j = K_i$, and ΔE_{ij}^{k} is the voltage magnitude derivation of DG i and j in the kth control period; ΔQ_{ij}^{k} is the reactive power sharing error.

Similarly, we can get (4.10) in the $(k+1)$th interval

$$\Delta V_{ij}^{k+1} = -n\Delta Q_{ij}^{k+1} - \sum_{n=1}^{k} K\Delta Q_{ij}^{n} \tag{4.11}$$

Combining (4.10) and (4.11), it yields

$$\Delta V_{ij}^{k+1} - \Delta V_{ij}^{k} = -n\Delta Q_{ij}^{k+1} + n\Delta Q_{ij}^{k} - K\Delta Q_{ij}^{k} \tag{4.12}$$

According to the feeder characteristic, as shown in (4.2), the following expressions can be obtained:

$$\Delta V_{ij}^{k+1} = \frac{1}{V_{\text{pcc}}^{k+1}}\left(Q_i^{k+1}X_{line-i} - Q_j^{k+1}X_{line-j}\right) \tag{4.13}$$

$$\Delta V_{ij}^{k} = \frac{1}{V_{\text{pcc}}^{k+1}}\left(Q_i^{k}X_{line-i} - Q_j^{k}X_{line-j}\right) \tag{4.14}$$

Assume that the PCC voltage value satisfies the following:

$$V_{\text{pcc}}^{k+1} \approx V_{\text{pcc}}^{k} \approx V \gg 1 \tag{4.15}$$

Subtracting (4.12) from (4.13), it yields

$$\Delta V_{ij}^{k+1} - \Delta V_{ij}^{k} = \frac{X_{line-j}}{V}(\Delta Q_{ij}^{k+1} - n\Delta Q_{ij}^{k}) + \frac{\Delta X}{V}(Q_i^{k+1} - Q_i^{k}) \tag{4.16}$$

where $\Delta X = X_{line-i} - X_{line-j}$.

Combining the expression (4.12) and (4.16), then

$$\Delta Q_{ij}^{k+1} = r\Delta Q_{ij}^{k} - \frac{\Delta X}{V(n + X_{line-j}/V)}[Q_i^{k+1} - Q_i^{k}] \tag{4.17}$$

where $r = \frac{n+X_{line-j}/V-K}{n+X_{line-j}/V} < 1$. According to the contraction mapping theorem, if $|r| < 1$ and $\Delta X = 0$, then the reactive power sharing error will converge to zero. Generally, $\Delta X = 0$, we should also consider the effect of the second term of (4.17).

According to the feeder characteristic, as shown in (4.1), we have

$$Q_i^{k+1} - Q_i^k = \frac{(V_i^{k+1} - V_i^k)V}{X_{line-i}} \tag{4.18}$$

Because of the voltage recovery mechanism, we can ensure $V_{\min} \le V_i{}^k \le V_{\max}$ for all k

$$\left|Q_i^{k+1} - Q_i^k\right| \le (V_{\max} - V_{\min})\frac{V}{X_{line-i}} \tag{4.19}$$

Therefore, the second term of (4.17) is bounded. According to aforesaid analysis, it can be concluded that the reactive power sharing error is also bounded.

4.3 Simulation Results

The improved reactive power sharing strategy is verified with MATLAB/Simulink. In the simulations, a microgrid with two DG systems, as shown in Fig. 4.1, is employed. The associated parameters for power stage and control of the DG unit are listed in Table 4.1. Also, in order to facilitate the observation of the reactive power sharing, the two DG units are designed with the same power rating and different line impedances. The detailed configuration of the single DG unit is depicted in Fig. 4.6, where an LCL filter is placed between the insulated gate bipolar transistor bridge output and the DG feeder. The DG line current and filter capacitor voltage are measured to calculate the real and reactive powers. In addition, the commonly used double closed-loop control is employed to track the reference voltage [8–10].

4.3.1 Case 1: Power Sharing Accuracy Improvement

Two identical DG units operate in parallel with the introduced voltage droop control. Figure 4.7 illustrates the reactive power sharing performance of the two DGs. Before

Table 4.1 Associated parameters for power stage and control of the DG unit

Parameters	Values	Parameters	Values
u_r (V)	220	k_{pu}	0.05
L_f (H)	1.5e−3	k_{pi}	50
r_f (Ω)	0.25	K_{pi}	0.2
C_f (μF)	20	Ω_c (rad/s)	31.4
L_{line1} (H)	0.6e−3	m (rad/s · w)	5e−5
L_{line2} (H)	0.3e−3	n (v/var)	5e−3
f_s (kHz)	12.8	K_e (v/var)	0.001
f_{rate} (Hz)	50	ΔE_0 (V)	5
T_s (s)	1/12.8e3	n (v/var)	0.1

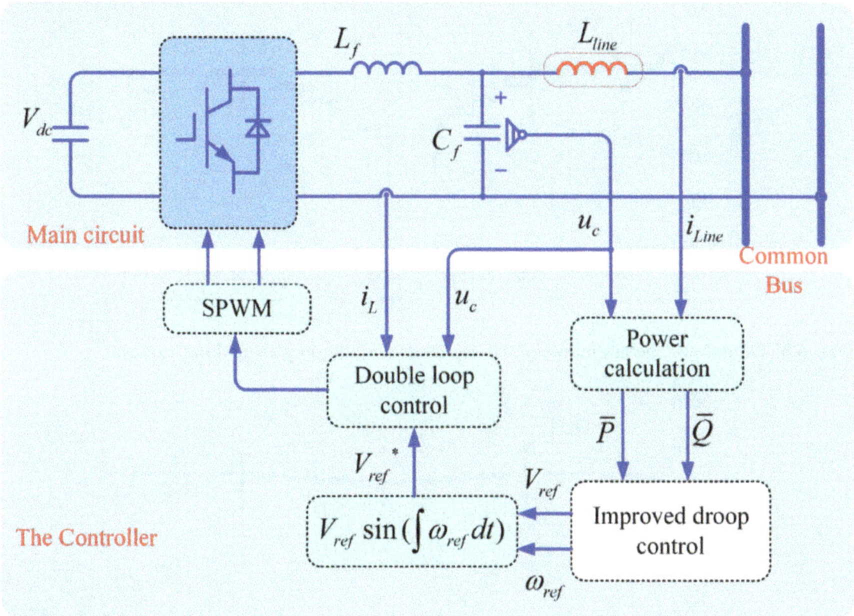

Fig. 4.6 Configuration of one single-phase DG unit

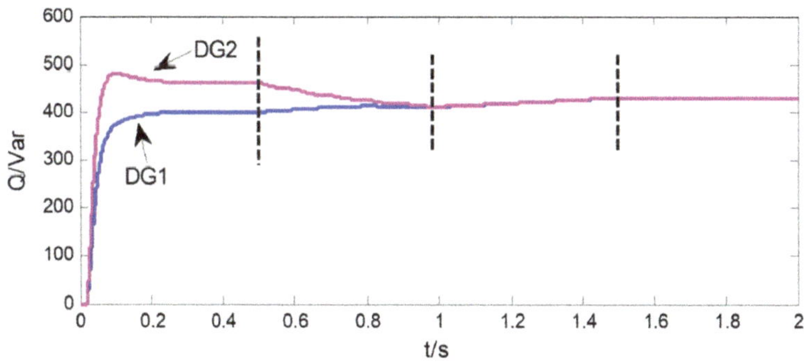

Fig. 4.7 Output reactive powers of two inverters with the improved droop control

$t = 0.5$ s, the sharing error reduction operation and voltage recovery operation are disabled, which is equivalent to the conventional droop control being in effect. There exists an obvious reactive power sharing error due to the unequal voltage drops on the feeders. After $t = 0.5$ s, the reactive power sharing error reduction operation is performed, and it is clear that the reactive power sharing error converges to zero gradually. After $t = 1$ s, the voltage recovery operation is performed. It can be observed that the output reactive power increases, but the reactive power sharing performance does not degrade. Figure 4.8 shows the corresponding output voltages. It can be observed that the output voltages decrease during the sharing

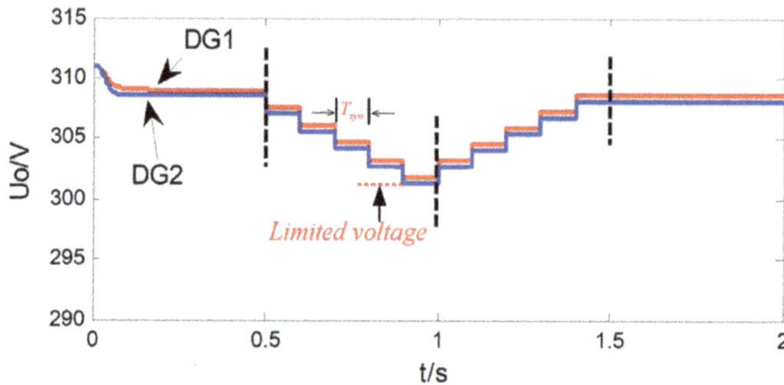

Fig. 4.8 Output voltage amplitude of two inverters with the improved droop control

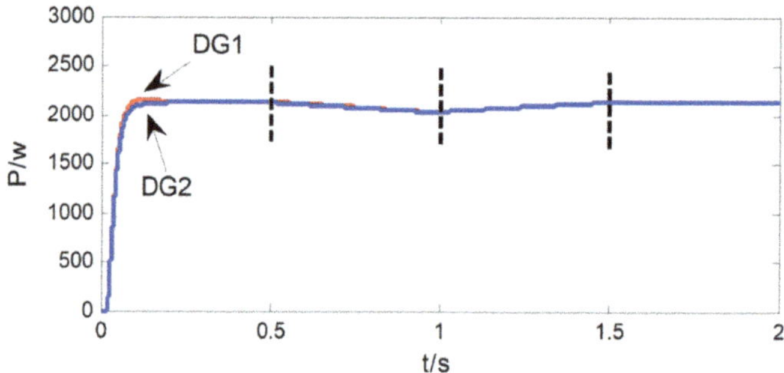

Fig. 4.9 Output active powers of two inverters with the improved droop control

error reduction operation, while the voltage recovery operation ensures that DG output voltage amplitudes can restore back nearby to the rated value. The whole process of adjustment can be done steadily in a relatively short period of time. Figure 4.9 illustrates active power sharing performance of the two DG units. It is obvious that the improved reactive power sharing strategy does not affect active power sharing performance.

4.3.2 *Case 2: Effect of Communication Delay*

To test the sensitivity of the improved droop control to the synchronized signal, a 0.02 s delay is intentionally added to the signal received by DG1 unit at $t = 0.5$ s as shown in Fig. 4.11, and the simulation results are shown in Figs. 4.10, 4.11, and 4.12. Compared to the Case 1 in Fig. 4.7 and Fig. 4.9, a small disturbance appears in both the reactive and active powers, while the voltage recovery operations are still able to ensure that the DG unit can deliver the expected reactive power. After $t = 2.0$ s, the active and reactive power sharing errors are almost zero. Therefore, the improved reactive power sharing strategy is not sensitive to the communication delay. Then, it is illustrated that it is robust to some small communication delays.

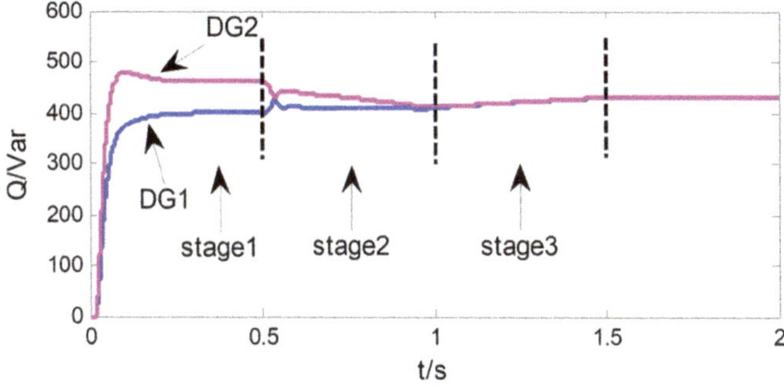

Fig. 4.10 Output reactive powers of the two inverters when 0.02 s time delay occurs in synchronization signal of DG1 unit

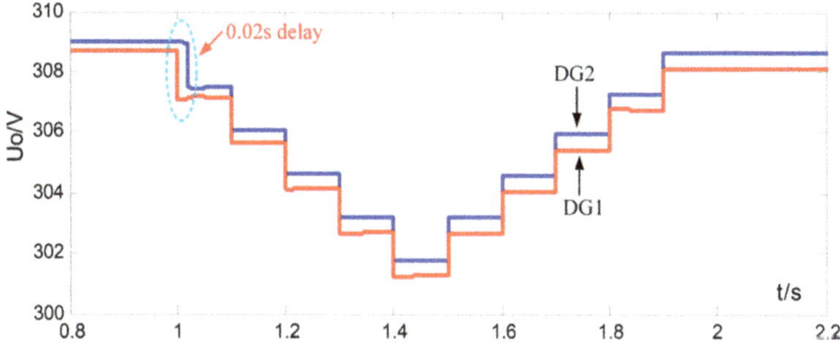

Fig. 4.11 DG output voltage of the inverters when 0.02 s time delay occurs in synchronization signal of DG1 unit

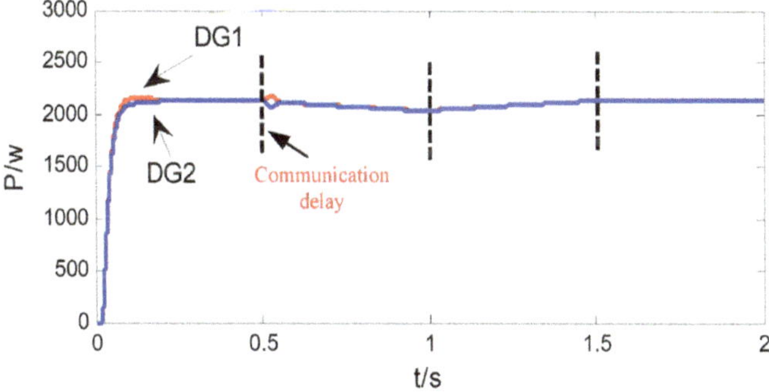

Fig. 4.12 Output active powers of the two inverters when 0.02 s time delay occurs in synchronization signal of DG1 unit

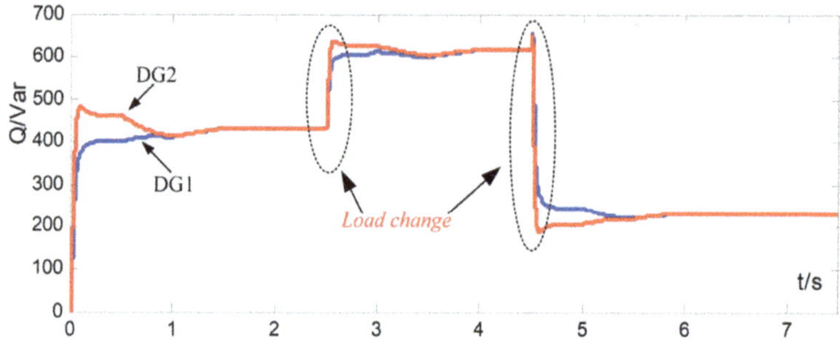

Fig. 4.13 Reactive power sharing performance of the improved droop control (with load changing)

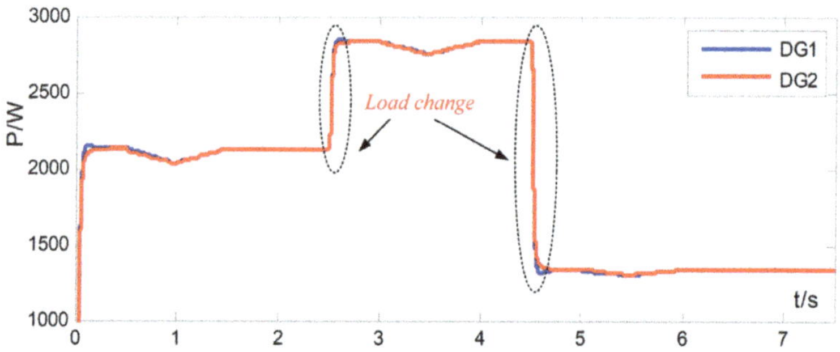

Fig. 4.14 Active power sharing performance of the improved droop control (with load changing)

4.3.3 *Case 3: Effect of Load Change*

In order to test the effect of load change with the method, the active load increases about 1.6 kW and the reactive load increases about 0.4 kVar at $t = 2.5$ s, and at $t = 4.5$ s, the active load decreases about 3.0 kW and the reactive load decreases about 0.8 kVar. The corresponding simulation results are shown in Figs. 4.13 and 4.14. As can be seen, a large reactive power sharing deviation appears at $t = 2.5$ s and $t = 4.5$ s. However, the deviation becomes almost zero after a while. Figure 4.15 illustrates the corresponding output voltage waveforms. It can be found that there exist obvious output voltage decrease and output voltage increase processes during each reactive power sharing error reduction process.

4.4 Experimental Results

The tested microgrid consists of two microsources based on the single-phase inverter. The parameters for output filter are the same as those in simulation. The

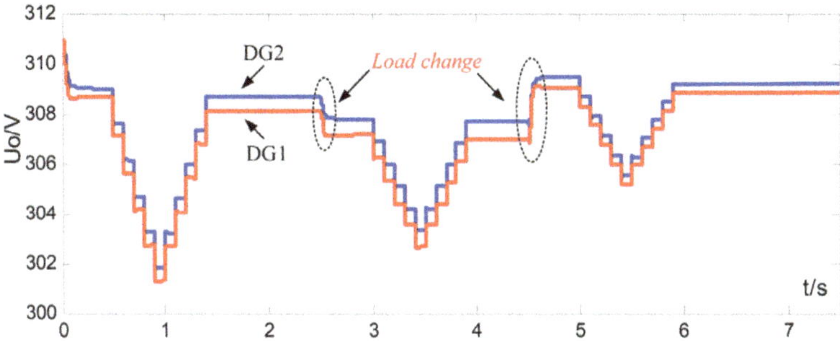

Fig. 4.15 DG output voltage of the improved droop control (with load changing)

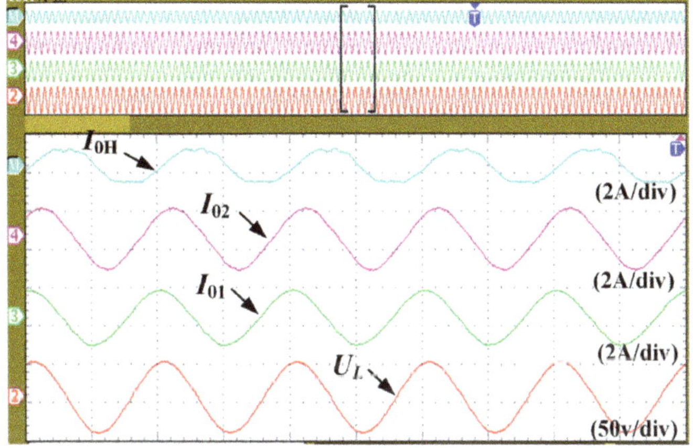

Fig. 4.16 Steady-state experimental waveforms with the conventional droop control

load consists of a resistor of 16 Ω and an inductor of 3 mH. The sample frequency is 12.8 kHz. A permissible minimum time interval between two consecutive synchronization events is 0.5 s. The permissible minimum output voltage does not less than the rated voltage by 90%.

Figures 4.16 and 4.17 show the measured waveforms with the conventional and improved droop control methods, respectively. The waveforms from top to down are circular current ($i_{0H} = i_{01}–i_{02}$), the output current of inverter 1 (i_{01}), the output current (i_{02}) of inverter 2, and PCC voltage (U_L), respectively. As can be seen from Fig. 4.16, there is a quite large phase difference between two output currents when the conventional droop control is applied. As a result, the circular current is pretty high and the peak value of circular current is up to 1.80 A. The main reason for it is the impedance difference of DG feeders. Compared with the circular current in Fig. 4.16, the circular current in Fig. 4.17 is very small, which indicates that the improved method is efficient in reducing the circular current mainly caused by the output reactive power difference between the inverters.

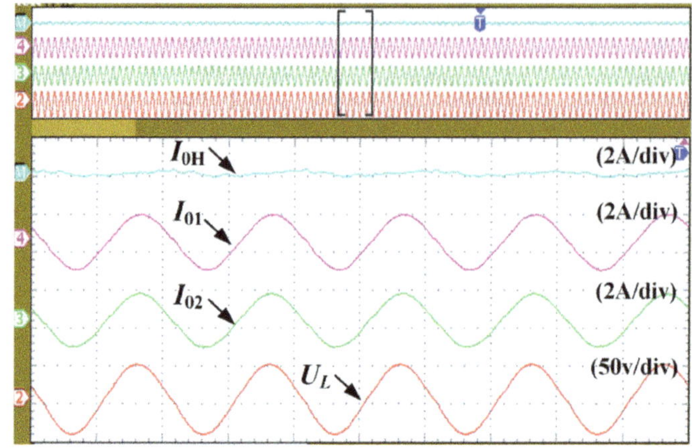

Fig. 4.17 Steady-state experimental waveforms with the improved droop control

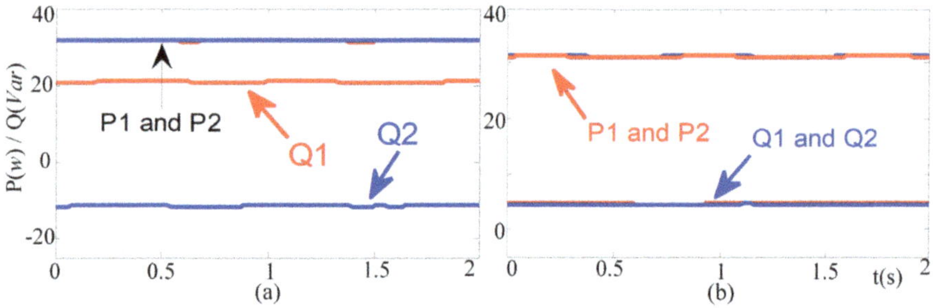

Fig. 4.18 Steady-state active power and reactive power: (**a**) with the conventional droop and (**b**) with the improved droop control

Figure 4.18 shows the steady-state output active and reactive power of each inverter with the conventional and the improved droop control. Figure 4.18a shows the results with the conventional droop. The steady-state output active powers of the inverters are 31.4 and 30 W, and the output reactive powers are 21.2 and −10.4 Var. When using conventional $P - f$ droop control, no active power divergence appears since frequency is a global variable, i.e., the same frequency can be measured along the microgrid; however, voltage may drop along the microgrid power lines, which produce the well-known reactive power divergence. Figure 4.18b shows the results with the improved droop. As can be seen, the output active powers of the inverters are 30.6 and 31.1 W, and the reactive powers are 3.9 and 4.4 Var. These results indicate that the improved droop control has no effect on the active power sharing performance but makes reactive power be shared precisely.

To verify the effectiveness of the sharing error reduction operation and voltage recovery operation of the improved method, the experiments with only one operation being continuously used are performed. As can be seen from Fig. 4.19, the circular current converges to a small value gradually when only the reactive power

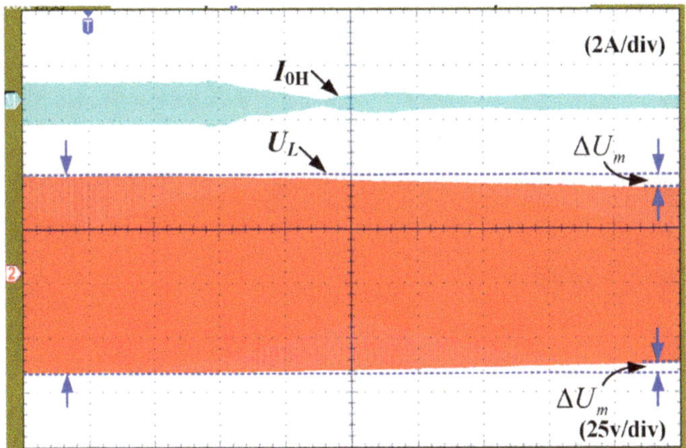

Fig. 4.19 Circulating current and PCC voltage waveforms of DGs with only sharing error reduction operation performed

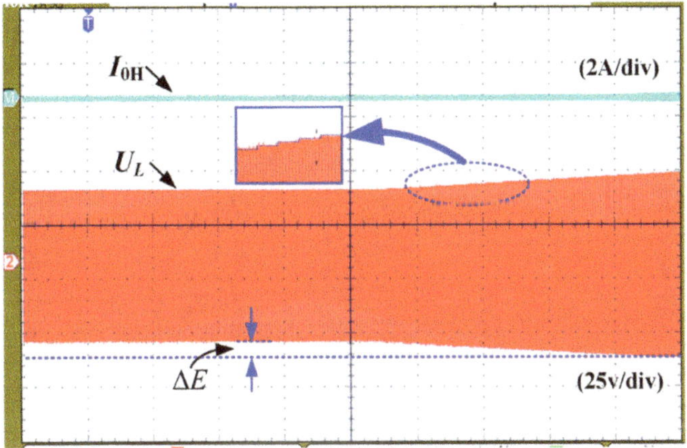

Fig. 4.20 Circulating current and PCC voltage waveforms of DGs with only voltage recovery operation performed

sharing error reduction operation is performed. In the meanwhile, a continuous decrease in PCC voltage could be found. Figure 4.20 shows the results when only the voltage recovery operation is performed. It can be seen that the PCC voltage increases linearly during this time, and the circular current is always small and almost kept constant. Figure 4.21 shows the results when the two operations are combined, i.e., the improved method is applied. The circular current is controlled to be a small value, and the quality of the PCC voltage is guaranteed successfully.

To test the sensitivity of the improved method to synchronization signal, a 0.2 s delay is intentionally added to the synchronization signal received by DG1 unit every time. The associated experimental results are shown in Fig. 4.22. Compared

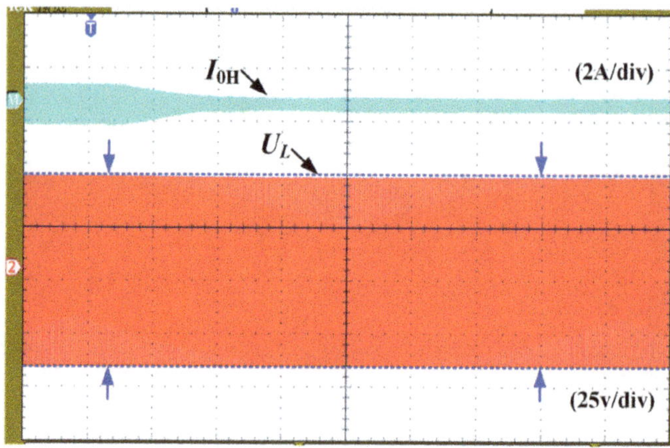

Fig. 4.21 Circulating current and PCC voltage waveforms of DGs with the improved droop

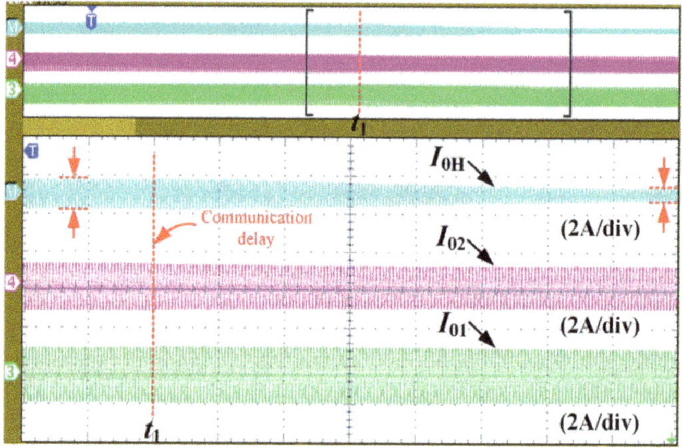

Fig. 4.22 Output current and circulating current waveforms when 0.2 s time delay occurs in synchronization signal of DG1 unit

to the normal case, there is no obvious difference between the two cases, and the reactive power sharing error can still reduce to a small value. Therefore, the improved method is robust to the communication delay because all the necessary operations only need to be completed in an interval, not a critical point.

Figure 4.23 shows the experimental results when the synchronization signal of DG1 unit fails, which is equivalent to the time delay is infinity. It is obvious that, before $t = t_1$, the circulating current is kept to be a small value because the improved droop control is in effect. After $t = t_1$, the sharing error reduction operation and voltage recovery operation are disabled due to the loss of the synchronization signal of DG1 unit. As a result, the peak value of the circulating current increases to about 2.8 A from a small value. In conclusion, the results in Figs. 4.22 and 4.23 indicate

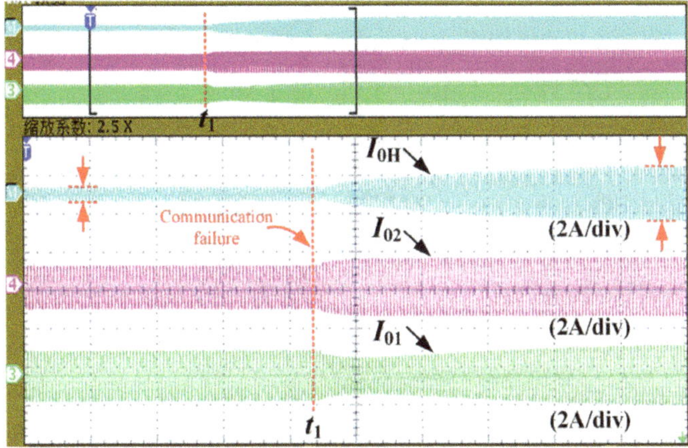

Fig. 4.23 Output current and circulating current waveforms when the synchronization signal is lost in DG1 unit

that the method only needs a low-bandwidth requirement, and it is robust to a small time delay of communication. However, once communication fails completely, the reactive power sharing accuracy performance may be worse.

4.5 Conclusion

In this chapter, a new reactive power control for improving the reactive sharing is presented for power electronics interfaced DG units in AC microgrids. The control strategy is realized through the following two operations: sharing error reduction operation and voltage recovery operation. The first operation changes the voltage bias of the conventional droop characteristic curve periodically, which is activated by the low-bandwidth synchronization signals. The second operation is performed to restore the output voltage to its rated value. The improved power sharing can be achieved with very simple communications among DG units. Furthermore, the plug-and-play feature of each DG unit will not be affected. Both simulations and experimental results are provided to verify the effectiveness of the improved control strategy.

References

1. A. Luna, F. Blaabjerg, P. Rodríguez, Control of power converters in AC microgrids. IEEE Trans. Power Electron. **27**(11), 4734–4749 (2012)
2. J.M. Guerrero, L.G. Vicuña, J. Matas, M. Castilla, J. Miret, Output impedance design of parallel-connected UPS inverters with wireless load-sharing control. IEEE Trans. Ind. Electron. **52**(4), 1126–1135 (2005)

3. J.M. Guerrero, J. Matas, L. Garcia de Vicuna, M. Castilla, J. Miret, Decentralized control for parallel operation of distributed generation inverters using resistive output impedance. IEEE Trans. Ind. Electron. **54**(2), 994–1004 (2007)
4. J.M. Guerrero, L.G. Vicuña, J. Matas, M. Castilla, J. Miret, A wireless controller to enhance dynamic performance of parallel inverters in distributed generation systems. IEEE Trans. Power Electron. **15**(9), 1205–1213 (2004)
5. A. Moawwad, V. Khadkikar, J.L. Kirtley, A new P-Q-V droop control method for an interline photovoltaic (I-PV) power system. IEEE Trans. Power Del. **28**(2), 658–668 (2013)
6. E.A.A. Coelho, P.C. Cortizo, P.F.D. Garcia, Small-signal stability for parallel-connected inverters in stand-alone AC supply systems. IEEE Trans. Ind. Appl. **38**(2), 533–542 (2002)
7. N. Pogaku, M. Prodanović, T.C. Green, Modeling, analysis and testing of autonomous operation of an inverter-based microgrid. IEEE Trans. Power Electron. **22**(2), 613–625 (2007)
8. Y. Mohamed, E.F. El-Saadany, Adaptive decentralized droop controller to preserve power sharing stability of paralleled inverters indistributed generation microgrids. IEEE Trans. Power Electron. **23**(6), 2806–2816 (2008)
9. J.M. Guerrero, J.C. Vásquez, J. Matas, L. Garcia de Vicuña, M. Castilla, Hierarchical control of droop-controlled AC and DC microgrids—a general approach towards standardization. IEEE Trans. Ind. Electron. **58**(1), 158–172 (2010)
10. Y.W. Li, C.-N. Kao, An accurate power control strategy for power-electronics-interfaced distributed generation units operating in a low-voltage multibus microgrid. IEEE Trans. Power Electron. **24**(12), 2977–2988 (2009)
11. H. Han, Y. Liu, Y. Sun, M. Su, J.M. Guerrero, An improved droop control strategy for reactive power sharing in islanded microgrid. IEEE Trans. Power Electron. **30**(6), 3133–3141 (2014)

Chapter 5
Droop-Based Economical Dispatch

5.1 Economical Dispatch Problems Formulation

Mathematically speaking, the objective of economical dispatch problem is to minimize the total generation cost subjected to the demand–supply constraints as well as the generator constraints. The optimization problem is formulated as

$$\begin{cases} \min \sum_{i=1}^{n} C_i (P_i) \\ s.t. \ \ \sum_{i=1}^{n} P_i = P_L; \ 0 \le P_i \le P_{i,\max}, \ i \in \{1, 2, \ldots, n\} \end{cases} \tag{5.1}$$

where P_i, $C_i(P_i)$, and $P_{i,\max}$ are the active power, cost function, and upper bound of P_i, respectively, and P_L is the total load. For an AC microgrid with inductive transmission network, P_L is equal to the sum of all loads. We use $P_{L,\max}$ to denote the upper bound of P_L, i.e., $P_L \le P_{L,\max}$. Because the constraints in (5.1) are compact and the cost functions are all continuous, it follows from the well-known extreme value theorem that the EDP has a global optimal solution (GOS) P^*.

Each P_L ($P_L \in [0, \ P_{L,\max}]$) is exactly related to the GOS $(P_1^*, P_2^*, \ldots, P_n^*)$ such that the cost is minimal. Thus, the optimal solution P_i^* can be regarded as the function of the load P_L, whose curve can be sketched by off-line calculation. For convenience, we name it as optimal solution function (OSF). Define $g_i (P_L) \triangleq P_i^* (P_L)$ and it inherently yields

$$\sum_{i=1}^{n} g_i (P_L) = P_L \tag{5.2}$$

In AC microgrid, the economical dispatch is usually achieved via the methods based on droop controls [1–5]. The droop-based control scheme $(P - f)$ is presented as

$$f_i = f^* - \varphi_i (P_i) \tag{5.3}$$

Y. Sun et al., *Series-Parallel Converter-Based Microgrids*, Power Systems,
https://doi.org/10.1007/978-3-030-91511-7_5

where f_i and f^* are the ith DG actual frequency and reference, respectively, and $\varphi_i(P_i)$ is the droop function. To guarantee the uniqueness of the steady operating point (SOP), the droop function $\varphi_i(P_i)$ should be monotonic. When the system frequency achieves synchronization, we have

$$\varphi_1(P_1) = \varphi_2(P_2) = \cdots = \varphi_n(P_n) \tag{5.4}$$

From the balance of power supply–demand, it yields

$$P_1 + P_2 + \cdots + P_n = P_L \quad \left(0 \le P_L \le P_{L,\max}\right) \tag{5.5}$$

Then, in steady state, the output power of each DG, i.e., SOP, is determined by (5.4) and (5.5). Given that all the $\varphi_i(P_i)$ have the same value in steady state, we assume they are equal to a common variable y, i.e.,

$$\varphi_1(P_1) = \varphi_2(P_2) = \cdots = \varphi_n(P_n) = y \tag{5.6}$$

Then, we obtain $P_i = \varphi_i^{-1}(y)$, where φ_i^{-1} is the inverse of φ_i. Substituting it into (5.5), we have $\sum_{i=1}^{n} \varphi_i^{-1}(y) = P_L$. Denote $F(\cdot) = \sum_{i=1}^{n} \varphi_i^{-1}(\cdot)$. Then, the SOP is obtained as

$$P_i = \varphi_i^{-1}(y) = \varphi_i^{-1} \circ F^{-1}(P_L) \tag{5.7}$$

where $f \circ g$ represents $f(g(x))$. Likewise, the system SOP also changes with variation of the load. It can be regarded as the function of P_L and we use $h_i(P_L) \triangleq \varphi_i^{-1}\left(F^{-1}(P_L)\right)$ to denote the SOP function (SOPF).

For the problem of droop-based economical dispatch, the key is to design the droop function $\varphi_i(P_i)$ to ensure that SOP is the GOS, $\forall P_L \in \left[0,\ P_{L,\max}\right]$. That is, the curve of $h_i(P_L)$ completely coincides with $g_i(P_L)$. Thus, the EDP can be formulated as

$$\min_{f_i} \int_0^{P_{L,\max}} \left(h_i(P_L) - g_i(P_L)\right)^2 dP_L \tag{5.8}$$

5.2 GOD Criterion and Decentralized Control Schemes

5.2.1 GOD Criterion Via Decentralized Manner

To solve the optimization problem (5.8), the following two questions are crucial:

1. Under what conditions there exists a series of droop functions φ_i such that $h_i(P_L) \equiv g_i(P_L)$?

2. How to design a suboptimal f_i if there do not exist droop functions such that $h_i(P_L) \equiv g_i(P_L)$?

The main results are listed as follows:

Theorem 5.1 *If all the OSFs are strictly monotonically increasing, the GOD can be achieved by constructing the following droop law:*

$$f_i = f^* - m{g_i}^{-1}(P_i) \tag{5.9}$$

where m is a positive constant to keep the frequency deviation in the acceptable range, i.e.,

$$m = \frac{f^* - f_{\min}}{\max_i \left\{ g_i^{-1}\left(P_{L,\max}\right)\right\}} \tag{5.10}$$

where f_{min} is the acceptable minimal value of frequency.

Proof When the system frequency reaches synchronization, according to (5.7), the SOPF can be obtained as

$$P_i = \left(m g_i^{-1}\right)^{-1} \circ F^{-1}(P_L) = g_i\left(m^{-1}F^{-1}(P_L)\right) \tag{5.11}$$

Considering $\varphi_i = m g_i^{-1}$, $F(P_L)$ becomes $\sum_{i=1}^{n} g_i\left(m^{-1}P_L\right)$. According to (5.2), we obtain $F(P_L) = m^{-1}P_L$. Substituting it into (5.10), we have

$$P_i = g_i\left(m^{-1}F^{-1}(P_L)\right) = g_i\left(m^{-1}mP_L\right) = g_i(P_L) \tag{5.12}$$

Thus, $h_i(P_L) \equiv g_i(P_L)$. That is, for any P_L $\left(P_L \le P_{L,\max}\right)$, the SOP is the GOS. Hence, when the system achieves the steady state, the GOD will be achieved by taking the droop law as $f_i = f^* - m g_i^{-1}(P_i)$. □

Theorem 5.2 *If any of the OSFs is not strictly monotonically increasing, GOD cannot be achieved in decentralized manners.*

Proof To illustrate Theorem 5.2, we first give two typical non-monotonic examples: saturation and dead-zone. Considering a DG with limited capacity, the DG optimal power may become saturated (reaching the limitation) when the load is too heavy. On this occasion, the DG should be set to current/power control mode to maintain the output power at the limiting value. However, only when the common frequency is available can current/power control be implemented, which means communication is necessary. Moreover, without knowing the real-time information of load, when to switch back to the droop mode cannot be determined. Likewise, in the case that DG OSF shows characteristics of dead-zone, communication is also necessary to generate a command signal. Therefore, in the above two cases, only with centralized communication can GOD be realized. Next, we give a generalized proof of Theorem 5.2.

It is assumed that there exists a series of function φ_i such that $h_i(P_L) \equiv g_i(P_L)$, where $g_i(P_L)$ is non-monotonic. Then, we obtain $F^{-1}(P_L) \equiv \varphi_i(g_i(P_L))$. Given that F^{-1} is the inverse of function F, F^{-1} must be monotonic, which is contradictory with the fact that $\varphi_i(g_i(P_L))$ is non-monotonic. Therefore, there is no droop function such that the SOP is the GOS for each load.

Overall, the optimization criterion is summarized as follows:

Optimization Criterion GOD can be achieved in decentralized manners if and only if OSFs are all strictly monotonically increasing for $P_L \in \left[0,\ P_{L,\max}\right]$.

□

When OSFs are all strictly monotonically increasing, the decentralized control law of GOD is presented in (5.9) according to optimization criterion. This statement will be verified by Case 1 in the simulation section.

5.2.2 *Decentralized Suboptimal Scheme*

The prerequisite in which all OSFs are strictly monotonically increasing in micro-grid might be too harsh to be realistic [6–13]. In fact, the OSFs of DGs may have saturation, dead, and hysteretic characteristics as shown in Fig. 5.1 when practical constraints are taken into account. According to Theorem 5.2, GOD cannot be achieved by decentralized methods. Then, how to construct a satisfying suboptimal method becomes a problem that needs to be solved. An intuitive method is to modify the OSF curves and make it monotonous. Define $\gamma_i(P_i)$ as the modified curve, and it can also be regarded as suboptimal solution function (SOSF). Thus, according to Theorem 5.1, the suboptimal droop-based method can be designed as

$$f_i = f^* - m\gamma_i^{-1}(P_i) \tag{5.13}$$

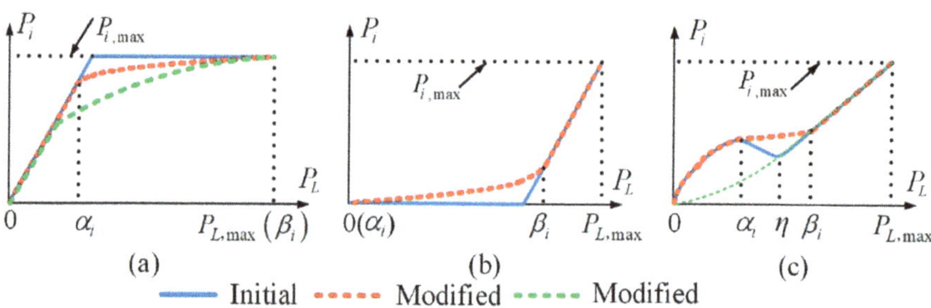

Fig. 5.1 Construction of suboptimal solution (the solid and dotted lines represent OSFs and SOSFs, respectively)

The dotted lines in Fig. 5.1 imply three typical modified smooth curves, respectively. In fact, there are numerous functions that satisfy the condition; is there a best one among them? The answer is no. On the one hand, the error between OSF and SOSF is smaller, and the result of SOSF is better. For example, the red line is better than green line in Fig. 5.1a, but there is not a best monotonically increasing SOSF. On the other hand, the results of SOSF depend on the characteristics of load. For example, when $P_L \in [0, \alpha_i]$ at most of the time, the red line may be better than the green line in Fig. 5.1c, and when $P_L \in [\alpha_i, \eta]$ at most of the time, the green line may be better than the red line. Therefore, there is no best SOSF.

To ensure satisfactory SOSFs, we design $\gamma_i(P_i)$ as

$$\gamma_i = \begin{cases} \sum_{k=0}^{N} a_{ik} P_L^k + \frac{c_i}{P_L + b_i}, & P_L \in [\alpha_i, \beta_i] \\ g_i, & P_L \notin [\alpha_i, \beta_i] \end{cases} \tag{5.14}$$

where $[\alpha_i, \beta_i]$ is the interval in which OSF curves need to be modified by polynomial fitting; the fractional item in (5.1) is introduced to reduce the order of the polynomial. The parameters a_{ik}, b_i, and c_i are determined by the following optimization problem:

$$\begin{cases} \min \int_{\alpha_i}^{\beta_i} (F_i - \gamma_i)^2 dP_L \\ s.t. \text{(I)}\, \gamma_i(\alpha_i) = F_i(\alpha_i), \gamma_i(\beta_i) = F_i(\beta_i); \text{(II)} \frac{d\gamma_i}{dP_L} > \varepsilon \end{cases} \tag{5.15}$$

where ε is a positive number. The constraints in (5.15) are to, respectively, guarantee continuity and monotonicity.

Although $\gamma_i(P_i)$ is suboptimal, it remains good performances. When PL belongs to the subintervals in which $g_i(P_L) = \gamma_i(P_i)$ for all DGs, GOD can still be achieved. When P_L belongs to the $[\alpha_i, \beta_i]$, (5.14) ensures that the operation point is close to the optimal point. Thus, the suboptimal droop-based method is also practical and economical.

Remark The droop strategies need to get the OSF $g_i(P_L)$, which comes from the off-line calculation, and the overall information (DGs' cost functions, maximal capacities, and the maximal loads) is needed. Generally, this information is time-invariant. Therefore, in system operation process after the off-line calculation, the presented control strategies do not need the other DG's real-time information, i.e., communication is not needed.

For an AC microgrid with inductive transmission network, P_L is equal to the sum of all loads because there is no transmission loss. Thus, the EDP can be formulated as (5.1) for any network with inductive cable. That is, the introduced methods can be used in network microgrids.

5.3 Simulation and Experimental Results

To verify the correctness and effectiveness of the method, two simulation cases (Cases 1 and 2) and one experiment case (Case 3) are tested. The general operation cost function is $C_i(P_i) = a_i P_i^3 + b_i P_i^2 + c_i P_i + d_i \exp(e_i P_i)$, and the coefficients and system parameters are listed in Table 5.1.

5.3.1 Case 1: Global Optimal Case

Firstly, according to (5.1), sketch the variation curve of $g_i(P_L)$ as P_L increases from 0 to $P_{L,\max}$, max by off-line calculation. As shown in Fig. 5.2a-1, since all OSFs are strictly monotonically increasing, we could directly find ${g_i}^{-1}$ and construct the droop law as $f_i = f^* - m g_i^{-1}(P_i)$. The $f - P$ droop curves and their approximate

Table 5.1 Cost coefficient for simulation and experiment

DGs	a/b/c/d/e (10-3/10-3/10-2/10-3/10-1)	$P_{\max}$ (kW)	$P_L, P_{L,\max}$ (kW)
DG1 of case 1	0/4/0.4/3/2.86	8	$10 \to 15 \to 20,\ 25$
DG1 of case 2	0.4/–5/6/0/0	8	$10 \to 15 \to 20,\ 25$
DG2 of cases 1,2	0/5.4/0.4/2/2.86	5	$10 \to 15 \to 20,\ 25$
DG3 of cases 1,2	0/3.3/1.1/1/2.86	8	$10 \to 15 \to 20,\ 25$
DG4 of cases 1,2	0/2.4/0.8/4/2.86	8	$10 \to 15 \to 20,\ 25$
DG1 of case 3	0/800/4/2/28.6	1	$0.8 \to 1.2 \to 1.5,\ 2$
DG2 of case 3	0/240/8/2/28.6	1	$0.8 \to 1.2 \to 1.5,\ 2$

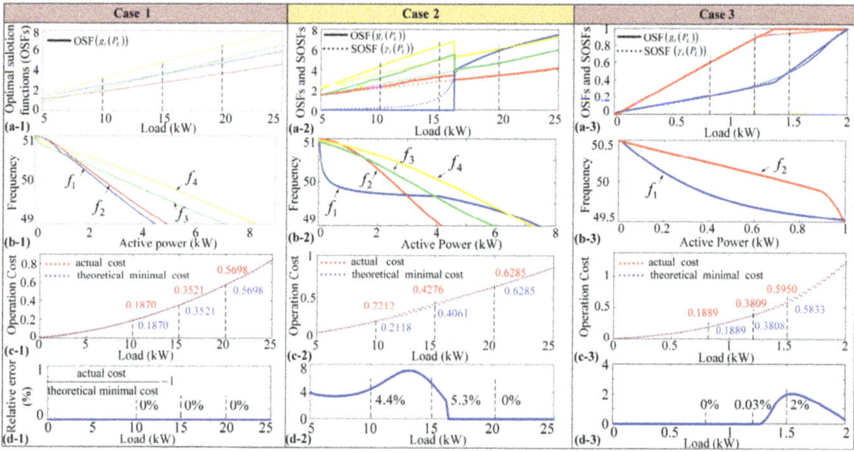

Fig. 5.2 (**a**) OSFs $g_i(P_i)$ and SOSFs $\gamma_i(P_i)$, (**b**) $f - P$ droop, (**c**) actual total operation costs, and (**d**) relative error in simulation and theoretical minimal total operation cost. The red and blue numbers are the actual total costs and the theoretical minimal total costs, respectively

Table 5.2 $f - P$ droop curves

Cases	Approximate polynomial of $\omega - P$	
1	$f_1 = 0.0058P_1^3 - 0.0388P_1^2 - 0.39P_1 + 51$	$P_1 \in [0, 4.5]$
	$f_2 = 0.006P_2^3 - 0.0.042P_2^2 - 0.357P_2 + 51$	$P_2 \in [0, 5]$
	$f_3 = 0.012P_3^2 - 0.31P_3 + 51$	$P_3 \in [0, 7.2]$
	$f_4 = -0.2132P_3 + 51$	$P_4 \in [0, 8.3]$
2	$f_1 = \begin{cases} 0.134(P_1 + 0.1)^{-1} + 50.866 \\ -0.0044P_1^3 + 0.0494P_1^2 - 0.3212P_1 + 50.47 \end{cases}$	$P_1 \in [0, 4.09]$ $P_1 \in (4.09, 7.5]$
	$f_2 = 0.025P_2^3 - 0.216P_2^2 - 0.088P_2 + 51$	$P_2 \in [0, 4.2]$
	$f_3 = 0.006P_3^3 - 0.0625P_3^2 - 0.1562P_3 + 51$	$P_3 \in [0, 6]$
	$f_4 = 0.002P_4^3 - 0.0365P_4^2 - 0.107P_3 + 51$	$P_4 \in [0, 7.3]$
3	$f_1 = \begin{cases} 4.634P_1^3 - 1.863P_1^2 - 1.812P_1 + 50.5 \\ -2.086P_2^3 + 4.46P_2^2 - 3.552P_2 + 50.66 \end{cases}$	$P_1 \in [0, 0.323]$ $P_2 \in (0.323, 1]$
	$f_2 = \begin{cases} -0.15P_2^3 + 0.1873P_2^2 - 0.7324P_2 + 50.5 \\ 40.36P_2^3 - 135.9P_2^2 + 145P_2 + 0.0352 \end{cases}$	$P_2 \in [0, 0.877]$ $P_2 \in (0.877, 1]$

polynomial are presented in Fig. 5.2b-1 and Table 5.2, respectively. The curves of the theoretical minimal total cost $(\sum_{i=1}^{n} C_i(g_i(P_L)))$ and the actual total cost of the method are sketched in Fig. 5.2c-1. Figure 5.2d-1 shows that the relative error $(\frac{actual\ cost}{theoretical\ minimal\ cost} - 1)$ is identically equal to zero over the whole load range, which means that the GOD can be achieved for each $P_L \in \left[0,\ P_{L,\max}\right]$.

5.3.2 *Case 2: Suboptimal Case*

Likewise, according to (5.1), the variation curve of $g_i(P_L)$ is sketched as P_L increases from 0 to $P_{L,\max}$. As shown in Fig. 5.2a-2, the OSFs are not all monotonically increasing, so it is necessary to fit an alternative monotonically increasing function by calculating γ_i from (5.14) and (5.15). The curve of SOSF γ_i is sketched in Fig. 5.2a-2, based on which we find the γ_i^{-1} and construct the droop law as $f_i = f^* - m\gamma_i^{-1}(P_i)$. The $f - P$ droop curves and their approximate polynomial are presented in Fig. 5.2b-2 and Table 5.2, respectively. Figure 5.2a-2 shows that $\gamma_i(P_L) \equiv g_i(P_L)$ when $P_L \in [16, 25]$. According to Theorem 1, GOD can be achieved. After fitting, $\gamma_i(P_L) \neq g_i(P_L)$, while $P_L \in [16, 25]$. According to Theorem 5.1, only a sub-optimization can be realized. The curves of the theoretical minimal total cost and the actual total cost of the method are sketched in Fig. 5.2c-2. Figure 5.2d-2 shows that the relative error is between 0 and 8% for $P_L \in [0, 16]$ and is identically equal to zero for $P_L \in [0, 16]$, which verifies Theorems 5.1 and 5.1. When the load is 10, 15, and 20 kW, the frequency and active power are presented in Fig. 5.3a-2 and b-2, respectively. The relative error in cost is 4.4%, 5.3%, and 0%, respectively.

5.3.3 Case 3: Suboptimal Case

The curve of $g_i(P_L)$ in Case 3 is shown in Fig. 5.2a-3. Here since $g_1(P_L)$ has saturation characteristics, it is necessary to fit an alternative monotonically increasing function by calculating γ_i from (5.14) and (5.15). The curve of SOSF γ_i is also sketched in Fig. 5.2a-3 with dotted lines, based on which we find the γ_i^{-1} and construct the droop law as $f_i = f^* - m\gamma_i^{-1}(P_i)$. The $f - P$ droop curves and their approximate polynomial are presented in Fig. 5.2b-3 and Table 5.2, respectively. Figure 5.2a-3 shows that $\gamma_i(P_L) \equiv g_i(P_L)$ when $P_L \in [0, 1.2)$. According to Theorem 5.1, GOD can be achieved. After fitting, $\gamma_i(P_L) \neq g_i(P_L)$, while $P_L \in [0, 16]$. According to Theorem 5.2, only a sub-optimization can be realized. The curves of the theoretical minimal total cost, the actual total cost and relative error in cost are sketched in Fig. 5.2c-3 and d-3, respectively. As shown in Fig. 5.2d-3, the relative error is identically equal to zero for $P_L \in [0, 1.2)$ and is between 0 and 2.1% for $P_L \in [1.2, 2]$.

Figures 5.3a-3, b-3, and 5.4 show the curves of frequency, active power, and output voltages when the load is 0.8, 1.2, and 1.5 kW and each load lasts for 20

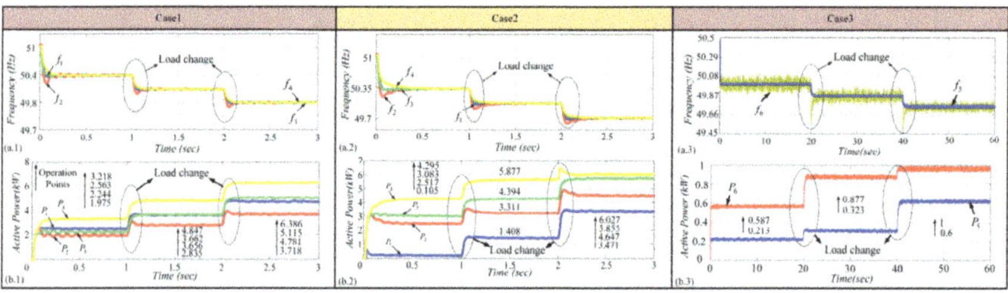

Fig. 5.3 Simulation and experiment results of (**a**) frequency and (**b**) active power under three cases

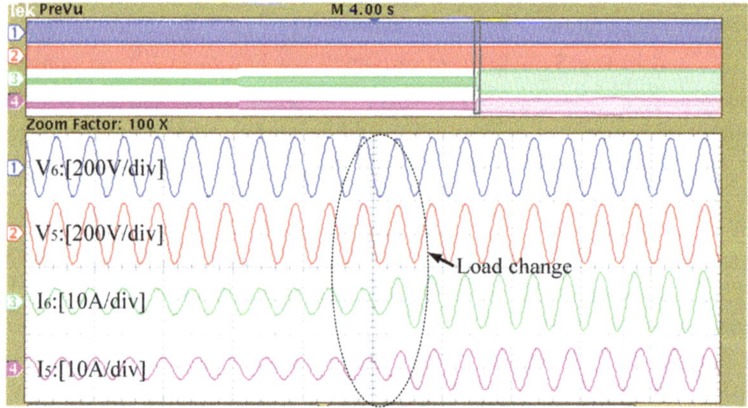

Fig. 5.4 Experimental results of the scheme

seconds. From Fig. 5.3b-3, the SOPs of DG1 and 2 are (0.209, 0.339, 0.545) and (0.591, 0.861, 0.955). The theoretical GOSs by solving (5.1) are (0.213, 0.323, 0.5) and (0.587, 0.877, 1), respectively. The SOPs almost coincide with the GOSs when PL is 0.8 kW and are close to the GOSs when P_L is 1.2 and 1.5 kW. And the relative error in cost is 0%, 0.03%, and 2%, respectively. Thus, the experiment results are consistent with the theoretical analysis.

Therefore, in Case 1, the SOPs of the point are identically equal to the GOS, i.e., GOD can be achieved by the method. In Case 2, when the load PL belongs to (0, 16], the system operation cost can reach the theoretical global minimum only if the output power of DG1 is zero. When the load P_L belongs to (16, 25], DG1 should work under droop control mode. However, when DG1 should switch the control mode cannot be determined without knowing the real-time information of load, i.e., it needs communication. Therefore, under these conditions, the global optimal dispatch cannot be realized via a decentralized manner.

In Cases 2 and 3, although GOD cannot be achieved, the suboptimal scheme is also acceptable to some extent. So, the simulation and experimental results verify the correctness and effectiveness of the method.

5.4 Conclusion

In this chapter, the problem that under what conditions and how can the system realize global optimal dispatch via decentralized control is studied. Three contributions are concluded as follows:

1. The criterion of decentralized GOD is that optimal solution functions (OSFs) are all strictly monotonically increasing.
2. If the system meets this criterion, a decentralized $P - f$ droop method is introduced to achieve GOD.
3. In particular, the saturation, dead, and hysteretic characteristics of OSFs will not obey the criterion, and then a sub-optimal droop control method is introduced.

Overall, this chapter presents a new method of decentralized optimal dispatch which provides a design guideline to build economical microgrids.

References

1. F. Guo, C. Wen, J. Mao, Y.-D. Song, Distributed economic dispatch for smart grids with random wind power, IEEE Trans. Smart Grid **7**(3), 1572–1583 (2016)
2. F. Guo, C. Wen, J. Mao, J. Chen, Y.-D. Song, Hierarchical decentralized optimization architecture for economic dispatch: a new approach for large-scale power system. IEEE Trans. Ind. Inf. **14**(2), 523–534 (2018)
3. F. Guo, C. Wen, Y.-D. Song, *Distributed Control and Optimization Technologies in Smart Grid Systems* (CRC Press, Boca Raton, 2017)

4. I.U. Nutkani, P.C. Loh, W. Peng, F. Blaabjerg, Cost-based droop scheme with lower generation costs for microgrids. IET Power Electron. **7**(5), 1171–1180 (2014)
5. I.U. Nutkani, P.C. Loh, W. Peng, F. Blaabjerg, Cost-prioritized droop schemes for autonomous AC microgrids. IEEE Trans. Power Electron. **30**(2),1109–1119 (2015)
6. I.U. Nutkani, P.C. Loh, W. Peng, F. Blaabjerg, Linear decentralized power sharing schemes for economic operation of AC microgrids. IEEE Trans. Ind. Electron. **63**(1), 225–234 (2016)
7. I.U. Nutkani, P.C. Loh, W. Peng, F. Blaabjerg, Decentralized economic dispatch scheme with online power reserve for microgrids. IEEE Trans. Smart Grid **8**(1), 139–148 (2017)
8. A. Elrayyah, F. Cingoz, Y. Sozer, Construction of nonlinear droop relations to optimize Islanded microgrid operation. IEEE Trans. Ind. Appl. **51**(4), 3404–3413 (2015)
9. A. Elrayyah, F. Cingoz, Y. Sozer, Plug-and-play nonlinear droop construction scheme to optimize Islanded microgrid operations. IEEE Trans. Power Electron. **32**(4), 2743–2756 (2017)
10. F. Chen, M. Chen, Q. Li, K. Meng, Y. Zheng, J.M. Guerrero, D. Abbott, Cost based droop schemes for economic dispatch in Islanded microgrids. IEEE Trans. Smart Grid **8**(1), 63–74 (2017)
11. Q. Xu, J. Xiao, P. Wang, C. Wen, A decentralized control strategy for economic operation of autonomous AC, DC, and hybrid AC/DC microgrids. IEEE Trans. Energy Convers. **32**(4), 1345–1355 (2017)
12. H. Hua, L. Li, L. Wang, M. Su, Y. Zhao, J.M. Guerrero, A novel decentralized economic operation in Islanded AC microgrids. Energies **10**(6),1–18 (2017)
13. Z. Liu, et al., Optimal criterion and global/sub-optimal control schemes of decentralized economical dispatch for AC microgrid. Int. J. Electr. Power Energy Syst. **104**, 38–42 (2019)

Chapter 6
Dynamic Distributed Consensus Control Strategy

6.1 Analysis of Modular UPS System

6.1.1 Configuration of Modular UPS System

Recently, the large installation of the critical loads for data center, communication network, and IT servers has propelled the fast-growing market of the uninterruptible power supply (UPS) system [1]. Meanwhile, the advanced technologies have been widely applied in the UPS system to achieve highly reliable and secure products [2].

According to the IEC standard 62040-3 [3], the UPS systems are classified into off-line, line-interactive, and on-line UPS systems. Among these three types of configuration, the on-line UPS system has been receiving increasing attention from both researchers and engineers, as this type of UPS system can provide the isolation between the grid and the load [4, 5]. In such a way, the critical load is not affected by the grid voltage irregularity, frequency variation, and other grid issues. Recently, the modular on-line UPS system has been popularized due to its flexible, reliable, and easy maintaining features [6]. As is seen from Fig. 6.1, in the modular UPS system, each module is comprised of an AC/DC rectifier, a DC/AC inverter, and a battery pack, and several modules are connected together to form the modular UPS system. In the normal mode of operation, the critical load is powered by the combination of these AC/DC rectifiers and DC/AC inverters. Meanwhile, the battery can be charged through the AC/DC rectifiers. However, under the abnormal situation, the on-line UPS system switches to the backup mode, where the battery is in charge of supplying the power to the critical loads. On the other hand, the bypass switch needs to be closed to directly support the load in case of overloading or the power failure of the UPS system.

In the normal operation of the parallel inverters in the modular on-line UPS system, master–slave control [7], droop control [8], average current sharing control [9], and circular chain control [10] have been proposed to regulate the power delivery of the modular UPS system. Among these control strategies, the droop control has

Y. Sun et al., *Series-Parallel Converter-Based Microgrids*, Power Systems,
https://doi.org/10.1007/978-3-030-91511-7_6

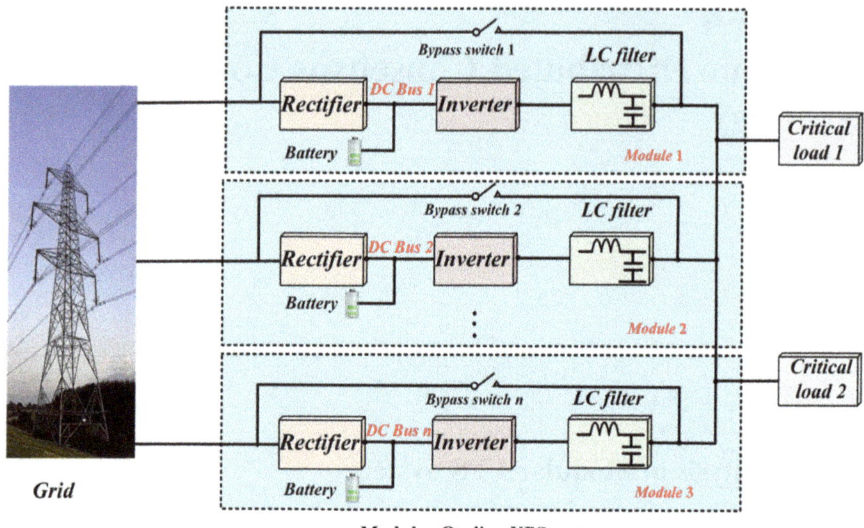

Fig. 6.1 The configuration of modular UPS system

been widely adopted for the UPS system, as the droop control is a decentralized control method, which is able to regulate the voltage amplitude and the frequency without communication among the UPS modules [11, 12]. Accordingly, flexible, expandable, and reliable features of the UPS modules are achieved by using this control strategy. However, the droop control strategy may cause the active power sharing error in the system when the line impedance is mismatched. It has been pointed out that in the low-voltage network, such as UPS system, the reactive power sharing is always accurate, while the active power sharing is dependent on the line impedance of UPS system [13]. As a result, the circulating current exists in the UPS system due to the unequal active power sharing. In order to enhance the power sharing capability and reduce the circulating current, the virtual impedance has been equipped in the system [14].

6.1.2 Operation Principle of Modular UPS System

Normally, the UPS system is connected to the 120/220 V low voltage distribution power system, where the line impedance shows the resistive characteristic. In order to achieve the wireless control strategy, the modified droop control strategy is applied in the system and expressed as [13]

$$V_i = V^* - D_p P_{LPF} \tag{6.1}$$

$$\omega = \omega^* + D_q Q_{LPF} \tag{6.2}$$

where ω^* and ω are the UPS's nominal and reference angular frequency. V^* and V_i are the nominal and reference voltage amplitude, respectively. D_p and D_q are the droop parameters to regulate the active power and reactive power, respectively. P_{LPF} and Q_{LPF} are the output active power and reactive power by a low-pass filter.

In the UPS system, the accuracy of the active power sharing is affected by the mismatch of the line resistance. In order to share the active power in the UPS system, the line resistance of the UPS system should be designed to be in inverse proportion to the rating of the inverter, such as

$$R_{line,1}P_{1,rated} = R_{line,2}P_{2,rated} = \cdots = R_{line,n}P_{n,rated} \tag{6.3}$$

where $R_{line,1}$ to $R_{line,n}$ are the physical line resistance, and $P_{1,rated}$ to $P_{n,rated}$ are the UPS modules' rated output active power. However, the power rating of each UPS module is normally the same from the view of modular production and the plug-and-play feature. Therefore, in order to achieve the active power sharing, the virtual resistance has to be added in the control strategy to accurately share the active power.

Figure 6.2 provides the details for the principle of the active power sharing. In Fig. 6.2, the parallel UPS modules are modeled as the controlled voltage source V_{ref}, $R_{Line,f}$ indicates the line resistance at the fundamental frequency, $R_{V,f}$ is the virtual resistance at the fundamental frequency, and PCC linear load is modeled as the passive RL load. Generally, when the power flows through a feeder that is comprised of inductance and resistance, the voltage drop across the feeder provides the following expression:

$$\Delta V = \frac{X \cdot Q + R \cdot P}{V^*} \tag{6.4}$$

where P and Q are the active and reactive instantaneous powers that flow out of the line feeder, and R and X are the resistance and inductance of the line feeder. V^* is the nominal voltage amplitude, and ΔV is the voltage magnitude drop on the feeder. When this equation is applied for the UPS system, the inductance is usually neglected as the line feeder shows the resistive characteristic. As a result, the power flow through the line resistance leads to the voltage drop, which is expressed as

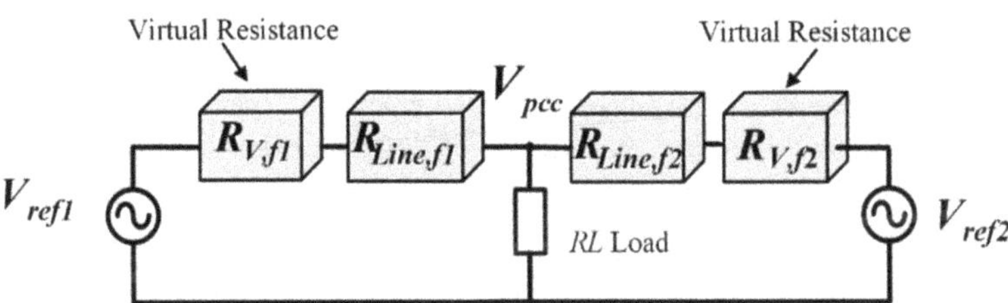

Fig. 6.2 Equivalent circuit for the UPS system with linear load

$$\Delta V = \frac{R \cdot P}{V*} \tag{6.5}$$

Therefore, when applying the voltage drop equation in (6.5) to the UPS system, the relationship between the UPS output voltage and the PCC voltage is expressed as

$$V_{ref1} = V_{pcc} + \frac{R_{e1} \cdot P_1}{V*} \tag{6.6}$$

$$V_{ref2} = V_{pcc} + \frac{R_{e2} \cdot P_2}{V*} \tag{6.7}$$

where $R_{e1} = R_{Line,f1} + R_{V,f1}$, and $R_{e2} = R_{Line,f2} + R_{V,f2}$, and P_1 and P_2 are the output active power of the two inverters.

By combing (6.6) and (6.7), it is found that the active power difference can be expressed in (6.8) and shown as

$$P_1 - P_2 = \frac{(V_{ref1} - V_{pcc})V^*}{R_{e1}} - \frac{(V_{ref2} - V_{pcc})V^*}{R_{e2}} \tag{6.8}$$

It is shown from (6.8) that the active power difference is mainly caused by two factors, namely the total resistance difference (R_{e1} and R_{e2}) and the voltage amplitude difference (V_{ref1} and V_{ref2}). In practical situation, the line resistances $R_{Line,f1}$ and $R_{Line,f2}$ have different values because of the stray parameters. As a result, without regulating the virtual resistance ($R_{V,f1}$ and $R_{V,f2}$), the circulating current inevitably exists in the UPS system. Therefore, the solution to eliminate the active power difference caused by circulating current is to adaptively adjust the virtual resistance ($R_{V,f1}$ and $R_{V,f2}$).

Meanwhile, the nonlinear load, such as sensitive load, may be supplied power by the modular on-line UPS system. Therefore, the harmonic power difference may also exist in the system as well due to the line resistance mismatch. Hence, the equivalent circuit at the harmonic frequency is illustrated in Fig. 6.3. The harmonic resistance is expressed as

$$R_H = R_{Line,H} + R_{v,H} \tag{6.9}$$

where $R_{Line,H}$ is the physical line resistance at the harmonic frequency, and $R_{v,H}$ is the virtual resistance at the harmonic frequency. It is noted that the voltage source at the harmonic frequency is not shown in Fig. 6.3, since the output harmonic voltage of the UPS system should be controlled to be zero in order to satisfy IEC 62020-3 standard.

The circulating harmonic current is defined as

$$I_{cir,H} = \frac{I_{H1} - I_{H2}}{2} \tag{6.10}$$

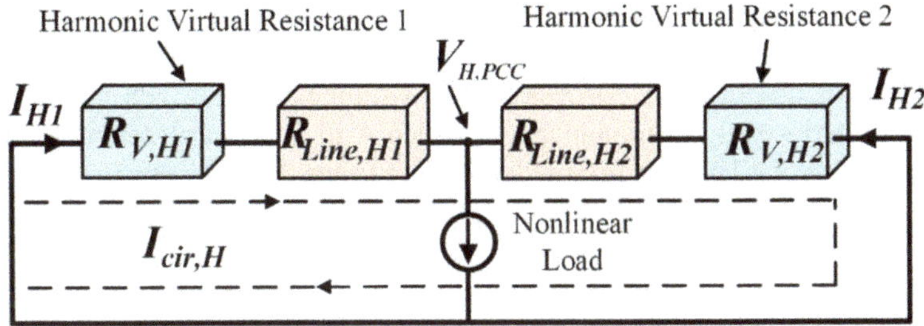

Fig. 6.3 Equivalent circuit for the UPS system with nonlinear load

According to Fig. 6.3, I_{H1} and I_{H2} are expressed as

$$I_{H1} = \frac{V_{H,pcc}}{R_{e,H1}} \tag{6.11}$$

$$I_{H2} = \frac{V_{H,pcc}}{R_{e,H2}} \tag{6.12}$$

where $R_{e,H1} = R_{Line,H1} + R_{v,H1}$ and $R_{e,H2} = R_{Line,H2} + R_{v,H2}$. As a result, by substituting (6.11) and (6.12) into (6.10), the circulating harmonic current can be expressed as

$$I_{cir,H} = \frac{V_{H,pcc}}{R_{e,H1} + R_{Line,H1}} - \frac{V_{H,pcc}}{R_{e,H2} + R_{Line,H2}} \tag{6.13}$$

From the above analysis, it is found that if physical line resistances are equal to each other, even without virtual harmonic resistance, the circulating harmonic current does not exist in the system. However, in the real system, the stray parameters lead to the different value of the line resistance. Therefore, the adaptive virtual harmonic resistance is added in the system to eliminate the circulating harmonic current.

6.2 Dynamic Consensus-Based Adaptive Virtual Resistance Control

In a centralized control system, the central controller plays a key role for the information sharing. As a consequence, the reliability of the system highly depends on this central controller. In addition, as the number of the UPS modules increases, an increasing amount of information needs to be exchanged as well, which may even lead to the communication jam. In order to increase the system's flexibility and reliability, a DCA-based adaptive virtual resistance for the power sharing is presented in this section, which can reduce the communication burden and eliminate the dependence on the only central controller. With DCA strategy, each UPS module only communicates its desired state with the adjacent UPS module, which effectively reduces the communication burden with the central controller. Finally, the states of all the UPS modules will converge to the desired average value.

The discrete expression of the consensus algorithm is written as

$$x_i(k+1) = x_i(k) + \varepsilon \cdot \sum_{j \in N} a_{ij}\left(x_j(k) - x_i(k)\right) \tag{6.14}$$

where $x_i(k)$ indicates the information status of agent i at iteration k, and a_{ij} is the connection status between node i and node j. ε is the consensus edge weight used for tuning the parameter of DCA. It is noted that if node i and node j are not adjacent, a_{ij}=0.

In order to guarantee that all of the agents' shared information can accurately converge to the consensus value, a modified type of the consensus algorithm, namely DCA, is applied and is shown in Fig. 6.4 and expressed as

$$\begin{cases} x_i(k+1) = x_i(0) + \varepsilon \cdot \sum\limits_{j \in N} \delta_{ij}(k+1) \\ \delta_{ij}(k+1) = \delta_{ij}(k) + \sum\limits_{j \in N} a_{ij}\left(x_j(k) - x_i(k)\right) \end{cases} \tag{6.15}$$

where $\delta_{ij}(k)$ stores the cumulative difference between the two agents.

Regarding the physical meaning, the x_i may indicate the i UPS module's active power $P_{i,LPF}$ at the fundamental frequency and the harmonic power $P_{i,H}$ for the modular on-line UPS system. Hence, (6.15) can be rewritten as (6.16) to deduce the DCA strategy of active power:

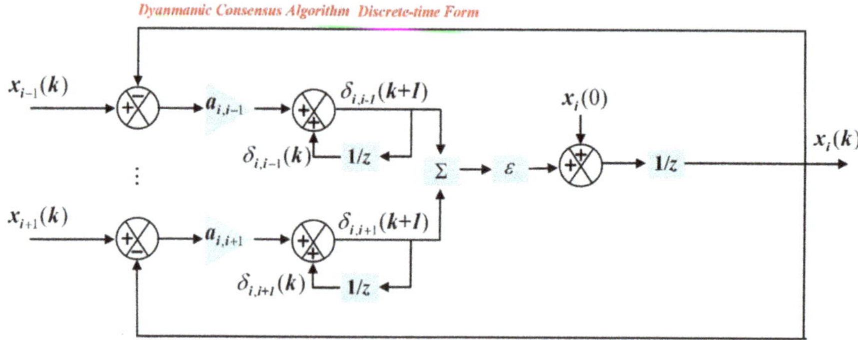

Fig. 6.4 Details of the dynamic consensus algorithm in discrete-time form

$$\begin{cases} P_{i,LPF}(k+1) = P_{i,LPF}(0) + \varepsilon_{LPF} \cdot \sum_{j \in N} \delta_{ij,LPF}(k+1) \\ \delta_{ij,LPF}(k+1) = \delta_{ij,LPF}(k) + \sum_{j \in N} a_{ij,LPF}\left(P_{i,LPF}(k) - P_{i,LPF}(k)\right) \end{cases} \tag{6.16}$$

It is noted that the final value of each module's active power will converge to the averaged active power $P_{ave,LPF}$ with DCA strategy. The proof of the convergence can be found in [15]. Meanwhile, the DCA strategy for the harmonic power can be expressed as

$$\begin{cases} P_{i,H}(k+1) = P_{i,H}(0) + \varepsilon_{H} \cdot \sum_{j \in N} \delta_{ij,H}(k+1) \\ \delta_{ij,H}(k+1) = \delta_{ij,H}(k) + \sum_{j \in N} a_{ij,H}\left(P_{i,H}(k) - P_{i,H}(k)\right) \end{cases} \tag{6.17}$$

It is noted that the final value of each module's harmonic power will converge to the averaged harmonic power $P_{ave,H}$ as well. In addition, the edge weight constants, ε_{LPF} and ε_{H}, affect the convergence stability and the converged dynamics, and the parameter selection for ε_{f} and ε_{H} is obtained by determining the following equation:

$$0 < \varepsilon_i < \frac{2}{\lambda_1(L)} \tag{6.18}$$

where ε_i can be ε_{LPF} or ε_{H}, L is the Laplacian matrix, which is defined as $L = A \cdot A^T$, and the element of matrix A is expressed as

$$A_{ij} = \begin{cases} 1, & if\ edg\ l\ starts\ from\ node\ i \\ -1, & if\ edg\ l\ ends\ at\ node \\ 0, & otherwise \end{cases} \tag{6.19}$$

Therefore, $\lambda_1(L)$ is chosen to be the largest eigenvalue of L. Hence, when the edge weight constant ε_i is between zero and $\frac{2}{\lambda_1(L)}$, the convergence is promised. In order to minimize the convergence time for the communication network, ε is suggested to be tuned as the following equation and expressed as

$$\varepsilon = \frac{2}{\lambda_1(L) + \lambda_{n-1}(L)} \tag{6.20}$$

Once the average fundamental active power and average harmonic active power are obtained by the DCA via neighboring information exchange. By comparing the average active power obtained by DCA and the respective measured local fundamental active power, the difference of these active powers goes through a PI controller to generate the adaptive virtual resistance. The adaptive fundamental virtual resistance is expressed as

$$\Delta R_{v,f} = \left(P_{ave,LPF} - P_{i,LPF}\right) \cdot \left(K_{p,f} + \frac{K_{i,f}}{s}\right) \quad (6.21)$$

where $K_{p,f}$ and $K_{p,i}$ are the proportional and integral gain of the controller. For each UPS module, the active power is expressed as

$$P_{LPF} = \frac{3\omega_c}{2\left(s + \omega_c\right)}\left(V_{c\alpha} \cdot I_{\alpha,f} + V_{c\beta} \cdot I_{\beta,f}\right) \quad (6.22)$$

where $V_{c\alpha}$ and $V_{c\beta}$ are the measured sinusoidal UPS voltage in stationary frame, and $I_{c\alpha}$ and $I_{c\beta}$ are the UPS fundamental positive sequence current. The ripples of active and reactive power are attenuated by the low-pass filter with cutoff frequency ω_c (Figs. 6.5 and 6.6).

Similarly, by comparing the average harmonic power obtained from DCA and respective local harmonic power in each UPS module, the harmonic power error will be used to regulate the adaptive harmonic virtual resistance at the selected harmonic frequency as

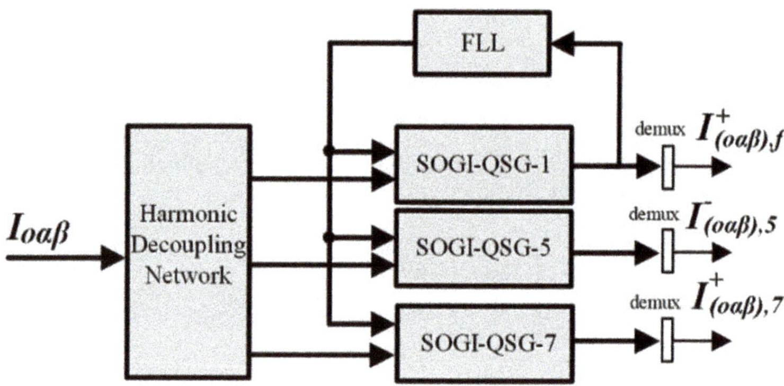

Fig. 6.5 SOGI-based current decomposition diagram

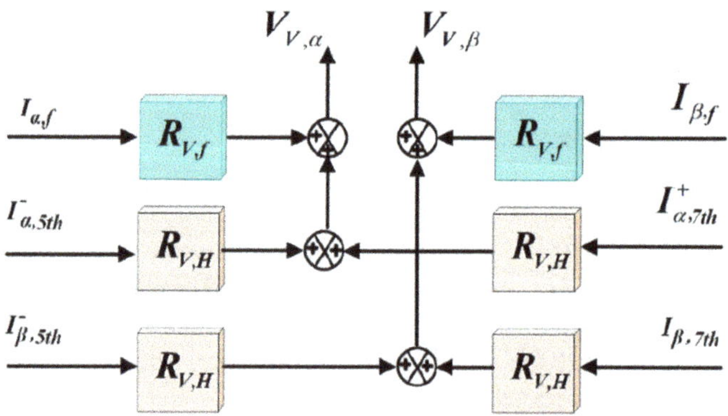

Fig. 6.6 Virtual resistance loop at fundamental and harmonic frequencies

$$\Delta R_{v,H} = \left(P_{ave,H} - P_{i,H}\right) \cdot \left(K_{p,H} + \frac{K_{i,H}}{s}\right) \tag{6.23}$$

where $P_{i,H}$ indicates the active power of the ith UPS module. $K_{p,H}$ and $K_{i,H}$ are the proportional and integral gain of the controller. The harmonic power in each UPS module is expressed as

$$P_H = \frac{3}{2} V^* \sqrt{\left(I_{\alpha,5}^{-}\right)^2 + \left(I_{\beta,5}^{-}\right)^2 + \left(I_{\alpha,7}^{+}\right)^2 + \left(I_{\beta,7}^{+}\right)^2} \tag{6.24}$$

where V^* is the nominal voltage of the UPS system, $I_{\alpha,5}^{-}$ and $I_{\beta,5}^{-}$ indicate the negative sequence of the UPS fifth harmonic current, while $I_{\alpha,7}^{+}$ and $I_{\beta,7}^{+}$ indicate the positive sequence of the UPS seventh harmonic current. It is noteworthy that only the low-order harmonic current component, such as fifth and seventh harmonic currents, is employed to calculate the harmonic power, as these two harmonic currents take the majority of the harmonic current in the system. In addition, the separation of the fundamental current and the harmonic current can be achieved by using a second-order generalized integrator (SOGI)-based sequence decomposition method, which is composed of a harmonic decoupling network, frequency locked-loop, and the multiple SOGI quadrature signal generators.

The completed control diagram for each UPS module is illustrated in Fig. 6.7, where it can be observed that it includes the primary and secondary control layers. In the primary control layer, droop control, voltage controller, and current controllers are adopted. Meanwhile, in the secondary control layer, the DCA-based adaptive virtual resistance control strategy is adopted to mitigate the fundamental and harmonic power error in the system. It is noteworthy that in order to achieve the voltage tracking in the inverter, a double loop controller is adopted, which is comprised of voltage control and current control strategy. The outer loop mainly regulates the output capacitor's voltage, while the inner loop current controller is to regulate the inverter side current. These two controllers are expressed as

$$G_v(s) = k_{pv} + \frac{k_{rV} s}{s^2 + (\omega_0)^2} + \sum\nolimits_{h=5,7} \frac{k_{vh} s}{s^2 + (h\omega_0)^2} \tag{6.25}$$

$$G_i(s) = k_{pi} + \frac{k_{ri} s}{s^2 + (\omega_0)^2} \tag{6.26}$$

where k_{pv} and k_{pi} are the proportional terms, and k_{rV} and k_{ri} are the resonant term coefficient at ω_o=50 Hz. k_{vh} is the resonant coefficient term for the harmonics h (5th and 7th). The inner current loop is designed to provide sufficient damping and protect the inductor's current from overcurrent.

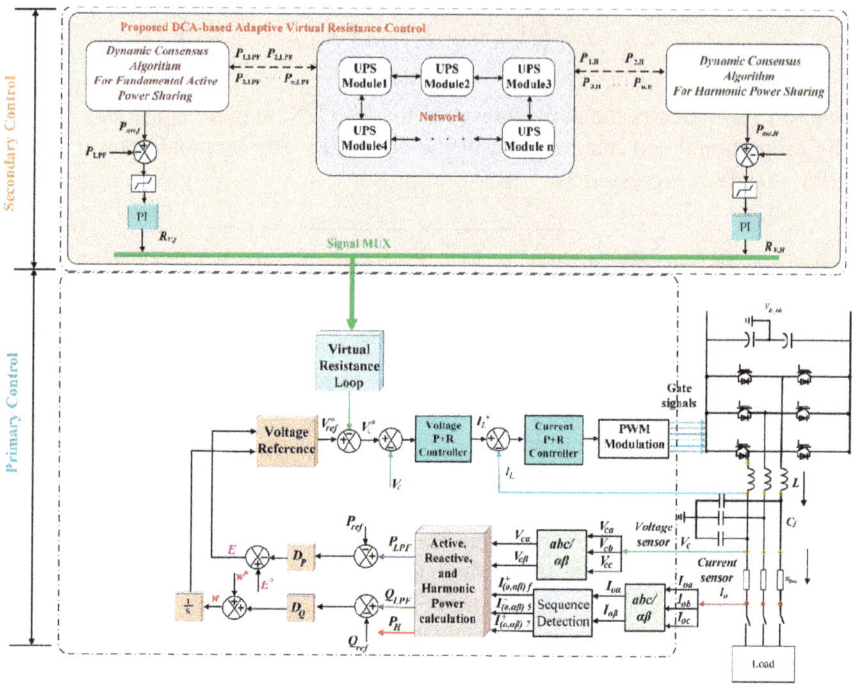

Fig. 6.7 Completed diagram of the UPS system with the DCA strategy for active and harmonic power sharing.

6.3 Simulation Results

In order to validate the feasibility of the control strategy, the module UPS system has been built up in MATLAB/Simulink environment. The UPS system consists of three modules and has two loads. The communication among each UPS module is shown in Fig. 6.8 as well, which is a ring-shaped topology.

6.3.1 *Case 1: Dynamic Performance Test with Linear Load and Mismatched Line Resistance*

In this test, the line resistance for UPS module 1, module 2, and module 3 is set to be 0.2 , 0.3, and 0.4 to emulate the mismatched line resistance. From the time range t=0.5 to t=1s, only the conventional method is adopted. It can be seen from Fig. 6.9 that due to the mismatched line resistance, the active power sharing is not equalized among these three modules. From 1s, the active power is accurately shared by adopting the consensus-based control strategy, where the adaptive virtual

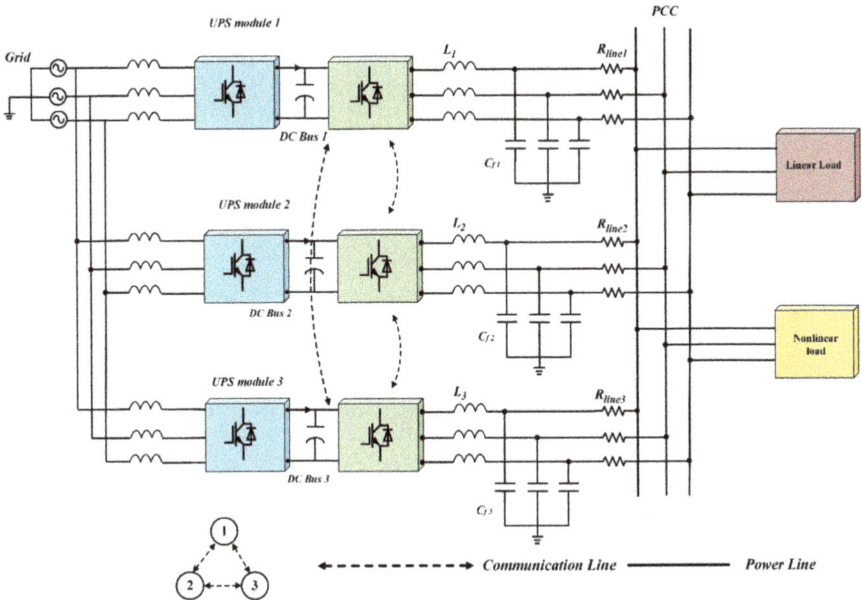

Fig. 6.8 The simulation test modular UPS system

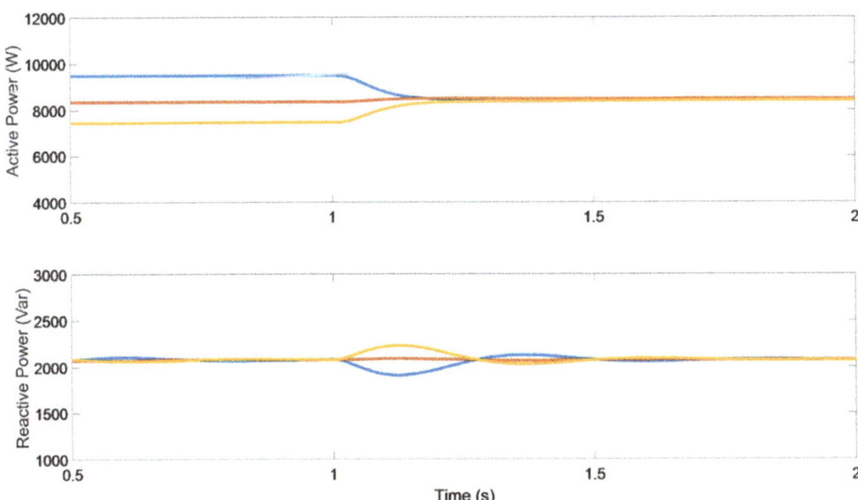

Fig. 6.9 Active and reactive power of three modules

resistance is regulated. As a result, all the UPS modules share the same amount of the active power. In addition, Fig. 6.10 shows that with the control strategy at 1s, the load current is properly shared by the three UPS modules.

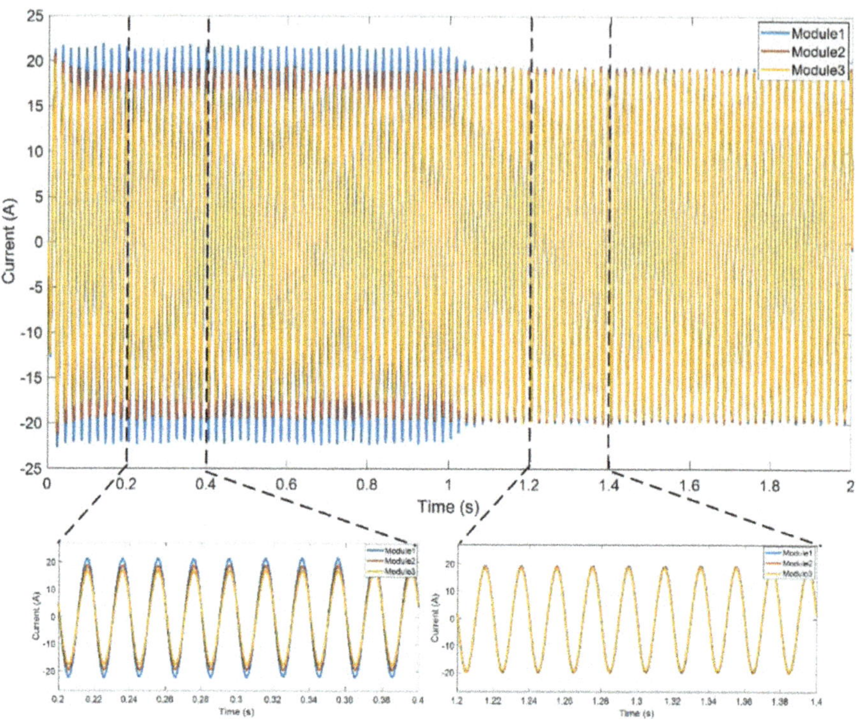

Fig. 6.10 Phase A output current of the UPS modules with linear load

6.3.2 *Case 2: Dynamic Performance Test with Both Linear and Nonlinear Loads*

In this test, both the linear load and the nonlinear load are connected at the PCC to test the effectiveness of the control strategies. The results are shown in Fig. 6.11. As illustrated in Fig. 6.11, without the control strategy, UPS system cannot share the linear load and nonlinear load. As a result, the active power and harmonic power are greatly diverse before 1s. However, after 1s, when the control strategy is activated, the active power and harmonic power are accurately shared among the three UPS modules as shown in Fig. 6.11.

6.4 Experimental Results

In order to validate the feasibility of the DCA-based virtual resistance control strategy, the control algorithm is implemented in dSPACE1006 platform for real-time control. The system parameters are listed in Table 6.1. Waveforms are captured by the oscilloscope.

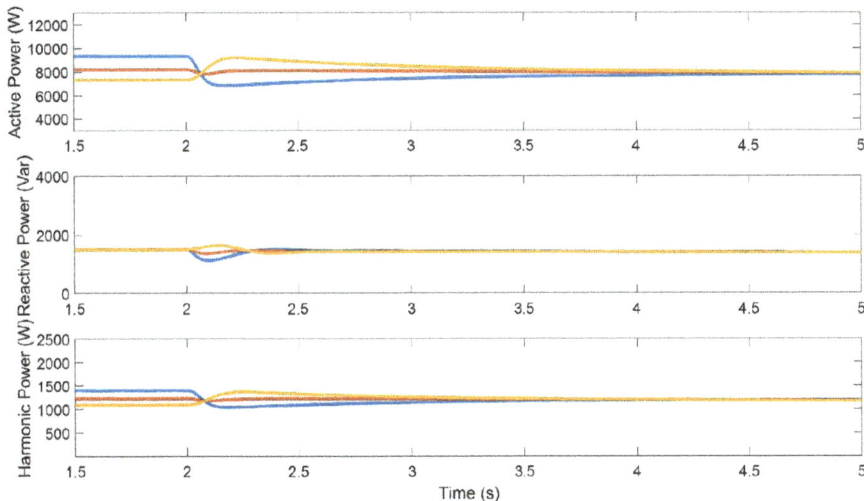

Fig. 6.11 Active, reactive, and harmonic power of the three UPS modules with general loads

Table 6.1 The experiment parameters

Parameter	Value
Filter inductor L_f	1.8 mH
ESR of inductor	0.02 Ω
Filter capacitor C_f	27 μF
Sampling frequency	10 kHz
Rated line-to-line RMS voltage	120 V
R-L Load	50 Ω−1.8 mH
Nonlinear Load (R-L-C type)	C=20 μF, L=1.8 mH, R=20 Ω
Proportional gain K_{p-f}	2e−4
Integral gain K_{i-f}	1e−4
Parameter A	35
Parameter B	10000
Frequency droop D_q	0.0001
Voltage droop D_p	0.00005
Proportional gain K_{pv}	0.2
Resonant gain K_{iv}	100
Proportional gain K_{p-i}	5.6
Resonant gain K_{i-i}	500

6.4.1 Case 1: Under Linear Load

In this case, the line resistances for the three UPS modules are mismatched, and in the following figures, the power sharing performance can be observed with conventional method and the DCA-based method. It can be seen from Fig. 6.12 that before the activation of the DCA-based method, the active power cannot be equally shared by the three UPS modules. After t_1, the DCA method is activated, so the active powers in the three UPS modules are beginning to share. It takes around 1.2s to equally share the active power. Figure 6.13 shows the circulating current in the system during the process of the active power sharing. It can be observed that before

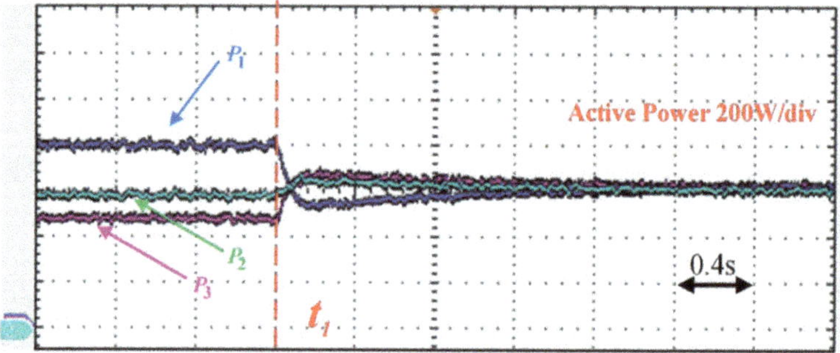

Fig. 6.12 Modular UPS active power sharing performance under linear load

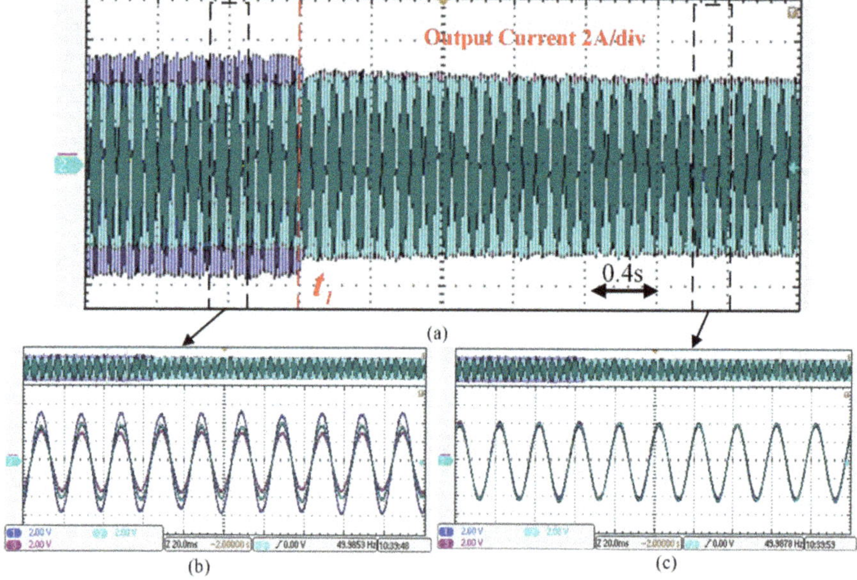

Fig. 6.13 Modular UPS current sharing performance under linear load

activating the control strategy, the Phase A current of the three UPS modules is not the same due to the unequal active power sharing, which may lead to the circulating current in the system. After t_1, when the DCA strategy is initiated, the Phase A current of the three UPS modules is almost the same, which can be observed from Fig. 6.13.

6.4.2 *Case 2: Under Generalized Load*

In this case, both linear and nonlinear loads are connected at the terminal with mismatched line resistances. From experimental results shown in Figs. 6.14, 6.15, and 6.16, it can be observed that fundamental active power sharing and harmonic power sharing are unequal due to the line resistance mismatch. To reduce the power sharing error, at t_2 instant, the DCA-based active power sharing and harmonic power sharing control strategies are both activated. By generating the virtual resistance through the DCA-based control strategy, the active power and the harmonic power are equally shared. And the current sharing performance is illustrated in Fig. 6.16, and compared to the performance before t_2 time instant, the control strategy can

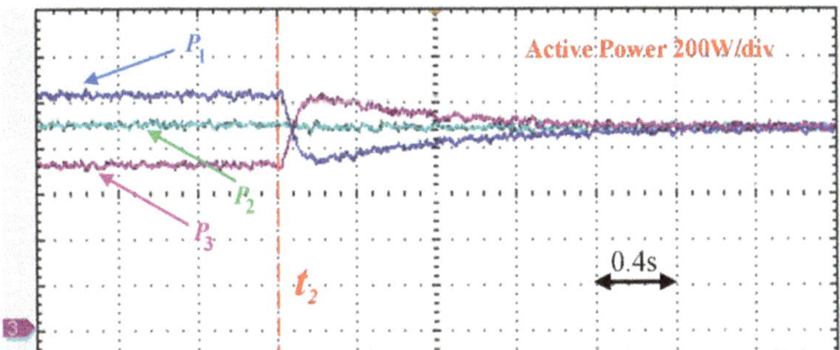

Fig. 6.14 Modular UPS active power sharing performance under generalized load

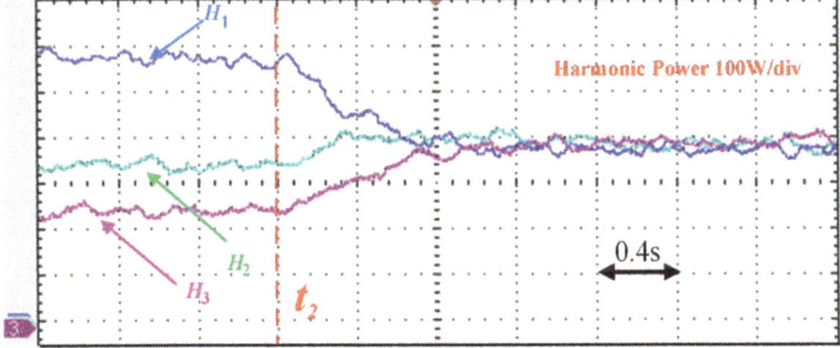

Fig. 6.15 Modular UPS harmonic active power sharing performance under generalized load

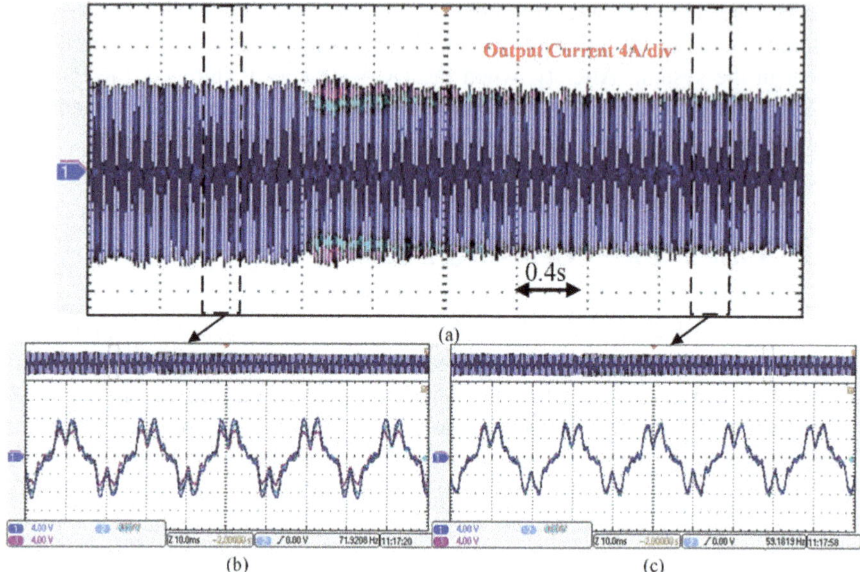

Fig. 6.16 Modular UPS current sharing performance under generalized load

effectively address the circulating current caused by the active power sharing and harmonic power sharing with generalized loads.

6.5 Conclusion

This chapter discusses the power sharing scheme for the modular UPS system. A DCA-based virtual resistance control strategy has been introduced to accurately control the power sharing of the modular UPS system. By adopting the distributed control strategy, the modular UPS system can conquer several drawbacks with the centralized control method. Simulation results including three UPS modules are obtained to show the effectiveness of the control strategy. The feasibility and effectiveness of the control strategy have also been demonstrated by experimental platform. Through the several dynamic tests, the circulating current among the parallel UPS modules can be effectively suppressed under different line resistances.

References

1. G.N. Rajani, Emerging trends in uninterrupted power supplies: patents view, in *2016 Biennial International Conference on Power and Energy Systems: Towards Sustainable Energy (PESTSE)* (2016), pp. 1–5

2. M.S. Racine, J.D. Parham, M.H. Rashid, An overview of uninterruptible power supplies, in *Proceedings of the 37th Annual North American Power Symposium* (2005), pp. 159–164
3. IEC-62040-3-2011 uninterruptible power systems (UPS)-part 3: method of specifying the performance and test requirements
4. J. Lu, et al., DC-link protection and control in modular uninterruptible power supply. IEEE Trans. Ind. Electron. **65**, 3942–3953 (2018)
5. J. Lu, M. Savaghebi, S. Golestan, J.C. Vasquez, J.M. Guerrero, A. Marzabal, Multi-mode operation for on-line uninterruptible power supply system. IEEE J. Emerg. Sel. Topics Power Electron. **7**, 1181–1196 (2019)
6. C. Zhang, J.M. Guerrero, J.C. Vasquez, E.A.A. Coelho, Control architecture for parallel-connected inverters in uninterruptible power systems. IEEE Trans. Power Electron. **31**(7), 5176–5188 (2016)
7. P. Yunqing, J. Guibin, Y. Xu, W. Zhaoan, Auto-master-slave control technique of parallel inverters in distributed AC power systems and UPS, in *2004 IEEE 35th Annual Power Electronics Specialists Conference, 2004. PESC04.* vol. 3 (2004), pp. 2050–2053
8. H. Shang-Hung, K. Chun-Yi, L. Tzung-Lin, J.M. Guerrero, Droop-controlled inverters with seamless transition between islanding and grid-connected operations, in *Energy Conversion Congress and Exposition (ECCE), 2011* (IEEE, Piscataway, 2011), pp. 2196–2201
9. S. Xiao, L. Yim-Shu, X. Dehong, Modeling, analysis, and implementation of parallel multi-inverter systems with instantaneous average-current-sharing scheme. IEEE Trans. Power Electron. **18**(3), 844–856 (2003)
10. K. Piboonwattanakit, W. Khan-ngern, Design of the two parallel inverter modules by circular chain control technique, in *2007 7th International Conference on Power Electronics and Drive Systems* (2007), pp. 1518–1522
11. J.M. Guerrero, J. Matas, L.G.D.V.D. Vicuna, M. Castilla, J. Miret, Wireless-control strategy for parallel operation of distributed-generation inverters. IEEE Trans. Ind. Electron. **53**(5), 1461–1470 (2006)
12. Y. Hu, J. Xiang, Y. Peng, P. Yang, W. Wei, Decentralised control for reactive power sharing using adaptive virtual impedance. IET Gener. Trans. Distrib. **12**(5), 1198–1205 (2018)
13. J.M. Guerrero, J.C. Vasquez, J. Matas, M. Castilla, L.G.D. Vicuna, Control strategy for flexible microgrid based on parallel line-interactive UPS systems. IEEE Trans. Ind. Electron. **56**(3), 726–736 (2009)
14. H. Shi, F. Zhuo, D. Zhang, Z. Geng, F. Wang, Adaptive implementation strategy of virtual impedance for paralleled inverters UPS, in *2014 IEEE Energy Conversion Congress and Exposition (ECCE)* (2014), pp. 158–162
15. J. Lu, X. Hou, A.M. Ghias, J. Ye, A dynamic consensus algorithm-based adaptive virtual resistance control strategy for modular on-line uninterruptible power supply system, in *2020 IEEE 9th International Power Electronics and Motion Control Conference (IPEMC2020-ECCE Asia)* (2020), pp. 1129–1136

Chapter 7
Distributed Event-Triggered Control with Less Communication

7.1 Islanded Microgrid Analysis

7.1.1 Reactive, Unbalanced, and Harmonic Power Sharing Analysis in Islanded AC Microgrids

In the islanded mode of microgrids, an essential requirement is the proper active and reactive power sharing between DG units. Traditionally, active power–frequency ($P - \omega$) and reactive power–voltage magnitude ($Q - V$) based droop control has been widely applied in the system to regulate the power delivery [1–3]. However, a small mismatch in the line impedance may cause a noticeable error in the reactive power sharing using the $Q - V$ droop [4]. Some efforts to address this challenge of the droop method have been made in recent years (for example, see [4, 5]). These methods, however, still suffer from some inaccuracies, such as sensitivity to complex loads, among others. In addition, a fully decentralized control of grid-connected series inverters is firstly proposed to achieve the autonomous power balance in [6]. On the other hand, with growing power quality issues such as harmonics and voltage imbalance, which are attributable to the proliferation of nonlinear loads and supplying single-phase and/or unbalanced three-phase loads, the DG units are expected to properly share unbalanced power and harmonic powers. The traditional droop control, however, may not be able to achieve this objective, as the traditional droop is derived from the fundamental voltage model and does not take the unbalanced and harmonic powers into account. This issue can be especially serious if a mismatch in the line impedances exists.

Figure 7.1 shows the structure of an islanded microgrid, where the DG units are interfaced to the microgrid through different feeders. Each DG unit is comprised of a DC source, an inverter, and an LC-type filter. The microgrid also includes linear, unbalanced, and nonlinear/harmonic loads placed at the PCC.

The power sharing of DG units in islanded microgrids, as mentioned before, is often based on the conventional $P-\omega$ and $Q-V$ droop controllers in DG units. The

Y. Sun et al., *Series-Parallel Converter-Based Microgrids*, Power Systems,
https://doi.org/10.1007/978-3-030-91511-7_7

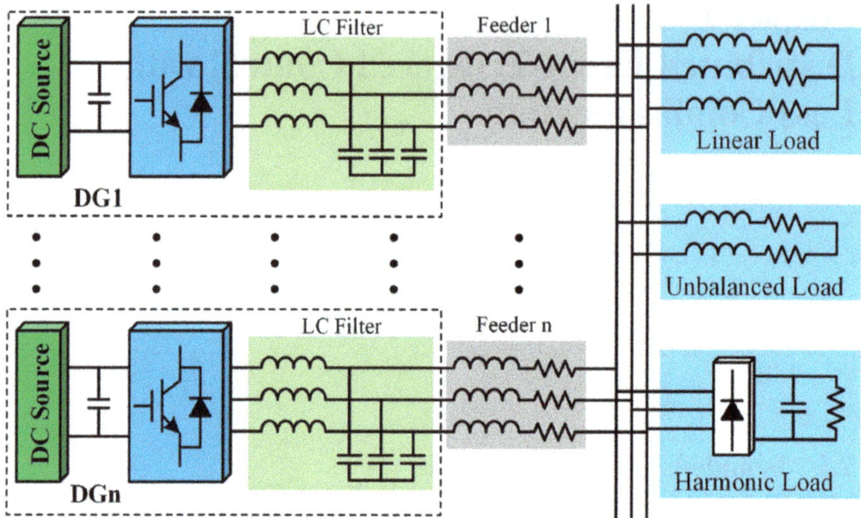

Fig. 7.1 Structure of islanded microgrid with n DG units

outputs of the droop control are fed to a reference generation unit, which generates the fundamental reference voltage $V_{droop,\alpha\beta}$ in the stationary reference frame for the DG unit.

A mismatch in DGs' feeders may adversely affect the stability and power sharing between DGs. To reduce the influence of feeder impedance, the virtual impedance (Z_{vir}) is introduced. Then, the voltage reference from the droop controller is modified as

$$V_{ref,\alpha\beta} = V_{droop,\alpha\beta} - Z_{vir} i_{\alpha\beta} \tag{7.1}$$

where $V_{ref,\alpha\beta}$ is the voltage reference considering virtual impedance and $i_{\alpha\beta}$ is the DG output current.

To simplify the analysis, an islanded microgrid including two DG units with the equal power rating is considered. In addition, it is assumed that the feeder impedances are inductive. Figures 7.2, 7.3, and 7.4 present an equivalent circuit with two DG units. It can be seen from Fig. 7.2 that DG units are modeled by controlled voltage sources at the fundamental positive sequence frequency (FPF). Meanwhile, the load is considered as a passive load. From Fig. 7.2, the DG equivalent impedance $L_{f,i}$ at the FPF becomes

$$L_{f,i} = L_{phyf,i} + L_{virf,i} \tag{7.2}$$

where $L_{phyf,i}$ and $L_{virf,i}$ are the physical feeder impedance and virtual impedance at the FPF.

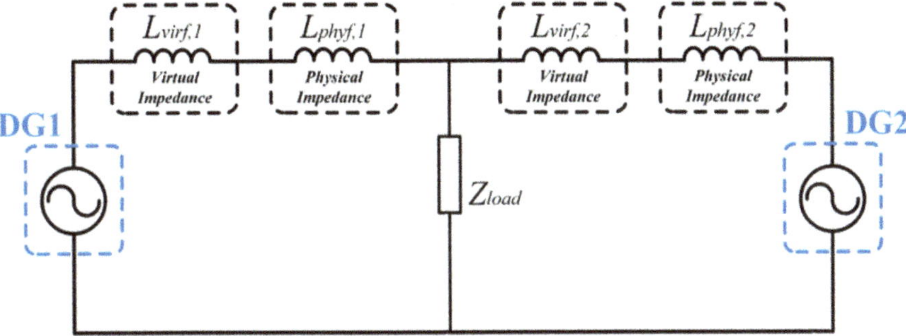

Fig. 7.2 Equivalent circuit at fundamental positive sequence frequency

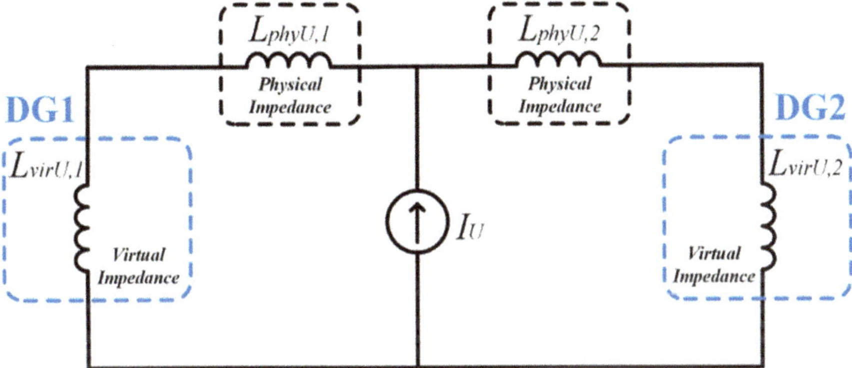

Fig. 7.3 Equivalent circuit at fundamental negative sequence frequency

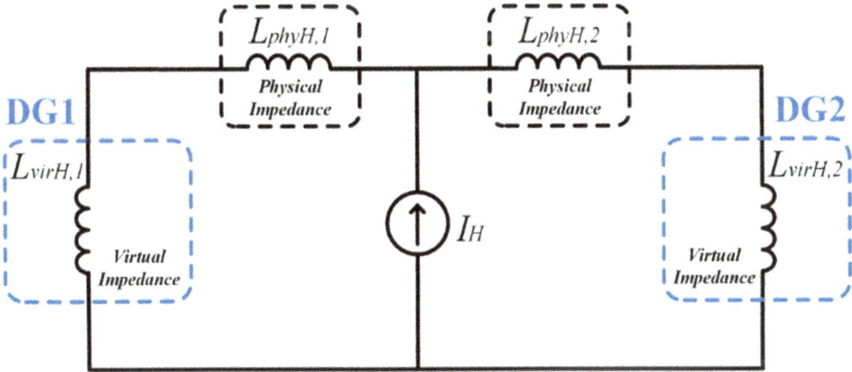

Fig. 7.4 Equivalent circuit at harmonic frequencies

In addition, the equivalent circuit at the fundamental negative sequence frequency (FNF) is illustrated in Fig. 7.3. The unbalanced load is modeled as a current source (IU) [7]. From Fig. 7.3, the DG equivalent fundamental negative sequence impedance $L_{U,i}$ becomes

$$L_{U,i} = L_{phyU,i} + L_{virU,i} \tag{7.3}$$

where $L_{phyU,i}$ and $L_{virU,i}$ are the physical feeder impedance and virtual impedance at the fundamental negative sequence frequency, respectively.

Similarly, the equivalent impedance at harmonic frequencies is given in Fig. 7.4 as

$$L_{H,i} = L_{phyH,i} + L_{virH,i} \tag{7.4}$$

where $L_{phyH,i}$ and $L_{virH,i}$ are, respectively, the physical feeder impedance and virtual impedance at harmonic frequencies.

To achieve accurate power sharing, the same equivalent impedance must be controlled in both DG units. Fortunately, as it can be seen in (7.2)–(7.4), the virtual impedance provides an efficient way for controlling the DG equivalent impedance at a given frequency. Therefore, by regulating the virtual impedances at fundamental positive and negative sequences and at harmonic frequencies, proper reactive/unbalanced/harmonic power sharing can be achieved.

Note that the DG units are assumed to be connected to the PCC with inductive physical feeders in the islanded microgrid. This assumption is reasonable because series coupling inductors are normally required in the DG units to ensure the stability of the power sharing. In addition, DG units are often interconnected to the distribution system with isolation transformers, which have highly inductive leakage impedance. Finally, even when DG units are interfaced to the PCC with resistive line impedance, the fixed value virtual inductive impedance can be added and preactivated in the control scheme. Therefore, if the preactivated virtual inductive impedance is properly designed, the DG equivalent impedance can be inductive.

7.1.2 Communication Network

The communication topology of a microgrid is depicted as a graph $G = (V, \Gamma, A)$, which consists of a set of nodes $V = (v_1, v_2, \ldots, v_N)$, where v_i represents DG i, a set of edges $\Gamma \subseteq V \times V$, and the adjacency matrix $A = \left[a_{ij}\right] \in R^{N\times N}$. $\left(v_j, v_i\right) \in \Gamma$ denotes an edge, which means that node j can transmit its own information to node i. The graph G is said to be undirected if for all edges $\left(v_j, v_i\right) \in \Gamma$. The neighboring set of i is presented as $N_i = \left\{v_j \in V | \left(v_j, v_i\right) \in \Gamma, i \neq j\right\}$. The elements $a_{ij} = 1$ if $\left(v_j, v_i\right) \in \Gamma$; otherwise, $a_{ij} = 0$. The degree matrix $D = diag d_1, \ldots, d_N$ is a diagonal matrix with $d_i = |N_i|$. The Laplacian matrix L is defined as $L = D - A$.

A path from node i to node j is a sequence of edges, belonging to Γ, which can be expressed as $\{(v_i, v_k), \cdots, (v_k, v_j)\}$. If there exists a path between any nodes, the graph is said to be connected [8].

7.2 Distributed Event-Triggered Control

In this section, a distributed event-triggered control scheme is presented to adaptively regulate the virtual impedance. The overall control diagram is shown in Fig. 7.5, which mainly includes the primary control, distributed adaptive virtual impedance controller as well as a communication layer. With the method, proper reactive, unbalanced, and harmonic power sharing can be achieved by, respectively, regulating the virtual impedances at fundamental positive sequence, fundamental negative sequence, and selected harmonic frequencies without any knowledge of line impedance. Note that the approach is fully distributed. Each controller requires only the local and its neighbors' information to achieve power sharing performance. Note also that the controller transmits its information to neighbors only at event-triggered times. Therefore, the communication burden is considerably reduced compared to the traditional periodic communication way.

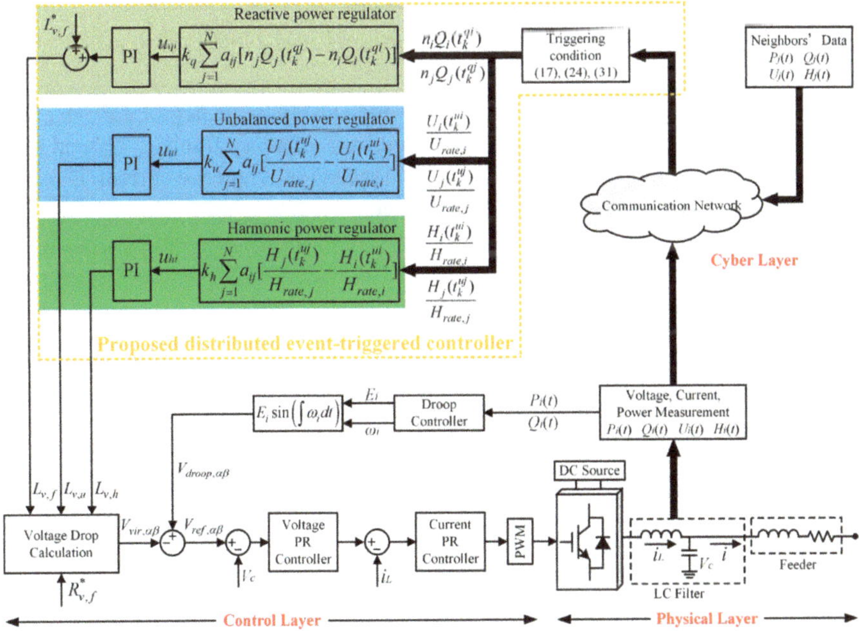

Fig. 7.5 Overall control diagram of the distributed control scheme

7.2.1 Power Calculation

The reactive, unbalanced, and harmonic power calculation involves extracting the fundamental-frequency positive- and negative sequence components and some concerned harmonic components. The multiple second-order generalized integrator (SOGI)-based frequency-locked loop is used for this purpose [9]. With the detected current and voltage components, the DG output active power P, reactive power Q, unbalanced power U, and harmonic power H can be calculated as

$$P = \frac{3\omega_c}{2(s+\omega_c)}\left(v_{c\alpha} \cdot i^{+}{}_{f\alpha} + v_{c\beta} \cdot i^{+}{}_{f\beta}\right) \tag{7.5}$$

$$Q = \frac{3\omega_c}{2(s+\omega_c)}\left(v_{c\beta} \cdot i^{+}{}_{f\alpha} - v_{c\alpha} \cdot i^{+}{}_{f\beta}\right) \tag{7.6}$$

$$U = \frac{3E_0\omega_c}{2(s+\omega_c)}\sqrt{\left(I^{-}{}_{f\alpha}\right)^2 + \left(I^{-}{}_{f\beta}\right)^2} \tag{7.7}$$

$$H = \frac{3E_0\omega_c}{2(s+\omega_c)}\sqrt{\left(I^{-}{}_{5\alpha}\right)^2 + \left(I^{-}{}_{5\beta}\right)^2 + \left(I^{+}{}_{7\alpha}\right)^2 + \left(I^{+}{}_{7\beta}\right)^2} \tag{7.8}$$

where E_0 is the nominal voltage and ω_c is the cut-off frequency of LPF. $v_{c\alpha}$ and $v_{c\beta}$ are the measured DG voltages in the stationary reference frame. $i^{+}{}_{f\alpha}$ and $i^{+}{}_{f\beta}$ ($i^{-}{}_{f\alpha}$ and $i^{-}{}_{f\beta}$) are the fundamental positive (negative) sequence current. $i^{+}{}_{7\alpha}$ and $i^{+}{}_{7\beta}$ are the 7-order positive sequence current.

7.2.2 Controller Design

A distributed event-triggered adaptive virtual impedance controller is designed in this subsection.

1. Reactive power sharing: Construct the state model of the reactive power sharing control as

$$n_i \dot{Q}_i = u_{qi} \tag{7.9}$$

where u_{qi} denotes the input for reactive power controller.

A distributed controller can be constructed as

$$u_{qi}(t) = k_q z_{qi}(t) \tag{7.10}$$

where k_q is a proportional gain and $z_{qi}(t)$ is defined as

$$z_{qi}(t) = \sum\nolimits_{j=1}^{N} a_{ij}\left[n_j Q_j(t) - n_i Q_i(t)\right] \tag{7.11}$$

Under the event-triggered control scheme, (7.11) is redefined as

$$z_{qi}(t) = \sum\nolimits_{j=1}^{N} a_{ij}\left[n_j Q_j\left(t_k^{qj}\right) - n_i Q_i\left(t_k^{qi}\right)\right] \tag{7.12}$$

The state measurement error is defined as

$$e_{qi}(t) = n_i Q_i\left(t_k^{qi}\right) - n_i Q_i(t),\ t \in \left[t_k^{qi}, t_{k+1}^{qi}\right) \tag{7.13}$$

Note that the measurement error $\left\|e_{qi}(t)\right\|$ is the deviation between the latest triggering time state and the real-time state. When $\left\|e_{qi}(t)\right\|$ reaches a predefined threshold, the event is triggered, the state estimate is equal to the actual value, and $\left\|e_{qi}(t)\right\|$ is reset to zero. During the event time interval, no communication is required.

Theorem 7.1 *Assume that the communication topology G is undirected and connected. Then, the controller in (7.10) and (7.12) can achieve reactive power sharing if the event-triggered time is defined as follows [10]:*

$$t_k^{qi} = \inf\left\{t > t_{k-1}^{qi} \middle| f_{qi}(t) = 0\right\} \tag{7.14}$$

where the triggering function $f_{qi}(t)$ can be defined as

$$f_{qi}(t) = \left\|e_{qi}(t)\right\|^2 - \frac{\sigma_q\left(1 - \alpha_q\left|N_i\right|\right)}{\left|N_i\right|/\alpha_q}\left\|z_{qi}(t)\right\|^2 \tag{7.15}$$

where $0 < \sigma_q < 1,\ 0 < \alpha_q < 1/\left|N_i\right|$.

If the reactive power is not shared accurately by each DG unit, the sharing error generated by the local controllers is utilized to adaptively regulate the virtual inductance as follows [11]:

$$L_{v,f} = L_{v,f}{}^{*} - G_q(s)\, u_{qi} \tag{7.16}$$

where $L_{v,f}^{*}$ is the static virtual impedance at the fundamental positive sequence, which is used to ensure that the fundamental equivalent impedance is inductive,

$L_{v,f}$ is adaptively regulated by an integral controller to eliminate the reactive power sharing error, and $G_q(s)$ is a proportional integral (PI) controller.

2. Unbalanced power sharing: Construct the state model of the unbalanced power sharing control as follows:

$$\frac{\dot{U}_i}{U_{rate,i}} = u_{ui} \tag{7.17}$$

where $U_{rate,i}$ is the unbalanced power rating of the ith DG units. u_{ui} denotes the input for unbalanced power controller. A distributed controller can be constructed as

$$u_{ui}(t) = k_u z_{ui}(t) \tag{7.18}$$

where k_u is a proportional gain and $z_{ui}(t)$ is defined as

$$z_{ui}(t) = \sum\nolimits_{j=1}^{N} a_{ij}\left[\frac{U_j\left(t_k^{uj}\right)}{U_{rate,j}} - \frac{U_i\left(t_k^{ui}\right)}{U_{rate,i}}\right] \tag{7.19}$$

The state measurement error is calculated from

$$e_{ui} = \frac{U_i\left(t_k^{ui}\right)}{U_{rate,i}} - \frac{U_i(t)}{U_{rate,i}}, \ t \in \left[t_k^{ui}, t_{k+1}^{ui}\right] \tag{7.20}$$

Theorem 7.2 *Assume that the communication topology G is undirected and connected. Then, the controller in (7.18) and (7.19) can achieve unbalanced power sharing if the event-triggered time is defined as follows:*

$$t_k^{ui} = \inf\left\{t > t_{k-1}^{ui} \middle| f_{ui}(t) = 0\right\} \tag{7.21}$$

where the triggering function fui(t) can be defined as

$$f_{ui}(t) = \|e_{ui}(t)\|^2 - \frac{\sigma_u\left(1 - \alpha_u |N_i|\right)}{|N_i| / \alpha_u}\|z_{ui}(t)\|^2 \tag{7.22}$$

where $0 < \sigma_u < 1, \ 0 < \alpha_u < 1/|N_i|$.

In a similar manner, the virtual impedance at fundamental negative sequence $L_{v,u}$ is adaptively regulated to remove the unbalanced power sharing error as

$$L_{v,u} = -G_u(s)\, u_{ui} \tag{7.23}$$

where $G_u(s)$ is a PI controller.

3. Harmonic power sharing: Construct the state model of the harmonic power sharing control as follows:

$$\frac{\dot{H}_i}{H_{rate,i}} = u_{hi} \tag{7.24}$$

where $H_{rate,i}$ is the harmonic power rating of the ith DG unit. u_{qi} denotes the input for harmonic power controller.

A distributed controller can be constructed as

$$u_{hi}(t) = k_h z_{hi}(t) \tag{7.25}$$

where k_h is a proportional gain and $z_{hi}(t)$ is defined as

$$z_{hi}(t) = \sum\nolimits_{j=1}^{N} a_{ij} \left[\frac{H_j\left(t_k^{uj}\right)}{H_{rate,j}} - \frac{H_i\left(t_k^{ui}\right)}{H_{rate,i}} \right] \tag{7.26}$$

The state measurement error is defined as

$$e_{hi} = \frac{H_i\left(t_k^{hi}\right)}{H_{rate,i}} - \frac{H_i(t)}{H_{rate,i}}, \; t \in \left[t_k^{hi}, t_{k+1}^{hi}\right] \tag{7.27}$$

Theorem 7.3 *Assume that the communication topology G is undirected and connected. Then, the controller in (7.25) and (7.26) can achieve harmonic power sharing if the event-triggered time is defined as follows:*

$$t_k^{hi} = \inf\left\{t > t_{k-1}^{hi} \middle| f_{hi}(t) = 0\right\} \tag{7.28}$$

where the triggering function $f_{hi}(t)$ can be defined as

$$f_{hi}(t) = \|e_{hi}(t)\|^2 - \frac{\sigma_h (1 - \alpha_h |N_i|)}{|N_i| / \alpha_h} \|z_{hi}(t)\|^2 \tag{7.29}$$

where $0 < \sigma_h < 1,\ 0 < \alpha_h < 1/|N_i|$.

Finally, the virtual impedance at harmonic frequency $L_{v,h}$ is adaptively regulated to remove the harmonic power sharing error as

$$L_{v,h} = -G_h(s)\, u_{hi} \tag{7.30}$$

where $G_h(s)$ is a PI controller.

Once the virtual impedance is determined, its corresponding voltage drops in the stationary reference frame can be calculated as follows:

$$V_{f,\alpha\beta} = R_{v,f}{}^{*}i_{f\alpha}{}^{+} - \omega L_{v,f} i_{f\beta}{}^{+} \tag{7.31}$$

$$V_{U,\alpha\beta} = -\omega L_{v,u} i_{f\beta}{}^{-} \tag{7.32}$$

$$V_{H,\alpha\beta} = -\omega L_{v,h} i_{h\beta} \tag{7.33}$$

Then, the voltage reference for the double-loop voltage controller is obtained as

$$V_{ref,\alpha\beta} = V_{droop,\alpha\beta} - V_{vir,\alpha\beta} \tag{7.34}$$

$$V_{ref,\alpha\beta} = V_{droop,\alpha\beta} - V_{f,\alpha\beta} - V_{U,\alpha\beta} - V_{H,\alpha\beta} \tag{7.35}$$

4. Double-loop voltage control: The outer loop voltage controller is implemented to regulate the output capacitor's voltage. The inner loop current controller is nested inside the voltage control loop to regulate the inverter side current. The controllers for voltage and current regulation are expressed as

$$G_V(s) = k_{pV} + \frac{k_{V1}s}{s^2 + (\omega_0)^2} + \sum_{h=5,7} \frac{k_{Vh}s}{s^2 + (h\omega_0)^2} \tag{7.36}$$

$$G_I(s) = k_{pI} + \frac{k_{rI}s}{s^2 + (\omega_0)^2} \tag{7.37}$$

where k_{pV} and k_{pI} are the proportional terms and k_{V1} and k_{rI} are the resonant term coefficients at $\omega_0 = 2\pi \times 50$ rad/s. k_{Vh} is the resonant coefficient term for the hth harmonics (5th, 7th). The inner current loop is designed to provide sufficient damping and protect the inductor's current from overcurrent.

7.3 Stability Analysis

7.3.1 Proof of Theorem

To simplify the proof, we omit the subscript Q, abbreviate $x(t)$ to x, and denote $q_i = n_i Q_i$. Combining (7.9), (7.10), (7.12), and (7.13), the overall system dynamics can be written as

$$\dot{q} = -kL(q+e) \tag{7.38}$$

where $q = [q_1, q_2, \ldots, q_N]^T$, $e = [e_1, e_2, \ldots, e_N]^T$. Similarly, we have

$$z = -L(q+e) \tag{7.39}$$

where $z = [z_1, z_2, \ldots, z_N]^T$.

Considering the Lyapunov function candidate

$$V = \frac{1}{2}q^T Lq \tag{7.40}$$

Then, the time derivative of (7.40) becomes

$$\dot{V} = q^T L\dot{q} \tag{7.41}$$

Combining (7.38), then (7.41) can be written as

$$\dot{V} = -kq^T L^2 (q+e) \tag{7.42}$$

Placing (7.39) into the upper equation in (7.42) yields

$$\dot{V} = -kz^T z - ke^T Lz = -k\|z\|^2 - ke^T Lz \tag{7.43}$$

Expanding (7.43), we get

$$\begin{aligned} V &= -k\sum_{i=1}^{N} z_i^2 - k\sum_{i=1}^{N}\sum_{j\in N_i} e_i\left(z_i - z_j\right) \\ &= -\mathrm{k}\sum_{i=1}^{N} z_i^2 - k\sum_{i=1}^{N}|N_i|e_i z_i + k\sum_{i=1}^{N}\sum_{j\in N_i} e_i z_i \end{aligned} \tag{7.44}$$

Using the inequality

$$|xy| \le \frac{1}{2\alpha}x^2 + \frac{\alpha}{2}y^2,\ \alpha > 0 \tag{7.45}$$

The equation in (7.46) can be bounded by

$$\dot{V} \le -\mathrm{k}\sum_{i=1}^{N} z_i^2 + k\sum_{i=1}^{N}\frac{1}{\alpha}|N_i|\, e_i^2 + k\sum_{i=1}^{N}\frac{\alpha}{2}|N_i|\, z_i^2 + \mathrm{k}\sum_{i=1}^{N}\sum_{j\in N_i}\frac{\alpha}{2}z_j^2 \tag{7.46}$$

Since the undirected graph G is symmetric, by interchanging the indices of the last term, we get

$$\sum_{i=1}^{N}\sum_{j\in N_i}\frac{\alpha}{2}z_j^2=\sum_{i=1}^{N}\sum_{j\in N_i}\frac{\alpha}{2}z_i^2=\sum_{i=1}^{N}\frac{\alpha}{2}|N_i|\,z_i^2 \tag{7.47}$$

So that

$$\dot{V}\le -\mathrm{k}\sum_{i=1}^{N}(1-\alpha\,|N_i|)z_i^2+k\sum_{i=1}^{N}\frac{1}{\alpha}\,|N_i|e_i^2 \tag{7.48}$$

Assuming that

$$0<\alpha<1/\,|N_i| \tag{7.49}$$

Then, if the following condition holds:

$$e_i{}^2\le\frac{\sigma\,(1-\alpha\,|N_i|)}{|N_i|\,/\alpha}z_i{}^2,\ 0<\sigma<1 \tag{7.50}$$

we get

$$\dot{V}\le 0 \tag{7.51}$$

Thus, the triggering function defined in (7.15) can ensure that $q(t)$ is stable. The proof is completed.

7.3.2 Inter-Event Interval Analysis

Theorem 7.4 *Assume that the communication topology G is undirected and connected. Consider the system in (7.9) with the controller in (7.10) and (7.12). If the triggering function is defined as (7.16), then the inter-event interval is lower bounded by a positive constant τ.*

Proof Considering the following time derivative:

$$\frac{d}{dt}(\frac{\|e\|}{\|z\|})=\frac{d(e^Te)^{\frac{1}{2}}}{dt(z^Tz)^{\frac{1}{2}}}=\frac{e^T\dot{e}}{\|e\|\,\|z\|}-\frac{\|e\|\,(z^T\dot{z})}{\|z\|^3} \tag{7.52}$$

According to (7.13), the derivative of state measurement error e can be written as

$$\dot{e}=-\dot{q} \tag{7.53}$$

Then the derivative of z in (7.39) is

$$\dot{z} = -L\left(\dot{q} + \dot{e}\right) = 0 \tag{7.54}$$

Placing (7.55), (7.56) into (7.54) yields

$$\frac{d}{dt}\left(\frac{\|e\|}{\|z\|}\right) = -\frac{e^T \dot{e}}{\|z\|^3} - \frac{\|e\| \left(z^T \dot{z}\right)}{\|z\|^3} \leq \frac{\|\dot{q}\|}{\|z\|} \tag{7.55}$$

Placing the derivative of q in (7.40) into (7.57), we get

$$\frac{d}{dt}(\frac{\|e\|}{\|z\|}) \leq k \frac{\|L\| \left(\|q\| + \|e\|\right)}{\|z\|} \tag{7.56}$$

Note that L is reversible; combining (7.58) with (7.41) yields

$$\frac{d}{dt}(\frac{\|e\|}{\|z\|}) \leq k \frac{\|L\| \left(\|q\| + \|e\|\right)}{\|z\|} \tag{7.57}$$

Thus, $\|e\| \,/\, \|z\|$ upper bounded by

$$\frac{\|e\|}{\|z\|} \leq \phi\left(t, \phi_0\right) \tag{7.58}$$

where $\phi\left(t, \phi_0\right)$ is the solution of the following differential equations:

$$\begin{cases} \dot{\phi} = k\left(2\|L\|\phi + 1\right) \\ \phi\left(0, \phi_0\right) = \phi_0 \end{cases} \tag{7.59}$$

□

Thus, we have

$$\phi\left(\tau, 0\right) = \frac{1}{2\|L\|}\left(e^{2k\|L\|\tau} - 1\right) \tag{7.60}$$

Denoting

$$h = \arg\max_i \|z_i\| \tag{7.61}$$

Because $\|e_i\| \leq \|e\|$ holds, we have

$$\frac{|e_h|}{|z_h|} \leq \frac{\|e\|}{|z_h|} \leq \frac{N\,\|e\|}{\|z\|} \tag{7.62}$$

Placing (7.59) and the triggering function in (7.15) into (7.62), we have

$$\sqrt{\frac{\sigma\left(1-\alpha\left|N_i\right|\right)}{\left|N_i\right|/\alpha}} \le \frac{N}{2\left\|L\right\|}\left(e^{2k\|L\|\tau}-1\right) \tag{7.63}$$

Then

$$\tau \ge \frac{1}{2k\left\|L\right\|}\ln\left[\frac{2\left\|L\right\|}{N}\sqrt{\frac{\sigma\left(1-\alpha\left|N_i\right|\right)}{\left|N_i\right|/\alpha}}+1\right] \tag{7.64}$$

Thus, the inter-event interval is lower bound, and Zeno behavior is avoided. The proof is completed.

7.4 Experimental Results

To demonstrate the effectiveness of the control technique, experiments are provided at the HIT-Shenzhen microgrid laboratory. The microgrid setup includes three DG units at the same power rating supplying unbalance/nonlinear loads connected to PCC. Each DG unit consists of a Danfoss FC302 converter with LC type filters. The dSPACE MicroLabBox is chosen to be the controller in the experiment. The communication links among three DG units are bidirectional. The physical system parameters and control ones are listed in Table 7.1.

Table 7.1 Experimental parameters

Symbol	Quantity	Nominal value
V_{DC}	DC voltage	400 V
f_{sw}	Switching frequency	10 kHz
ω_0	Nominal frequency	$2\pi \times 50$ rad/s
E_0	Nominal voltage	120 Vrms
L_0/C_f	LC filter	1.8 mH/25 μF
P_U/Q_U	Unbalanced load	5 kW/4 kVar
C_{HL}/R_{HL}	Nonlinear load	4700 µF/72 Ω
$R^*_{v,f}/L^*_{v,f}$	Static virtual impedance	0.5 Ω/1.5 mH
m	$P-\omega$ droop coefficient	0.0002
n	$Q-E$ droop coefficient	0.0005
ω_c	LPF cut-off frequency	4π rad/s
K_{pV}/K_{V1}	Voltage PR controller	0.04/100
K_{V5}/K_{V7}		50/50
K_{pI}/K_{r1}	Current PR controller	5.7/500
K_{pQ}/K_{IQ}	Reactive PI controller	0.001/0.03
K_{pU}/K_{IU}	Unbalance PI controller	0.001/0.02
K_{pH}/K_{IH}	Harmonic PI controller	0.002/0.05

7.4.1 Case 1: Unbalanced Load

In this section, only an unbalanced RL load is connected to PCC. The PCC voltage is 120 V rms, as shown in Fig. 7.6. Figure 7.7 shows the power sharing performance of the method. To save space, the output active powers are not shown. At $t < t_1$ (i.e., stage 1), the traditional droop controller is applied. It can be observed that the reactive and unbalanced powers are shared improperly among parallel units. This result was expected as there is a mismatch in line impedances. At $t = t_1$, the control mechanism is activated. Thanks to the action of the proposed control strategy, the reactive and unbalanced powers are converged to a common value after a short while. Recall that the controller adaptively regulates the virtual impedances

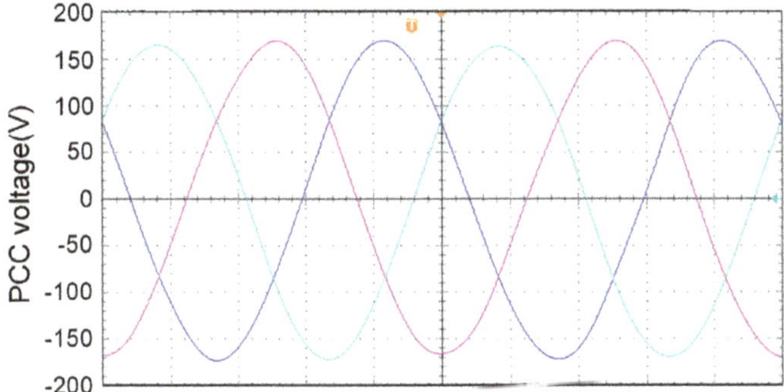

Fig. 7.6 PCC voltage with unbalanced loads

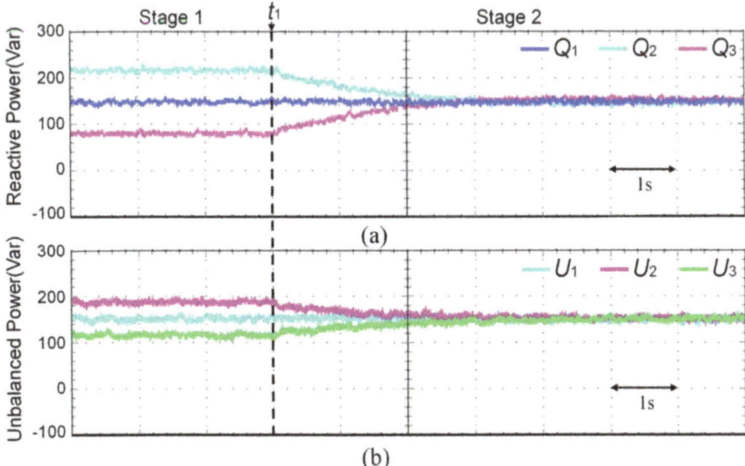

Fig. 7.7 Power sharing performance in the presence of unbalanced loads: (**a**) reactive power and (**b**) unbalanced power

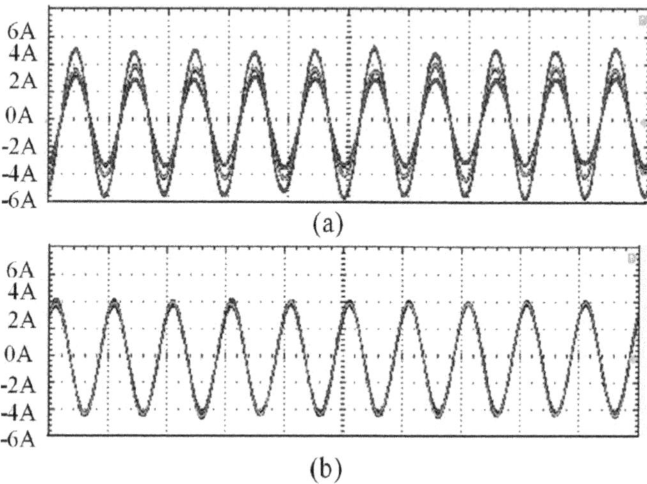

Fig. 7.8 Current performance with unbalanced loads. (**a**) Phase-A currents of DG units without the method. (**b**) Phase-A currents of DG units with the method (20 ms/div)

to compensate for the line impedance mismatches and achieve proper power sharing among parallel units.

The corresponding phase-A current waveforms of the three DG units are shown in Fig. 7.8. Figure 7.8a shows phase-A current of three DG units using the traditional droop method. It obviously illustrates that the current magnitudes of phase-A in three DG units are not the same. This is because of the unequal power sharing in the microgrid. After the method is activated, the phase-A currents of the three DG units are almost identical, as shown in Fig. 7.8b.

Figure 7.9 validates the plug-and-play ability of the method. At $t < t_1$ (i.e., stage 1), all three DG units supply the power to the unbalanced load connecting to the PCC. Figure 7.9a shows that the initial reactive power output of all DG units is around 150 Var. When DG2 is disconnected at t_1, the reactive power outputs of DG1 and DG3 are increased to around 225 Var, which is reasonable because the total reactive power is 450 Var. When DG2 is reconnected to the microgrid at t_2, three DG units equally share the total reactive power again. Meanwhile, the unbalanced power sharing process during the plug-and-play is shown Fig. 7.9b. It is clearly observed that unbalanced power is equally shared by three DG units during the stage 1 and then is equally shared by DG 1 and DG3 due to the disconnection of the DG2. Finally, at t_2, when DG2 is plugged into the microgrid, the three DG units cooperatively share the unbalanced power.

Figure 7.10 shows the power sharing performance under the load change. At $t < t_1$ (i.e., stage 1), all three DGs operate in the steady state and accurately share the reactive and unbalanced powers. At $t = t_1$, an extra 50Ω unbalanced load is connected to the microgrid. At $t = t_2$, this load is disconnected from the microgrid. As can be seen in Fig. 7.10, the control system efficiently handles the load change

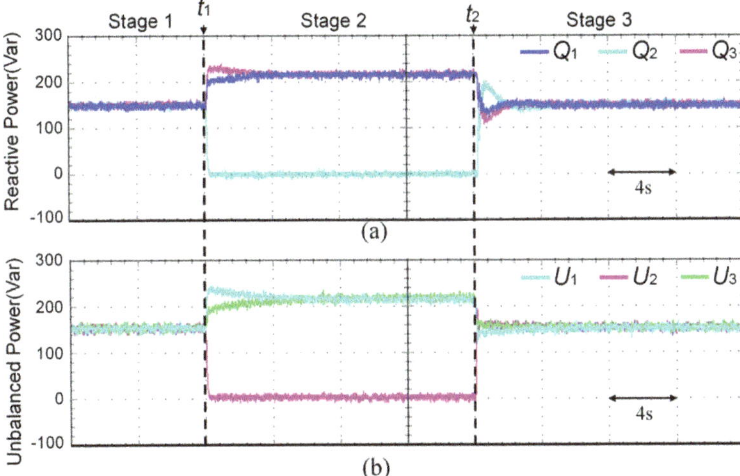

Fig. 7.9 Plug-and-play performance with unbalanced loads: (**a**) reactive power and (**b**) unbalanced power

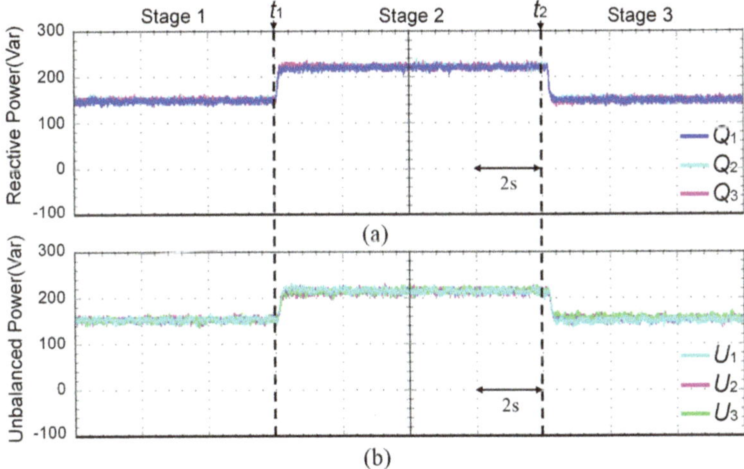

Fig. 7.10 Power sharing performance under unbalanced loads change: (**a**) reactive power and (**b**) unbalanced power

perfectly, and the reactive and unbalanced powers are accurately shared among three DGs all the time.

The control methods are also studied under communication link failure conditions, as shown in Fig. 7.11. At $t < t_1$ (i.e., stage 1), the communication network is intact, and all three DG units share the same power. At $t = t_1$, the communication link between DG1 and DG2 is disconnected. Finally, at $t = t_2$, all the communication link failures are cleared. As seen from Fig. 7.12, despite

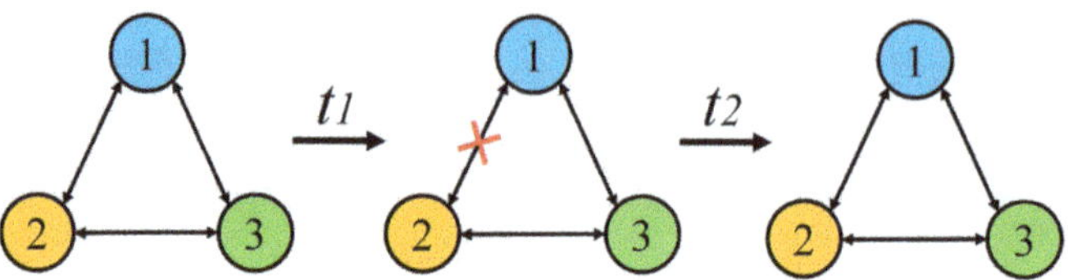

Fig. 7.11 Communication link failure process

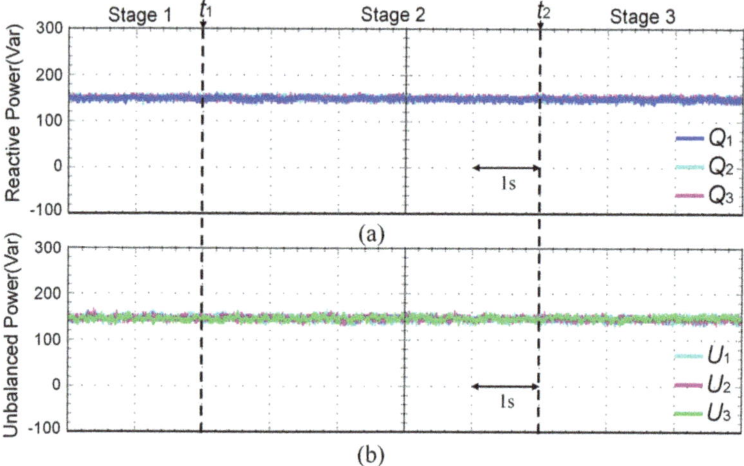

Fig. 7.12 Power sharing performance under communication link failure conditions: (**a**) reactive power and (**b**) unbalanced power

the changes of the communication network, the power sharing performance are not affected at the steady state.

7.4.2 *Case 2: Nonlinear Load*

To verify the harmonic power sharing performance, a three-phase diode rectifier load is connected to PCC. Before $t = t_1$, only the traditional droop method is adopted. The experimental results are presented in Fig. 7.13. As it can be observed, the traditional droop results in a poor power sharing. At $t = t_1$, when the method is activated, the control scheme enables DGs to share reactive and harmonic powers properly among each other.

Figure 7.14a shows the phase-A of the output current of DGs when the traditional droop method is active. It is observed that there is a noticeable magnitude error among output currents of DGs, which implies that there is a power sharing mismatch among them. With the activation of the method, however, as shown in Fig. 7.14b,

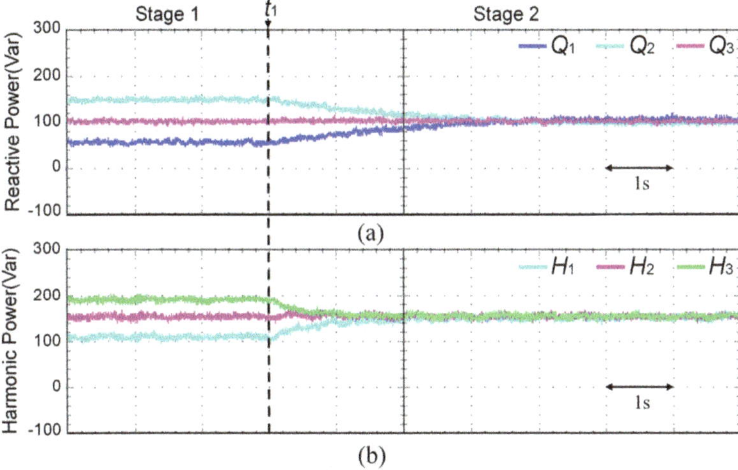

Fig. 7.13 Power sharing performance in the presence of the nonlinear load: (**a**) reactive power and (**b**) harmonic power

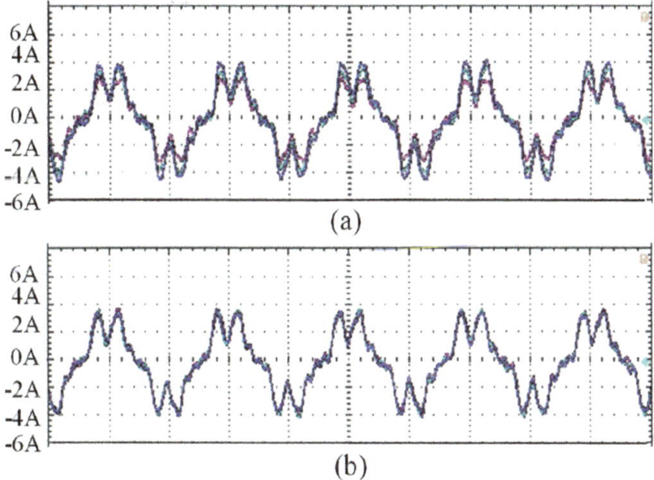

Fig. 7.14 Current sharing performance with nonlinear loads. (**a**) Phase-A currents of DG units without the method. (**b**) Phase-A currents of DG units with the method (10 ms/div)

the output currents of DGs are almost identical, which implies that the unbalance and harmonic power are equally shared.

The plug-and-play capability of DGs in the presence of a nonlinear load is investigated in Fig. 7.15. The initial reactive power and harmonic power output of all DG units are around 100 and 155 Var, and when DG2 is disconnected at t_1, the reactive and harmonic power output of DG1 and DG3 is around 150 and 233 Var.

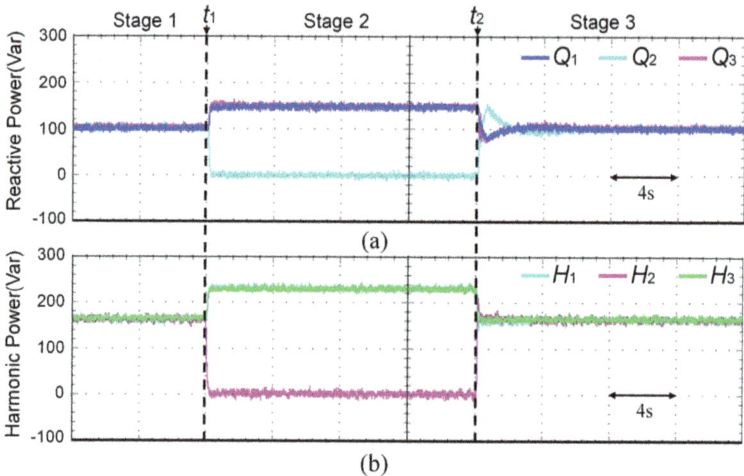

Fig. 7.15 Plug-and-play performance with nonlinear loads: (**a**) reactive power and (**b**) harmonic power

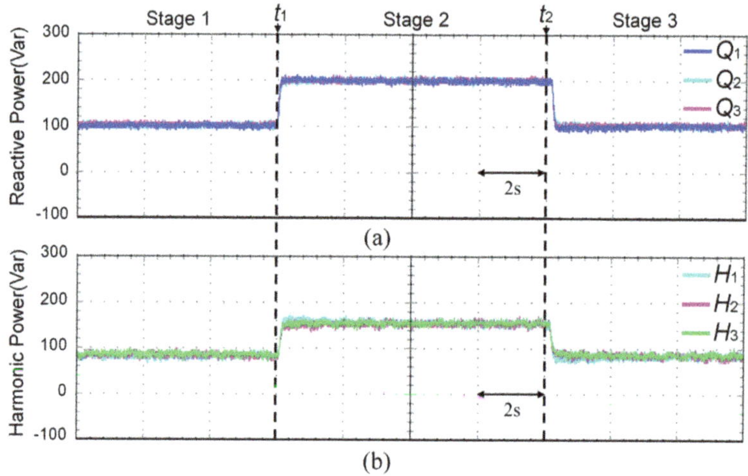

Fig. 7.16 Power sharing performance under nonlinear loads change: (**a**) reactive power and (**b**) harmonic power

Then, DG2 is reconnected to the microgrid at t_2, and the total 300 Var reactive power and 465 Var harmonic power are equally shared by these three DG units again.

The power sharing performance under the load change with nonlinear load is validated in Fig. 7.16. Similar to Fig. 7.9, the power can be accurately shared whenever a load is connected or disconnected from the microgrid.

The communication link failures with nonlinear loads are also studied in Fig. 7.17. The changes of the communication network are the same as in Fig. 7.10.

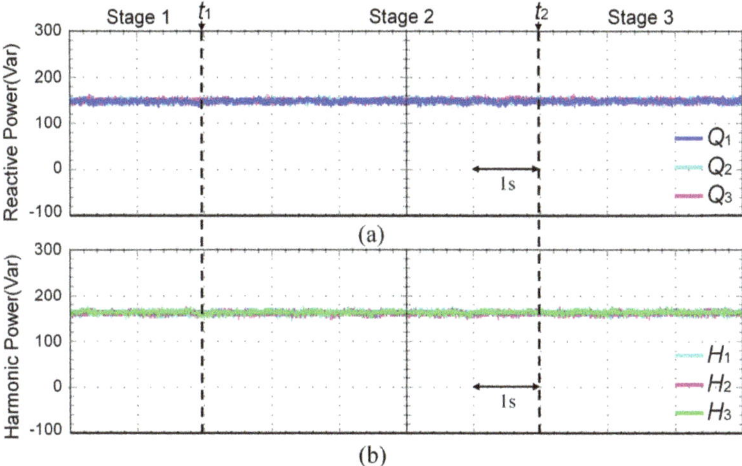

Fig. 7.17 Power sharing performance under communication link failure conditions: (**a**) reactive power and (**b**) harmonic power

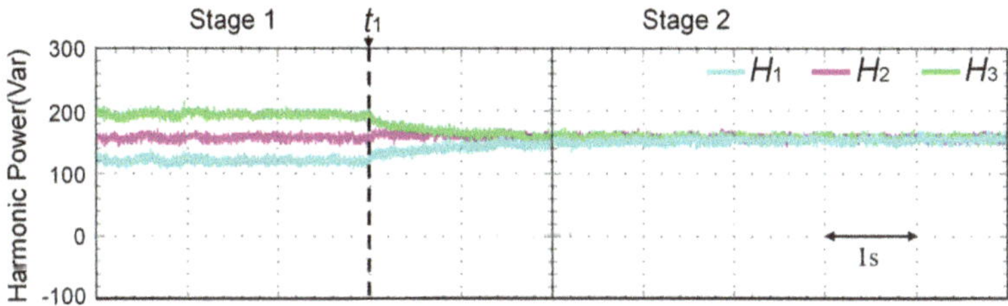

Fig. 7.18 Harmonic power sharing performance with nonlinear loads in continuous communication way

Similar to the case of Fig. 7.11, even though the communication link changes, the power sharing performance is almost unaffected.

7.4.3 *Case 3: Comparison with Periodic Communication*

In this subsection, the event-triggered control is compared to the periodic communication, where its sampling frequency is set as $f_c = 1$ kHz. The experimental results of harmonic power sharing with periodic communication is shown in Fig. 7.18. Compared to the event-triggered control method in Fig. 7.13b, they nearly achieve the same control performance. However, they are achieved at a different number of communication updates. The event-triggered time instant of DG1 harmonic power sharing is shown in Fig. 7.19. It is shown that the controller updates their communication in an aperiodic way. In addition, the communication triggering

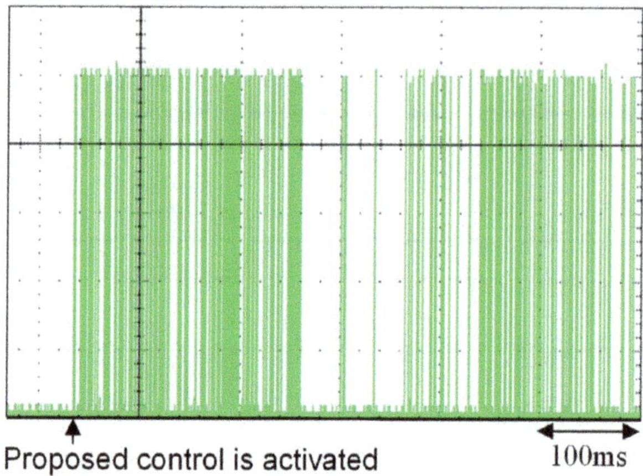

Fig. 7.19 Event-triggered time instant with DG1 harmonic power sharing

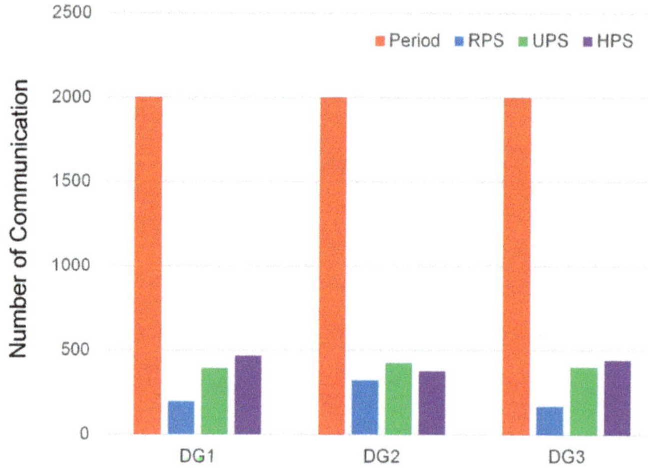

Fig. 7.20 Comparison for the number of communications. (RPS: reactive power sharing, UPS: unbalanced power sharing, HPS: harmonic power sharing)

times of these two approaches for reactive power, unbalanced power, and harmonic power sharing are calculated during 2 s period with the controllers activated, as shown in Fig. 7.20. From Fig. 7.20, it is concluded that the event-triggered control method has few triggering times, which can highly reduce the communication burden among DG units.

7.5 Conclusion

This chapter presented an event-triggered distributed control strategy for the reactive, unbalanced, and harmonic power sharing in an islanded microgrid. The introduced control strategy could realize the accurate power sharing while highly reducing communication data exchange and achieving the plug-and-play feature among the DG units. The stability of the control strategy was proved with Lyapunov function and the Zeno behavior can be excluded. Experimental results from microgrid laboratory demonstrated the effectiveness of the scheme.

References

1. K.D. Brabandere, B. Bolsens, J.V.D. Keybus, A. Woyte, J. Driesen, R. Belmans, A voltage and frequency droop control method for parallel inverters. IEEE Trans. Power Electron. **22**(4), 1107–1115 (2007)
2. J.C. Vasquez, R.A. Mastromauro, J.M. Guerrero, M. Liserre, Voltage support provided by a droop-controlled multifunctional inverter. IEEE Trans. Ind. Electron. **56**(11), 4510–4519 (2009)
3. J. Rocabert, A. Luna, F. Blaabjerg, P. Rodríguez, Control of power converters in AC microgrids. IEEE Trans. Power Electron. **27**(11), 4734–4749 (2012)
4. H. Mahmood, D. Michaelson, J. Jiang, Reactive power sharing in islanded microgrids using adaptive voltage droop control. IEEE Trans. Smart Grid **6**(6), 3052–3060 (2015)
5. J. He, Y.W. Li, An enhanced microgrid load demand sharing strategy. IEEE Trans. Power Electron. **27**(9), 3984–3995 (2012)
6. X. Hou, Y. Sun, H. Han, Z. Liu, W. Yuan, M. Su, A fully decentralized control of grid-connected cascaded inverters. IEEE Trans. Sustain. Energy **10**(1), 315–317 (2019). https://doi.org/10.1109/TPWRD.2018.2816813
7. J. He, Y.W. Li, F. Blaabjerg, An enhanced islanding microgrid reactive power, imbalance power, and harmonic power sharing scheme. IEEE Trans. Power Electron. **30**(6), 3389–3401 (2015). https://doi.org/10.1109/TPEL.2014.2332998
8. Y. Su, J. Huang, Cooperative output regulation of linear multi-agent systems. IEEE Tran. Autom. Control **57**(4), 1062–1066 (2012)
9. P. Rodríguez, A. Luna, I. Candela, R. Mujal, R. Teodorescu, F. Blaabjerg, Multiresonant frequency-locked loop for grid synchronization of power converters under distorted grid conditions. IEEE Trans. Ind. Electron. **58**(1), 127–138 (2011). https://doi.org/10.1109/TIE.2010.2042420
10. M. Chen, X. Xiao, J.M. Guerrero, Secondary restoration control of islanded microgrids with a decentralized event-triggered strategy. IEEE Trans. Ind. Inf. **14**(9), 3870–3880 (2018)
11. J. Lu, M. Zhao, S. Golestan, T. Dragicevic, X. Pan, J.M. Guerrero, Distributed event-triggered control for reactive, unbalanced and harmonic power sharing in islanded AC microgrids. IEEE Trans. Ind. Electron. **69**(2), 1548–1560 (2022).

Part II
Series-Type Microgrid Systems

Chapter 8
Decentralized Method for Islanded Operation Mode

8.1 Series-Type Microgrid Configuration

Figure 8.1 shows the schematic diagram of an islanded series-type microgrid. These DG interfacing inverters are cascaded and supply power for loads together. The output real power P_i and reactive power Q_i of the i-th DG are derived as follows:

$$P_i + jQ_i = V_i e^{j\delta_i} \cdot \left(V_{\text{pcc}} e^{j\delta_{\text{pcc}}} / \left| Z_{load} \right| e^{j\delta_{load}} \right)^*, \tag{8.1}$$

where V_i and δ_i represent the output voltage amplitude and phase angle of the i-th DG. The AC bus voltage is the sum of the each DG voltage:

$$V_{\text{pcc}} e^{j\delta_{\text{pcc}}} = \sum_{i=1}^{n} V_i e^{j\delta_i} \tag{8.2}$$

According to (8.1)–(8.2), the power transmission characteristics of the i-th DG are obtained as

$$P_i = V_i \sum_{j=1}^{n} V_j \cos\left(\delta_i - \delta_j + \theta_{load}\right) / \left| Z_{load} \right| \tag{8.3}$$

$$Q_i = V_i \sum_{j=1}^{n} V_j \sin\left(\delta_i - \delta_j + \theta_{load}\right) / \left| Z_{load} \right| \tag{8.4}$$

where Z_{load} is the load impedance. The subscript j represents the serial number of the j-th DG.

Y. Sun et al., *Series-Parallel Converter-Based Microgrids*, Power Systems,
https://doi.org/10.1007/978-3-030-91511-7_8

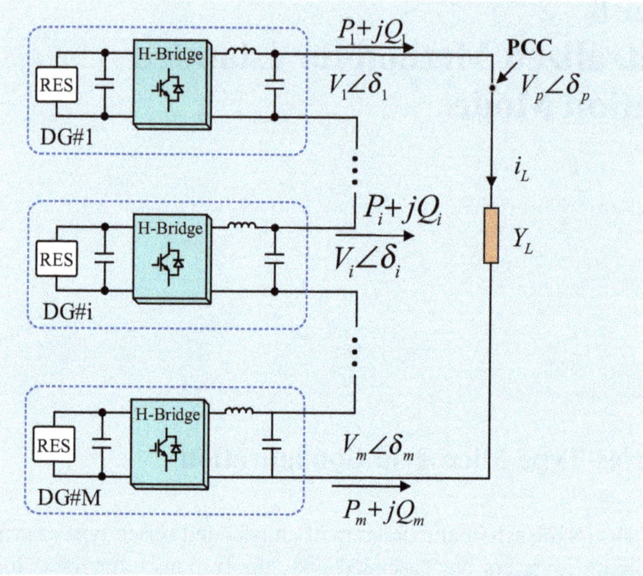

Fig. 8.1 Structure of islanded series-type microgrid

8.2 Traditional Operation Mode

The series converter is originally applied to multilevel inverters [1] and initially extended into microgrid applications [2, 3] for attaining higher voltage level and better utilization, especially for PV grid-connected application [2] and battery management [3]. The traditional operation mode of series-type microgrid is introduced in Fig. 8.2. The centralized synchronization control relies heavily on high bandwidth communication, that is, the same synchronization clock is used. The synchronization command is transmitted to each power electronic device in real time through the high bandwidth communication network. When the communication network fails or the synchronous clock deviates, the system will lose stability.

8.3 Decentralized Control Method Design

8.3.1 *An $f - P/Q$ Droop Control Scheme*

To synchronize each DG in the series-type microgrid without communication, a decentralized control scheme [4] is designed as follows:

$$\omega_i = \omega^* + m_i P_i / Q_i \tag{8.5}$$

$$V_i = V^* \tag{8.6}$$

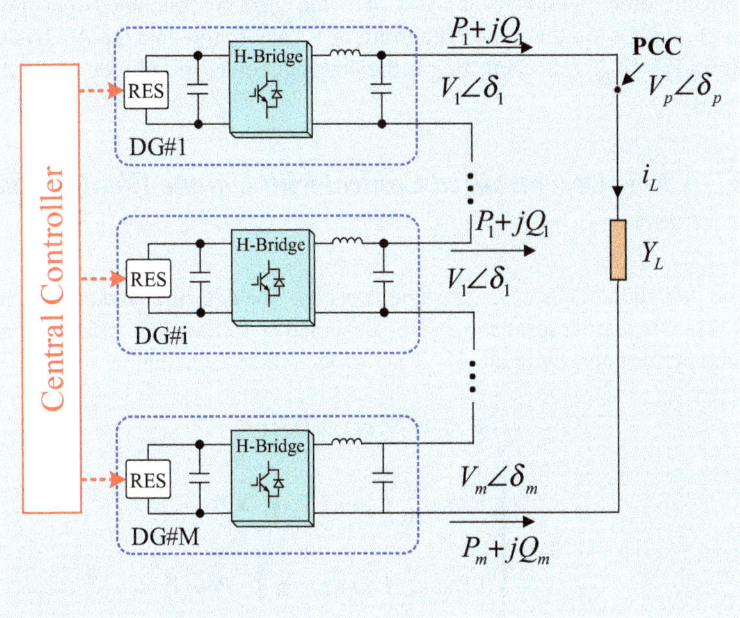

Fig. 8.2 Traditional operation mode

where ω_i and V_i are the angular frequency and voltage amplitude references of the i-th DG, respectively. ω^* represents the value of ω at no load. V^* is the nominal voltage value. Note that there is a singularity in (8.5) when $Q_i = 0$. To avoid the singularity and guarantee a reasonable frequency deviation, m_i is designed as follows:

$$m_i = \begin{cases} k_i Q_{\min}, |Q_i| \geqslant Q_{\min} \\ k_i |Q_i|, |Q_i| < Q_{\min} \end{cases} \tag{8.7}$$

where $Q_{\min}$ is a small positive constant, which is determined by practical requirements. k_i is a positive constant, which is equal to $\left|\Delta\omega_{\max}/P^*_{\max\_i}\right|$. $\Delta\omega_{\max}$ is the allowable maximum frequency deviation, and $P^*_{\max\_i}$ is the nominal rated power of the i-th DG.

For simplicity, we assume that the rated capacities of all DGs are the same and the system is in steady state. Because the voltage amplitude references are the same for all DGs and each DG shares the same load current, the apparent powers of all DGs are equal.

$$S_1 = S_2 = \cdots = S_n \tag{8.8}$$

For the purpose of power balance, let $k_i = K; i \in \{1, 2 \cdots n\}$. Then (8.9) is derived from (8.5), (8.7), and (8.8).

$$P_1 = P_2 = \cdots = P_n \tag{8.9}$$

From the aforementioned analysis, it could also be concluded that the phase angles of all DGs are the same. Therefore, it is easy to regulate the AC bus voltage by setting $V_i = V^*_{\mathrm{pcc}}/n$, where V^*_{pcc} is the nominal voltage amplitude of the AC bus.

8.3.2 A New Decentralized Control with Unique Equilibrium Point

Assume that all the DGs have the same capacity. The decentralized control strategy of the series-type microgrid can also be designed as follows [5], which can achieve the same performance with an $f - P/Q$ droop control scheme:

$$\omega_i = \omega^* + m\,\mathrm{sgn}\,(Q_i)\,P_i \tag{8.10}$$

$$\mathbf{v_i} = \begin{cases} V^* \sin\left(\int \omega_i dt\right), & P_i > 0 \\ V^* \sin\left(\int \omega_i dt + \pi\right), & P_i < 0 \end{cases} \tag{8.11}$$

where ω_i is the angular frequency. ω^*, V^* is the nominal angular frequency and voltage amplitude. $sgn(.)$ is a signum function. m is a positive coefficient. $\mathbf{v}_i$ is the voltage vector. Equation (8.11) can be rewritten as $\mathbf{v_i} = V_i \sin\left(\int \omega_i dt\right)$, and

$$V_i = V^*\,\mathrm{sgn}\,(P_i) \tag{8.12}$$

Cleary, the scheme in (8.10) and (8.11) only needs the local information of each DG, thus the decentralized manner is realized.

8.3.3 Power Factor Angle Droop Control

The power factor angle droop control strategy [6] of the single-phase series inverters is expressed as

$$\omega_i = \omega^* - m\left(\varphi_i - \varphi^*\right) \tag{8.13}$$

$$V_i = V^* \tag{8.14}$$

where ω_i is the angular frequency. ω^*, V^*, φ^* are the nominal angular frequency, voltage amplitude, and power factor angle, and its design is related to the actual application requirements of system. m is a positive coefficient. φ_i is the power factor angle. The output voltage of the i-th DG is regulated by the inner dual-loop controller. The control diagram is depicted in Fig. 8.3.

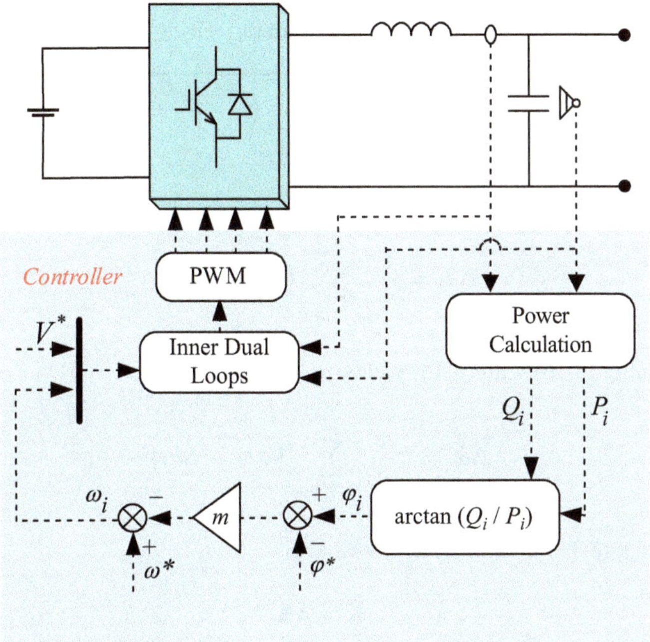

Fig. 8.3 Control diagram of the scheme

As seen, the scheme in (8.13) and (8.14) only needs the local information of each module, thus it is a decentralized approach.

8.4 Stability Analysis

To prove the stability of the introduced method in the islanded and grid-connected modes, the small signal analysis near the equilibrium point is carried out [7].

Assume that δ_s is the synchronous phase angle of series inverters in the steady state, and denote $\tilde{\delta}_i = \delta_i - \delta_s$. Since $\dot{\tilde{\delta}}_i = \omega_i$, (8.13) is rewritten as

$$\dot{\tilde{\delta}}_i = \omega^* - m\left(\varphi_i - \varphi^*\right) \tag{8.15}$$

Linearizing (8.15) around the equilibrium point [8], we have

$$\Delta\dot{\tilde{\delta}}_i = -m\Delta\varphi_i \tag{8.16}$$

In the islanded mode, combining (8.3) and (8.4) yields

$$\varphi_i = \operatorname{atan}\frac{Q_i}{P_i} = \operatorname{atan}\frac{\sum_{j=1}^{n}\sin\left(\delta_i - \delta_j + \theta'_{load}\right)}{\sum_{j=1}^{n}\cos\left(\delta_i - \delta_j + \theta'_{load}\right)} \tag{8.17}$$

Then, linearizing (8.17), we have

$$\Delta\varphi_i = \frac{1}{n}\sum_{j=1, j\neq i}^{n}\left(\Delta\delta_i - \Delta\delta_j\right) \tag{8.18}$$

Combining (8.16) with (8.18) yields

$$\Delta\dot{\delta}_i = -\frac{m}{n}\sum_{j=1, j\neq i}^{n}\left(\Delta\delta_i - \Delta\delta_j\right) \tag{8.19}$$

Rewriting (8.19) in matrix form, we have

$$\dot{\mathbf{X}} = \mathbf{AX} \tag{8.20}$$

where $\mathbf{X} = [\Delta\delta_1, \ldots, \Delta\delta_n]^T$, $\mathbf{A} = -(m/n)\,\mathbf{L}$. $\mathbf{L}$ is a Laplacian matrix [9], and it is expressed as

$$\mathbf{L} = n\mathbf{E} - \mathbf{1_n} \tag{8.21}$$

where $\mathbf{E} = diag\,(1, \ldots, 1)$, and $\mathbf{1_n}$ is a matrix where all elements are 1.

The eigenvalues of $\mathbf{A}$ are expressed as

$$\lambda_1(\mathbf{A}) = 0, \lambda_2(\mathbf{A}) = \cdots = \lambda_n(\mathbf{A}) = -m \tag{8.22}$$

Clearly, the system in the islanded mode is stable [10]. Moreover, the stability does not depend on load parameters.

According to the analysis above, the main results are summarized as follows:

1. From (8.22), the eigenvalues are independent of the load and transmission line impedance. Thus, the scheme is suitable for all types of load and transmission line impedance.
2. The system always holds a unique equilibrium point with implementation of the introduced scheme.
3. Due to $\varphi^* \in (-\pi, \pi]$ in (8.13), the scheme can realize four-quadrant operation of inverters.

Table 8.1 Parameters of Simulations

Parameters	Values	Parameters	Values
V_g (V)	315	Z_{line} (Ω)	$j0.314$
f^*/f_i (Hz)	50/[49,51]	V^* (V)	315/4
m	0.5	φ^*	0.2

8.5 Case Study

Simulations are performed on Matlab/Simulink platform. The related parameters of the tested system comprised of four DGs are listed in Table 8.1.

8.5.1 Case 1: Suited for All Types of Loads

In this case, the system in islanded mode is tested under pure resistance, RL and RC loads in the interval [0 s, 6 s], [6 s, 12 s], and [12 s,18 s]. Figure 8.4 show the waveforms of frequency, active power, and reactive power from top to bottom. The frequencies are always synchronous no matter how the loads change. Thus, the scheme is suited for all types of loads.

8.5.2 Case 2: Unique Equilibrium Point

In case 2, the series inverters operate in the islanded mode. The initial phase angle of inverter #1 is set in I, II, III, and IV quadrants in the interval [0 s, 5 s], [5 s, 10 s], [10 s, 15 s], and [15 s, 20 s], respectively, and the others are set as zero. The waveforms of frequency, active power, and reactive power are illustrated in Fig. 8.5a–c. Accordingly, the control scheme always has a unique equilibrium point regardless of the initial states.

8.6 Conclusion

This chapter introduced several decentralized control strategy of the single-phase series inverters, i.e., power factor angle droop control. It has the following pros: (1) suitable for all types of loads, (2) unique equilibrium point, (3) feasible for all types of transmission line impedance, and (4) four-quadrant operation of inverters. Based on the idea behind this study, the decentralized operation of series converters for applications such as PV, storage, and STATCOM will be realized.

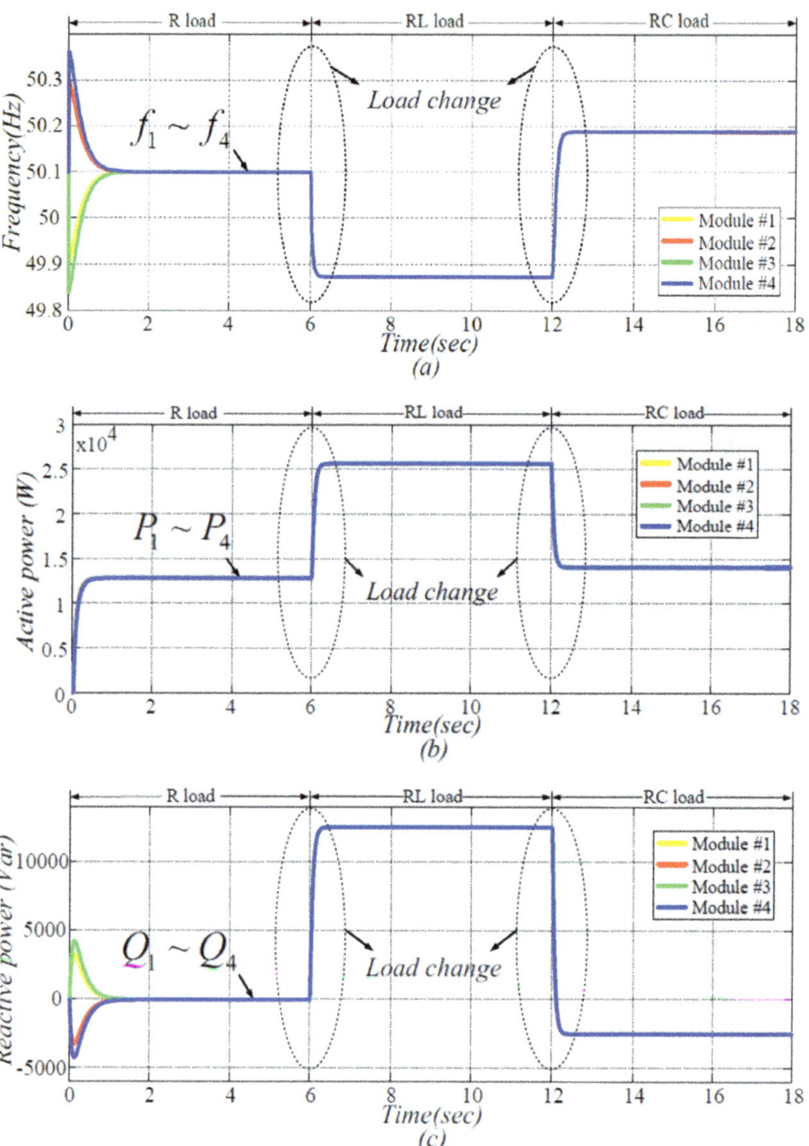

Fig. 8.4 Simulation results of case 1

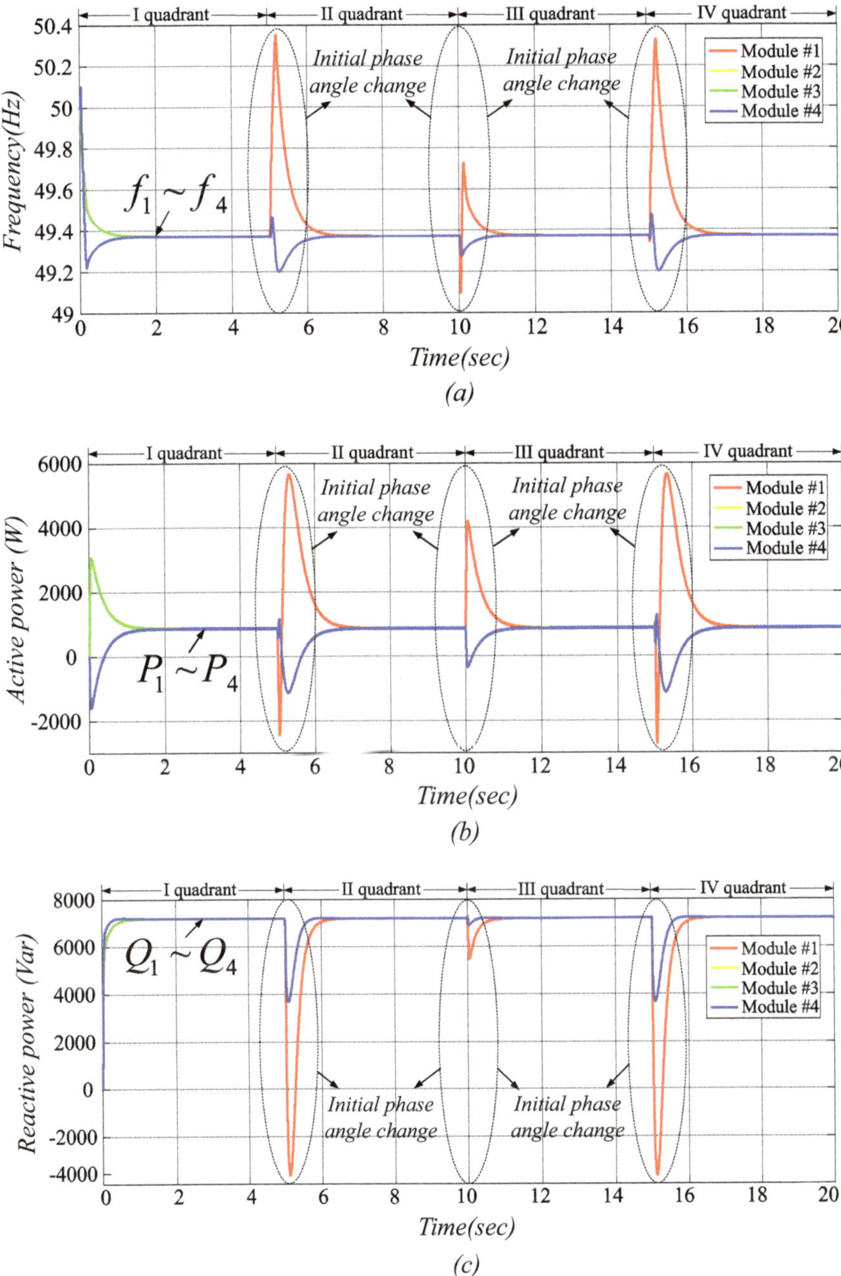

Fig. 8.5 Simulation results of case 2

References

1. H. Akagi, Classification, terminology, and application of the modular multilevel cascade converter (MMCC). IEEE Trans. Power Electron. **26**(11), 3119–3130 (2011)
2. S.B. Kjaer, J.K. Pedersen, F. Blaabjerg, A review of single-phase grid-connected inverters for photovoltaic modules. IEEE Trans. Ind. Appl. **41**(5), 1292–1306 (2005)
3. L. Maharjan, S. Inoue, H. Akagi, J. Asakura, State-of-charge (SOC)-balancing control of a battery energy storage system based on a cascade PWM converter. IEEE Trans. Power Electron. **24**(6), 1628–1636 (2009)
4. Y. Sun, G. Shi, X. Li, W. Yuan, M. Su, H. Han, X. Hou, An f-P/Q droop control in cascaded-type microgrid. IEEE Trans. Power Syst. **33**(1), 1136–1138 (2018)
5. L. Li, Y. Sun, Z. Liu, X. Hou, G. Shi, M. Su., A decentralized control with unique equilibrium point for cascaded-type microgrid. IEEE Trans. Sustain. Energy **10**(1), 324–326 (2019)
6. Y. Sun, L. Li, G. Shi, X. Hou, M. Su, Power factor angle droop control—a general decentralized control of cascaded inverters. IEEE Trans. Power Delivery **36**(1), 465–468 (2021)
7. E.A.A. Coelho, P.C. Cortizo, P.F.D. Garcia, Small-signal stability for parallel-connected inverters in stand-alone AC supply systems. IEEE Trans. Ind. Appl. **38**(2), 533–542 (2002)
8. D. Mondal, *Power System Small Signal Stability Analysis and Control*, Chap. 5 (Academic, London, 2014), pp. 119–143
9. F.L. Lewis, H. Zhang, K. Hengster-Movric, et al., *Cooperative Control of Multi-Agent Systems: Optimal and Adaptive Design Approaches* (Springer, London, 2014)
10. J.W. Simpson-Porco, F. Dörfler, F. Bullo, Synchronization and power sharing for droop-controlled inverters in islanded microgrids. Automatica **49**(9), 2603–2611 (2013)

Chapter 9
Decentralized Optimal Economical Dispatch Scheme

9.1 Economical Optimization of Series-Type Microgrids

Figure 9.1 presents the structure of islanded series-type microgrids, which comprises of n series micro-inverters. Then, the equivalent circuit is demonstrated in Fig. 9.2. In Figs. 9.1 and 9.2, $V_i e^{j\delta_i}$ represents the output voltage vector of the i-th DG, $V_{PCC} e^{j\delta_{PCC}}$ is the voltage vector at the PCC. V_i, and V_{PCC} denotes the corresponding voltage amplitudes, δ_i and δ_{PCC} are the corresponding voltage phase angles. L_f and C_f express the filter inductance and filter capacitance, respectively. R_f and R_d are the corresponding series resistances. Z_{load} is the load impedance, and Z_{line} denotes the line impedance.

According to Kirchhoff laws, $V_{PCC} e^{j\delta_{PCC}}$ is expressed as

$$V_{PCC} e^{j\delta_{PCC}} = y' Z_{load} \sum_{i=1}^{n} V_i e^{j\delta_i} \tag{9.1}$$

$$y' = 1 \Bigg/ \left(Z_{load} + \sum_{i=1}^{n} Z_{line} \right) \tag{9.2}$$

where y' is the equivalent admittance. For simplification, y' is rewritten as

$$y' = \left| Y' \right| e^{j\theta'} \tag{9.3}$$

where $\left| Y' \right|$ and θ' denote its corresponding modulus and phase angle of y'. Then, the instantaneous active power p_i and reactive power q_i are written as

$$p_i = V_i \left| Y' \right| \sum_{j=1}^{n} V_j \cos\left(\delta_i - \delta_j - \theta'\right) \tag{9.4}$$

Y. Sun et al., *Series-Parallel Converter-Based Microgrids*, Power Systems,
https://doi.org/10.1007/978-3-030-91511-7_9

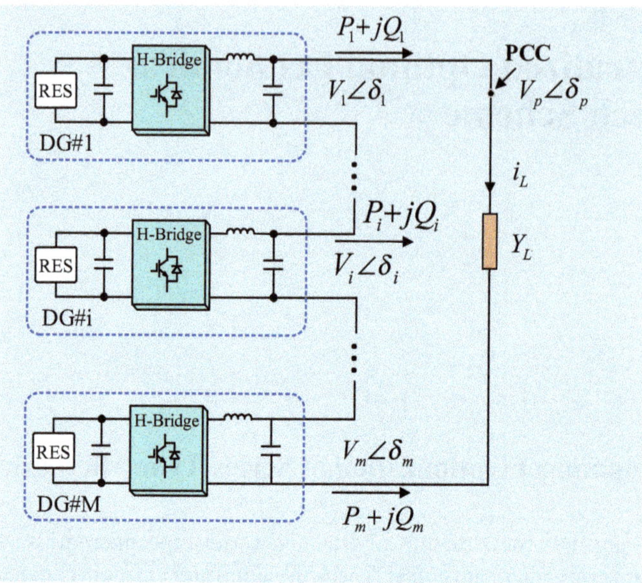

Fig. 9.1 Series-type microgrids

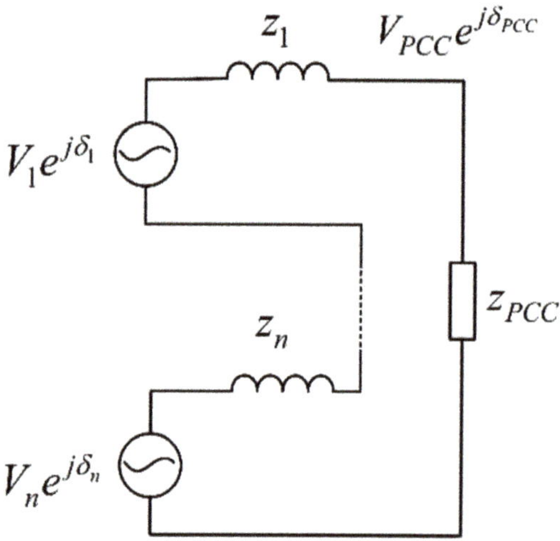

Fig. 9.2 Equivalent circuit of series-type microgrids

$$q_i = V_i \left|Y'\right| \sum_{j=1}^{n} V_j \sin\left(\delta_i - \delta_j - \theta'\right) \tag{9.5}$$

After passing a first-order low-pass filter, the filtered active power P_i and reactive power Q_i are expressed as

$$\dot{P}_i = \omega_c p_i - \omega_c P_i \tag{9.6}$$

$$\dot{Q}_i = \omega_c q_i - \omega_c Q_i \tag{9.7}$$

where ω_c is the cutoff frequency of the low-pass filter.

From (9.4)–(9.7), it can be concluded that P_i and Q_i are controlled by regulating the amplitude differences and phase angle differences of DGs output voltages.

9.1.1 Economical Optimization Problem Formulation

Usually, the economical optimization problem of power systems could be formulated as

$$\begin{aligned} & \min\left(\sum C_i\left(P_i\right)\right) \\ s.t.\ & \sum P_i = P_L \\ & P_{i,\min} \leqslant P_i \leqslant P_{i,\max} \end{aligned} \tag{9.8}$$

where P_L is the active power load demand including the transmission line losses. $P_{i,min}$, $P_{i,max}$ are the allowed output active power of the i-th DG. $C_i\left(P_i\right)$ is the general comprehensive operational costs consisting of fuel costs, maintenance costs, and so on, $i \in \{1, 2, \cdots, n\}$.

Assume that $C_i\left(P_i\right)$ is continuous, it follows from the extreme value theorem such that the economical optimization problem has a global optimal solution $\left(P_1^*, P_2^*, \cdots, P_n^*\right)$. And the optimal solution P_i^* is a map of P_L, which is expressed as

$$P_i^* = g_i\left(P_L\right) \tag{9.9}$$

where $g_i\left(P_L\right)$ is the optimal economical operation function (OEOF) of P_L. Usually, $g_i\left(P_L\right)$ can be obtained by off-line calculation in advance. It is worth noting that the focus of this study is not about how to solving the optimal solution but how to design a controller only based on the local information to realize the optimal economical operation of the studied system.

9.2 Communication-Free Economical Operation Control Scheme

To implement the optimal solution formula (9.9) and maintain the desired load voltage amplitudes, the communication-free $\varphi - f/P - V$ control is introduced in this section.

9.2.1 Control Scheme

The $\varphi - f/P - V$ control scheme of the i-th DG in the series-type microgrids is expressed as

$$f_i = f^* + m\,\mathrm{sgn}\,(Q_i)\cos\varphi_i \tag{9.10}$$

$$V_i = \frac{g_i\left(\bar{P}_i\right)}{\sum\limits_{j=1}^{n} g_j\left(\bar{P}_i\right)} V_{PCC}^* \tag{9.11}$$

where

$$\bar{P}_i = V_{PCC}^* I_i \cos\varphi_i \tag{9.12}$$

$$\cos\varphi_i = \frac{P_i}{\sqrt{P_i^2 + Q_i^2}} \tag{9.13}$$

where f^* is the nominal frequency of the microgrids and f_i is the reference frequency of the i-th DG. φ_i is the power factor angle of the i-th DG. V_{PCC}^* is the reference voltage at PCC. sgn(·) is a signum function. m is a certain positive coefficient determined by the feasible frequency ranges $[f_{\min}, f_{\max}]$. $f_{\max}$ and $f_{\min}$ are the given maximum and minimum permissible frequencies. Usually, the nominal frequency takes the value of $(f_{\max} + f_{\min})/2$. According to (9.10), m should be designed to satisfy the constraint: $0 < m \leqslant (f_{\max} - f_{\min})/2$.

9.2.2 Steady-State Analysis

In the steady state, the following equality for two different DGs is obtained from (9.10):

$$\cos\varphi_i = \cos\varphi_j \tag{9.14}$$

Equation (9.14) means that those power factors of each DG in the microgrids are the same under the scheme.

Remark Keeping the same power factor is very crucial, because it ensures: (1) the power sharing proportional to its output voltage amplitudes, (2) the same voltage phase of each DG to add them algebraically, and (3) acquiring the load information locally.

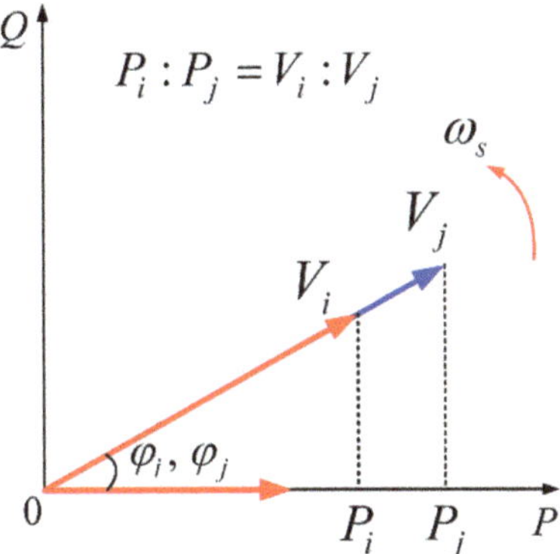

Fig. 9.3 Power operation control of the scheme

Since all the DGs have the same power factor, combine $I_i = I_j$ and $P_i = V_i I_i \cos \varphi_i$, then the following equalities hold:

$$P_i : P_j = V_i : V_j \tag{9.15}$$

$$\bar{P}_i = \bar{P}_j \tag{9.16}$$

Clearly, the active power allocations depend on the choice of the output voltage amplitudes shown in Fig. 9.3, in which ω_s is the synchronous frequency of the series-type microgrids in the steady state.

Remark Output power of each DG is proportional to its output voltage amplitudes, which is the foundation of economical allocation. We could regulate the voltage reference to realize the economical operation, which is totally different from the idea in [1].

Substituting (9.11) into (9.15) and combining (9.16), we have

$$P_i : P_j = g_i \left(\bar{P}_i\right) : g_j \left(\bar{P}_j\right) \tag{9.17}$$

In fact, according to the definition of $\bar{P}_i$ in (9.12), $\bar{P}_i$ in the steady state is equal to P_L, which can be obtained only based on the local measurements. Thus, (9.17) is rewritten as follows:

$$P_i : P_j = g_i \left(P_L\right) : g_j \left(P_L\right) \tag{9.18}$$

Combining (9.18) with (9.9) yields

$$P_i : P_j = P_i^* : P_j^* \tag{9.19}$$

Since $\sum P_i = \sum P_i^* = P_L$, then we have

$$P_i = P_i^* \tag{9.20}$$

Remark From (9.20), the output power of each DG under the scheme (9.10) and (9.11) is the optimal solution of the problem (9.8). That is to say, the first limitation in [1] is solved. From (9.10) and (9.11), the control scheme only depends on each DG's output voltage and current, so $\varphi - f/P - V$ is a communication-free control approach.

Usually, the feeder impedance is much less than the load impedance. Consequently, the voltage drop on feeder impedance can be neglected. Since the same voltage phase of each DG is achieved according to (9.10), then $\left|\sum V_i e^{j\delta_i}\right| = V_{PCC}^*$. Thus, the expected load voltage amplitudes are obtained while satisfying the optimality. And the second limitation in [1] is overcame.

Overall, it can be summarized as follows: the scheme can perform the optimal solution via the decentralized manner while satisfying the desired load voltage amplitudes.

9.3 Stability Analysis

As the economical operation is closely related with the underlying control of the microgrids, the stability of systems with the consideration of economical operation is a critical issue. Thus, the stability and sensibility analysis of parameters will be investigated in this section through the small-signal analysis method [2–4].

Equation (9.10) is rewritten as

$$\omega_i = \omega^* + 2\pi m \operatorname{sgn}(Q_i) \cos\varphi_i \tag{9.21}$$

where $\omega^* = 2\pi f^*$. Let $\delta_s = \int \omega_s dt$, and denote $\tilde{\delta_i} = \delta_i - \delta_s$, then we have

$$\dot{\tilde{\delta_i}} = \omega^* - \omega_s + 2\pi m \operatorname{sgn}(Q_i) \cos\varphi_i \tag{9.22}$$

Linearization of (9.6) and (9.7) yields

$$\begin{aligned} \Delta\dot{P}_i = {} & \omega_c \frac{\partial p_i}{\partial V_i} \Delta V_i + \omega_c \sum_{j=1, i\neq j}^{n} \frac{\partial p_i}{\partial V_j} \Delta V_j \\ & + \omega_c \frac{\partial p_i}{\partial \tilde{\delta}_i} \Delta\tilde{\delta}_i + \sum_{j=1, i\neq j}^{n} \frac{\partial p_i}{\partial \tilde{\delta}_j} \Delta\tilde{\delta}_j - \omega_c \Delta P_i \end{aligned} \tag{9.23}$$

$$\begin{aligned} \Delta\dot{Q}_i = {} & \omega_c \frac{\partial q_i}{\partial V_i} \Delta V_i + \omega_c \sum_{j=1, i\neq j}^{n} \frac{\partial q_i}{\partial V_j} \Delta V_j \\ & + \omega_c \frac{\partial q_i}{\partial \tilde{\delta}_i} \Delta\tilde{\delta}_i + \sum_{j=1, i\neq j}^{n} \frac{\partial q_i}{\partial \tilde{\delta}_j} \Delta\tilde{\delta}_j - \omega_c \Delta Q_i \end{aligned} \tag{9.24}$$

Combining $I_i = \sqrt{P_i^2 + Q_i^2}/V_i$ with (9.13) and (9.12) is rewritten as

$$\bar{P}_i = V_{PCC}^* P_i / V_i \tag{9.25}$$

By substituting (9.25) into (9.11), then it is rewritten as

$$F_i \left(V_i, P_i \right) = 0 \tag{9.26}$$

where $F_i(\cdot)$ is the function of V_i and P_i. Linearization of (9.26) yields

$$\Delta V_i = a_i \Delta P_i \tag{9.27}$$

where $a_i = -\left(\frac{\partial F_i}{\partial P_i} \Big/ \frac{\partial F_i}{\partial V_i} \right)$. Combining (9.22), (9.23), (9.24), and (9.27), the small-signal model of the i-th DG is

$$\Delta \dot{\tilde{\delta}}_i = -2\pi m \operatorname{sgn}\left(Q_i^o \right) \sin \varphi_i^o \left(\frac{\partial \varphi_i}{\partial P_i} \Delta P_i + \frac{\partial \varphi_i}{\partial Q_i} \Delta Q_i \right) \tag{9.28}$$

$$\begin{aligned} \Delta \dot{P}_i = \omega_c \left(\frac{\partial p_i}{\partial V_i} a_i - 1 \right) \Delta P_i + \omega_c \sum_{j=1, i \neq j}^{n} \frac{\partial p_i}{\partial V_j} a_j \Delta P_j \\ + \omega_c \frac{\partial p_i}{\partial \tilde{\delta}_i} \Delta \tilde{\delta}_i + \omega_c \sum_{j=1, i \neq j}^{n} \frac{\partial p_i}{\partial \tilde{\delta}_j} \Delta \tilde{\delta}_j \end{aligned} \tag{9.29}$$

$$\begin{aligned} \Delta \dot{Q}_i = \omega_c \frac{\partial q_i}{\partial V_i} a_i \Delta P_i + \omega_c \sum_{j=1, i \neq j}^{n} \frac{\partial q_i}{\partial V_j} a_j \Delta P_j \\ + \omega_c \frac{\partial q_i}{\partial \tilde{\delta}_i} \Delta \tilde{\delta}_i + \omega_c \sum_{j=1, i \neq j}^{n} \frac{\partial q_i}{\partial \tilde{\delta}_j} \Delta \tilde{\delta}_j - \omega_c \Delta Q_i \end{aligned} \tag{9.30}$$

Write (9.28)–(9.30) in the matrix form:

$$\dot{\mathbf{X}} = \mathbf{A}\mathbf{X} \tag{9.31}$$

where $\mathbf{X}$ is the state variable vector and $\mathbf{A}$ is the system matrix. Both of them are shown in the following.

The variable vectors are $\mathbf{X} = [\mathbf{\Delta P}\ \mathbf{\Delta Q}\ \mathbf{\Delta \tilde{\delta}}]^T$, $\mathbf{\Delta P} = [\Delta P_1\ \cdots\ \Delta P_n]$, $\mathbf{\Delta Q} = [\Delta Q_1\ \cdots\ \Delta Q_n]$, $\mathbf{\Delta \tilde{\delta}} = \left[\Delta \tilde{\delta}_1\ \cdots\ \Delta \tilde{\delta}_n \right]$.

The system matrix **A** is written as

$$\mathbf{A} = \begin{bmatrix} \mathbf{A_{11}} & 0 & \mathbf{A_{13}} \\ \mathbf{A_{21}} & \mathbf{A_{22}} & \mathbf{A_{23}} \\ \mathbf{A_{31}} & \mathbf{A_{32}} & 0 \end{bmatrix} \tag{9.32}$$

where

$$\mathbf{A}_1 1 = \omega_c \begin{bmatrix} \left(\frac{\partial p_1}{\partial V_1} a_1 - 1\right) & \cdots & \frac{\partial p_1}{\partial V_n} a_n \\ \vdots & \ddots & \vdots \\ \frac{\partial p_n}{\partial V_1} a_1 & \cdots & \left(\frac{\partial p_n}{\partial V_n} a_n - 1\right) \end{bmatrix} \tag{9.33}$$

$$\mathbf{A}_{13} = \omega_c \begin{bmatrix} \frac{\partial p_1}{\partial \tilde{\delta}_1} & \cdots & \frac{\partial p_1}{\partial \tilde{\delta}_n} \\ \vdots & \ddots & \vdots \\ \frac{\partial p_n}{\partial \tilde{\delta}_1} & \cdots & \frac{\partial p_n}{\partial \tilde{\delta}_n} \end{bmatrix} \tag{9.34}$$

$$\mathbf{A}_2 1 = \omega_c \begin{bmatrix} \frac{\partial p_1}{\partial \tilde{\delta}_1} & \cdots & \frac{\partial p_1}{\partial \tilde{\delta}_n} \\ \vdots & \ddots & \vdots \\ \frac{\partial p_n}{\partial \tilde{\delta}_1} & \cdots & \frac{\partial p_n}{\partial \tilde{\delta}_n} \end{bmatrix} \tag{9.35}$$

$$\mathbf{A}_{22} = -\omega_c \mathbf{I} \tag{9.36}$$

$$\mathbf{I} = diag[1 \ \cdots \ 1]_{n \times n} \tag{9.37}$$

$$\mathbf{A}_{23} = \omega_c \begin{bmatrix} \frac{\partial q_1}{\partial \tilde{\delta}_i} & \cdots & \frac{\partial q_1}{\partial \tilde{\delta}_n} \\ \vdots & \ddots & \vdots \\ \frac{\partial q_n}{\partial \tilde{\delta}_1} & \cdots & \frac{\partial q_n}{\partial \tilde{\delta}_n} \end{bmatrix} \tag{9.38}$$

$$\mathbf{A}_{31} = -2\pi m \begin{bmatrix} \operatorname{sgn}\left(Q_1^o\right) \sin \varphi_1^o \frac{\partial \varphi_1}{\partial P_1} & & \\ & \ddots & \\ & & \operatorname{sgn}\left(Q_n^o\right) \sin \varphi_n^o \frac{\partial \varphi_n}{\partial P_n} \end{bmatrix} \tag{9.39}$$

$$\mathbf{A}_{32} = -2\pi m \begin{bmatrix} \operatorname{sgn}\left(Q_1^o\right) \sin \varphi_1^o \frac{\partial \varphi_1}{\partial Q_1} & & \\ & \ddots & \\ & & \operatorname{sgn}\left(Q_n^o\right) \sin \varphi_n^o \frac{\partial \varphi_n}{\partial Q_n} \end{bmatrix} \tag{9.40}$$

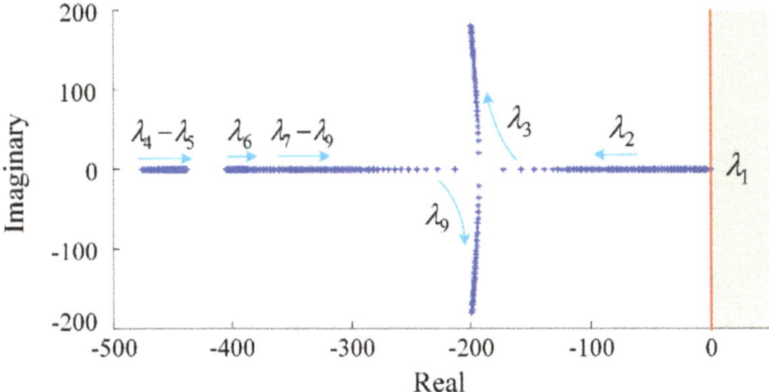

Fig. 9.4 Root locus as m increases from 0.01 to 0.5

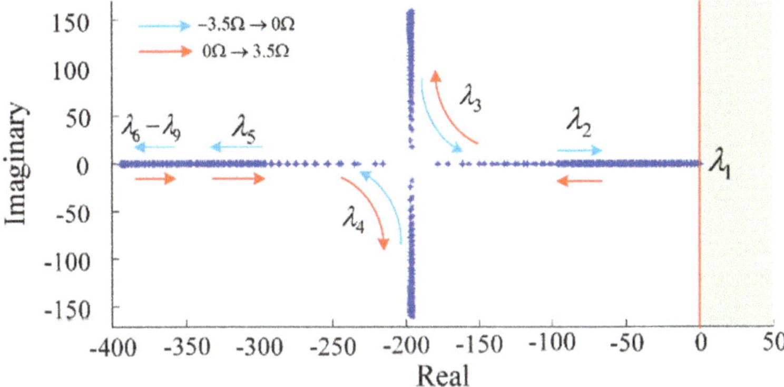

Fig. 9.5 Root locus as $X_{load} \in [-3.5,\ 3.5]$

The root locus method is used to investigate the system's stability around the operating point. Based on the simulation system described in this section, the root locus diagrams under different m and the load reactance X_{load} are studied.

Figure 9.4 shows the root locus diagram as m increases from 0.01 to 0.5, with the load resistance $R_{load} = 12.5\,\Omega$, $X_{load} = 3.14\,\Omega$. As seen, the matrix **A** has one zero eigenvalue as to rotational invariance, which is depicted in [5]. The remaining eigenvalues are in the left half-plane. As seen, the system is stable when $m \in [0.01, 0.5]$.

When $R_{load} = 12.5\,\Omega$ and $m = 0.3$, then let X_{load} change from $-3.5\,\Omega$ to $3.5\,\Omega$ and the root locus diagram is depicted in Fig. 9.5. As seen, the scheme can maintain the system's stable operation under both the resistance–capacitance loads and resistance–inductance loads.

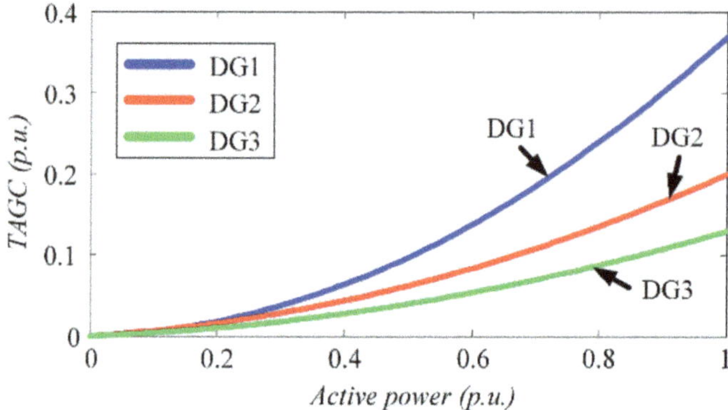

Fig. 9.6 TAGC of DG1, DG2, and DG3

Table 9.1 Parameters for simulations

Parameters	Values	Parameters	Values
f(Hz)	[49,51]	$R_d(\Omega)$	3.3
f^*(Hz)	50	L_{Line1}(H)	1.5e−3
m	0.3	L_{Line2}(H)	1.6e−3
V_{PCC}^*(V)	110	L_{Line3}(H)	1.2e−3
L_f(H)	1.5e−3	$P_{\max}$(W)	1000
$R_f(\Omega)$	0.4	$Q_{\max}$(Var)	1000
$C_f(\mu F)$	20	$P_i(p.u.)$	[0,1]

9.4 Simulation Results

The $\cos\varphi - f/P - V$ scheme is verified in MATLAB/Simulink platform. The series-type microgrid in the simulation model includes three DGs (see Fig. 9.1). The generation costs of DGs from the literature [6, 7] are depicted in Fig. 9.6. The parameters of simulation are listed in Table 9.1. The detailed control block diagram of the DG unit is shown in Fig. 9.7.

9.4.1 Case 1: Switch Between the RL and RC Load

This simulation is carried out under both the resistance–inductance (RL) and resistance–capacitance (RC) loads. The load demands are scheduled as follows: in the interval [0 s, 1 s] resistance–capacitance load, in [1 s, 2 s] resistance–inductance load, and resistance–capacitance load in the interval [2 s, 3 s]. The frequencies over time are depicted in Fig. 9.8a. The reactive power allocations among DGs are shown in Fig. 9.8b. Therefore, this scheme can realize the stable operation under the two types of load.

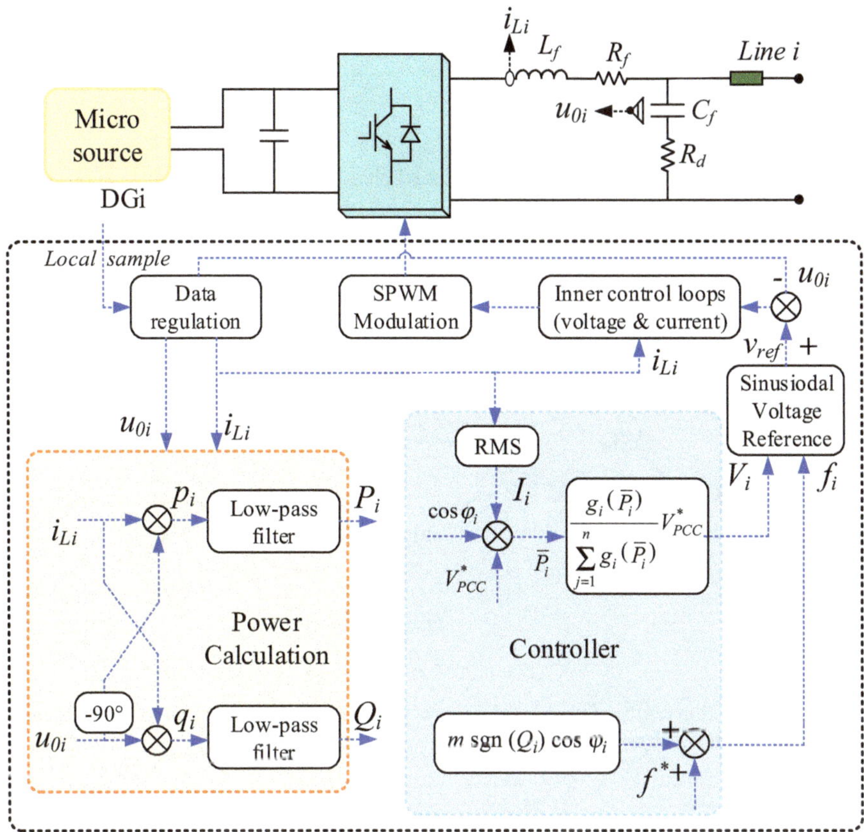

Fig. 9.7 Control block diagram of the single-phase DG unit

9.4.2 Case 2: Optimal Economical Operation Under RL Load

To verify the optimal economical operation under the resistance–inductance load, the load demands shown in Fig. 9.9a are scheduled as 0.5, 1, and 1.5$p.u.$ in the interval [0 s, 1 s], [1 s, 2 s], [2 s, 3 s], respectively. The frequencies are shown in Fig. 9.9b, which is higher than 50 Hz due to the feature of inverse droop. The active power sharing results are shown in Fig. 9.9c. The OEOF solved by interior point method [8] is depicted in Fig. 9.9d. Clearly, the operation results in Fig. 9.9c agree with Fig. 9.9d. It is concluded that the scheme can obtain the optimal economical operation under the resistance–inductance load.

Figure 9.10a shows the voltage waveform of each DG in the interval [1.4 s, 1.6 s]. The voltage phase angles of DGs are controlled to be a same value, but the voltage amplitudes are different. The load voltage at PCC is shown in Fig. 9.10b, in which

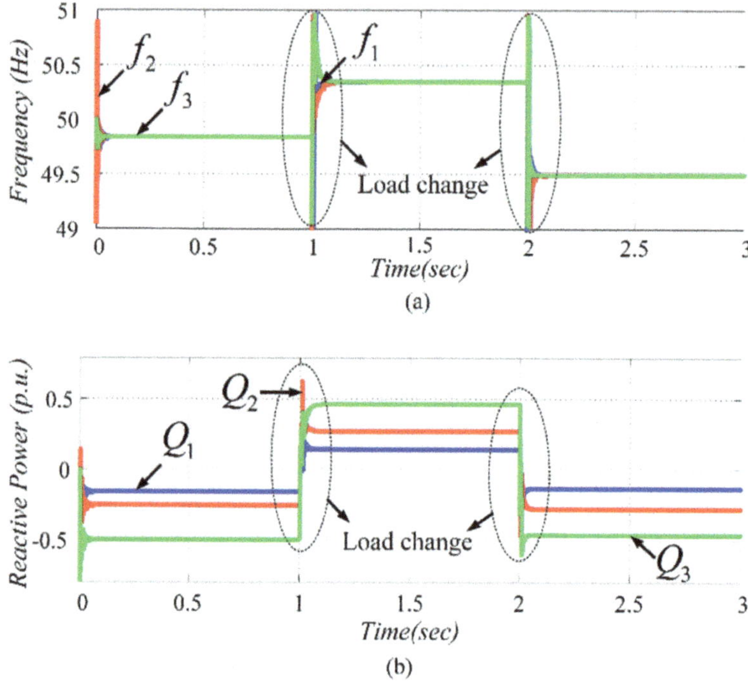

Fig. 9.8 Simulation results of case 1: (**a**) frequency and (**b**) reactive power under the resistance–inductance and resistance–capacitance loads

its amplitude stays at 110V regardless of the load changes. Therefore, the scheme can obtain an excellent load voltage quality [9, 10].

9.4.3 Case 3: Optimal Economical Operation Under RC Load

This simulation is carried out to verify the performance of the scheme under the resistance–capacitance load. The active power load schedules are the same as case 2 (see Fig. 9.9a). The waveform of frequencies is shown in Fig. 9.11a. It is lower than 50Hz because the droop control is performed. The active power allocations among DGs are shown in Fig. 9.11b, which agree with Fig. 9.9d. Therefore, the optimal economical operation is also obtained under the resistance–capacitance load.

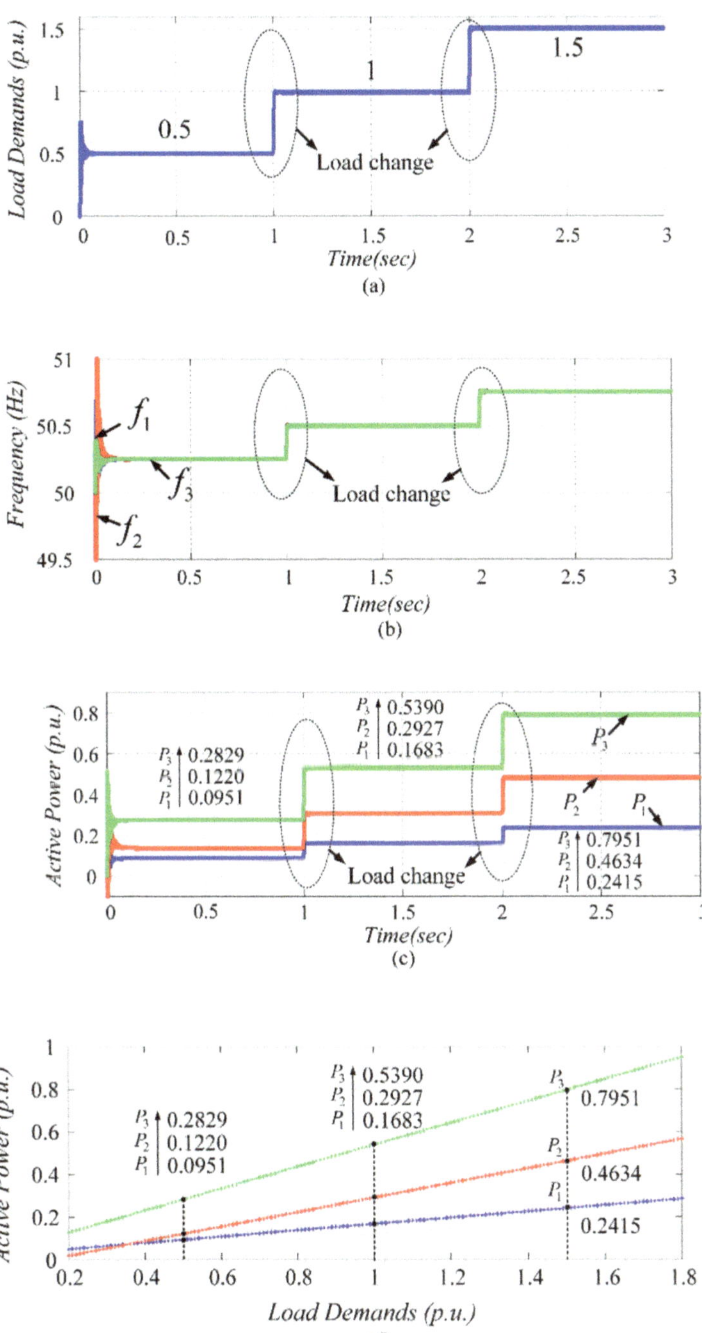

Fig. 9.9 Simulation results of case 2: (**a**) the active power load demands, (**b**) frequency, (**c**) active power allocations, and (**d**) OEOF

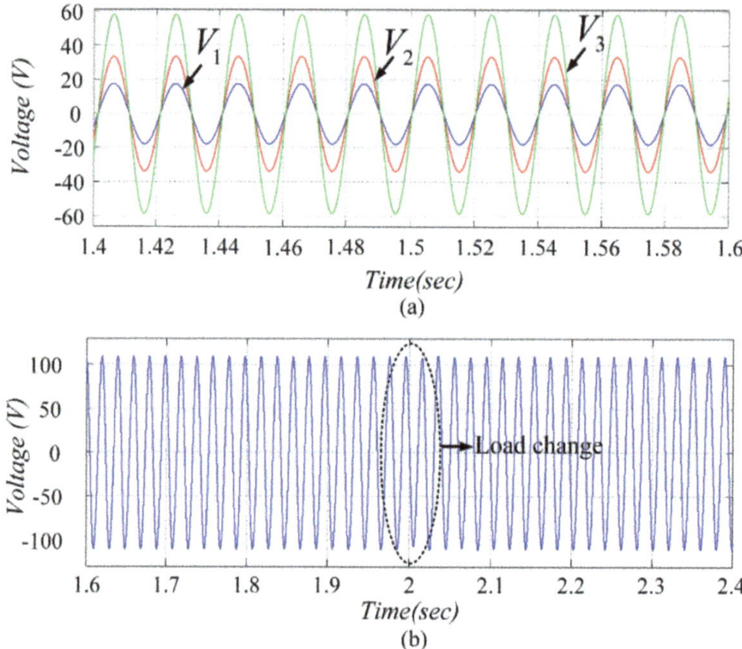

Fig. 9.10 Simulation voltage waveforms: (**a**) DGs in the steady state and (**b**) PCC during load changing

9.4.4 Case 4: Capacity Constraints

As shown in Fig. 9.12a, the load demands are scheduled as 1.8, 2, and 2.2$p.u.$ in the interval [0 s, 1 s], [1 s, 2 s], [2 s, 3 s], respectively. The frequencies are depicted in Fig. 9.12b. The active power allocations among DGs are shown in Fig. 9.12c, in which its maximum output power (1$p.u.$) in the second and third intervals are maintained. The OEOF curve is depicted in Fig. 9.12d with $P_L \in$ [1.6$p.u.$, 2.4$p.u.$]. Obviously, the simulation results in Fig. 9.12c agree with the theoretical results in Fig. 9.12d. Thus, the scheme is capable of capacity constraints for DGs.

9.4.5 Case 5: Comparisons Between the Scheme and Existing Method

In this case, the load demands are set as 2 and 2.2$p.u.$ in the interval [0 s, 1 s] and [1 s, 2 s], respectively. From the comparative results shown in Fig. 9.13a, the TAGC

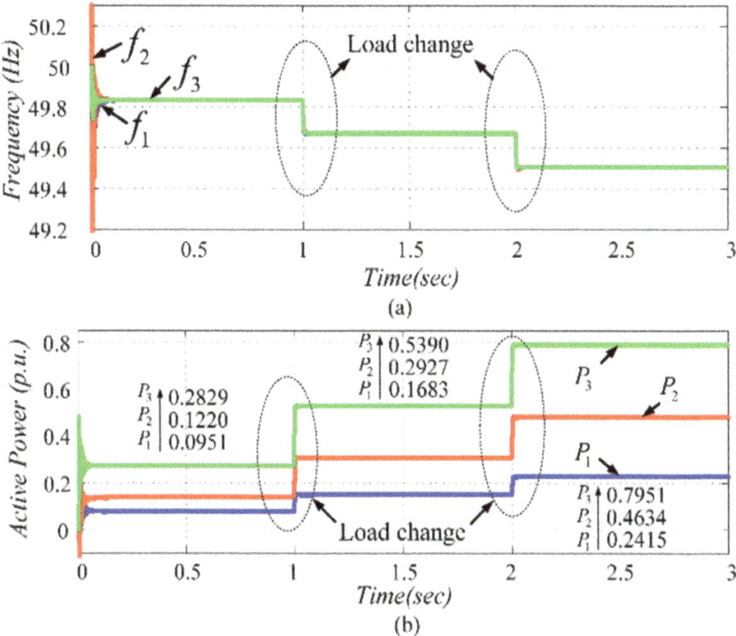

Fig. 9.11 Simulation results of case 3: (**a**) frequency and (**b**) active power allocations

of the scheme is lower than it based on the method in [1]. Therefore, the scheme is preferable to the method [1] from the perspective of economy. The load voltage amplitudes are shown in Fig. 9.13b, in which the scheme can maintain at the desired values. On the contrary, the load voltage amplitudes of the method in [1] deviate from its nominal values. As seen, the scheme is superior to the method [1] in the point of economy and load voltage quality.

9.4.6 Case 6: Performance of the Scheme Under the Feeder Impedance Variation

This test with feeder impedance variation (from 1.5 to 3 mH) at t=1 s is performed. The load voltage amplitudes are shown in Fig. 9.14a, which almost stay at a constant level. The active power allocations are depicted in Fig. 9.14b, which also almost maintains at a constant. Therefore, the effect of feeder impedance variation on performance of the method is almost negligible. That is to say, the scheme is robust to the feeder impedance variation to some extent.

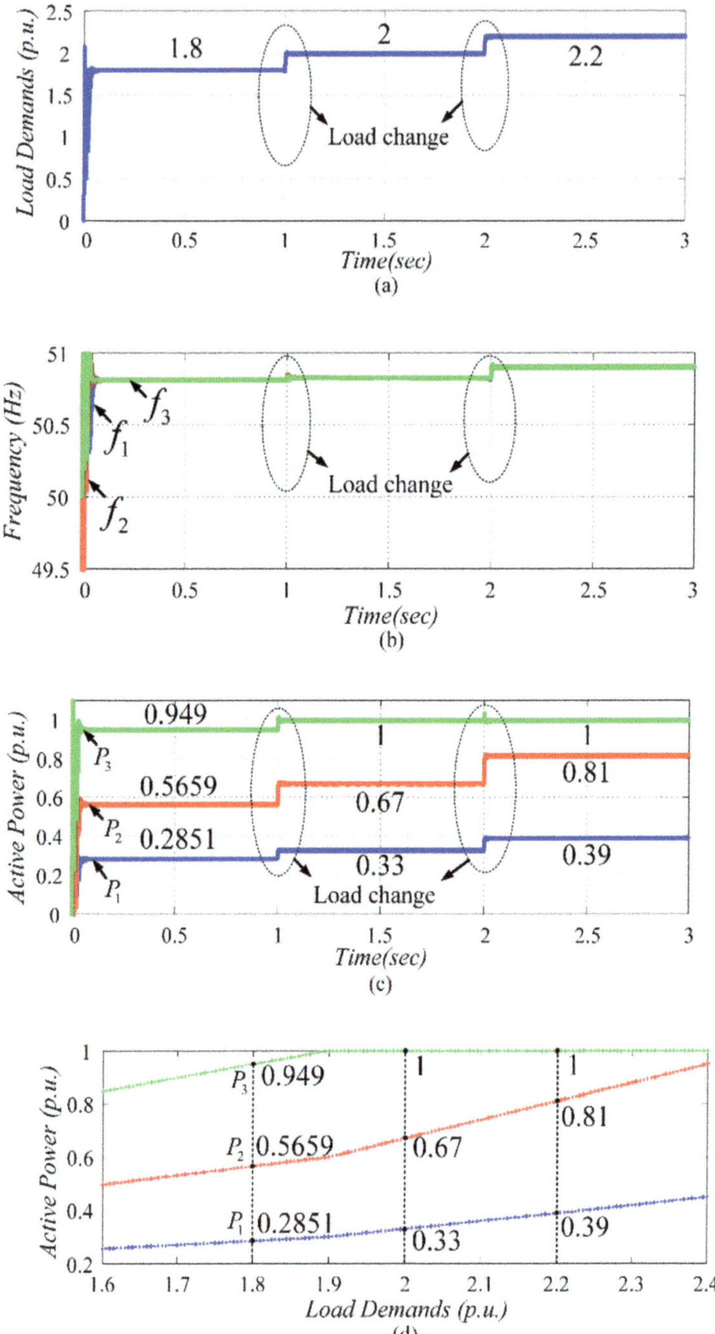

Fig. 9.12 Simulation results of case 4: (**a**) the active power load demands, (**b**) frequency, (**c**) active power allocations, and (**d**) OEOF

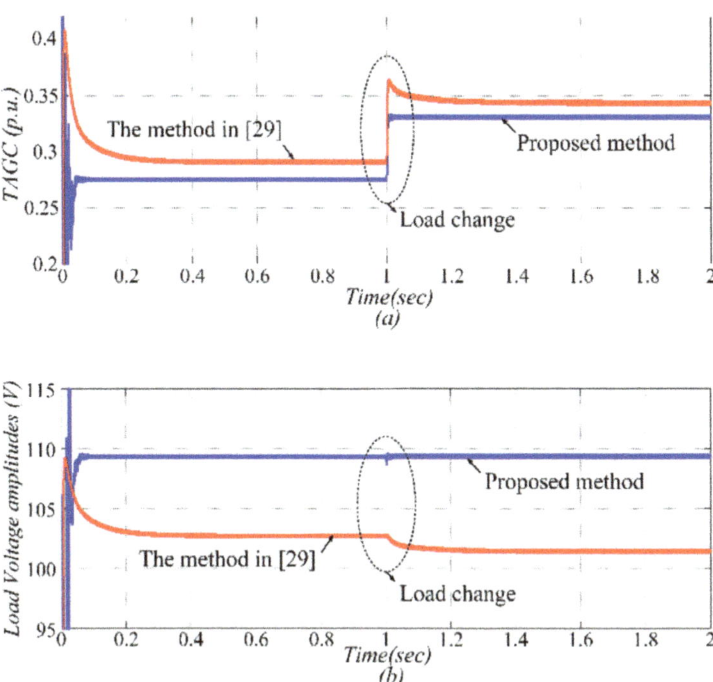

Fig. 9.13 Comparison results: (**a**) TAGC and (**b**) load voltage amplitudes

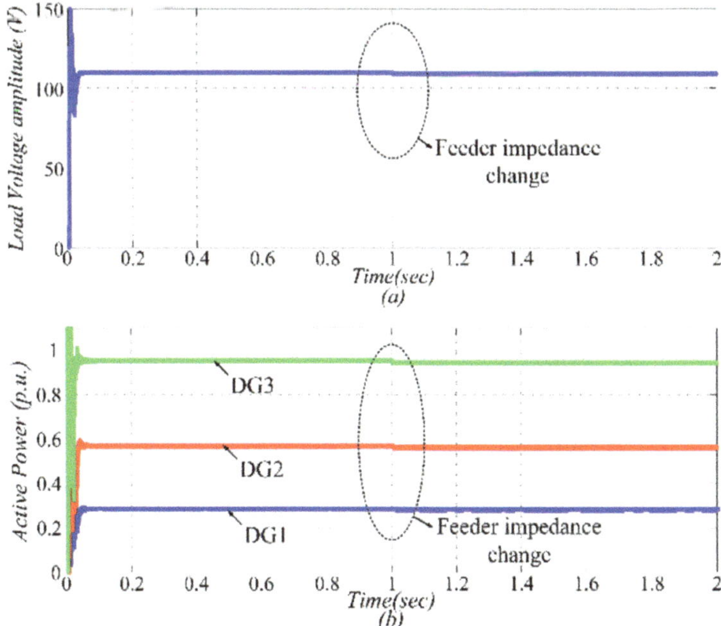

Fig. 9.14 Simulation results of case 6: (**a**) load voltage amplitudes and (**b**) active power allocations

9.5 Experimental Results

In order to verify the performance of the $\cos\varphi - f/P - V$ scheme, a microgrid prototype is built including two DGs, which is simulated by a single-phase voltage source inverter. The corresponding generation costs of the two DGs are the same as those of DG1 and DG2 in the simulation models, respectively. The parameters of this experiment are listed in Table 9.2.

The considered microgrid comprises two DGs. Even though, it could still meet the experimental requirements. The experiment is implemented under the resistance–inductance load. The active power load schedules are 0.4, 0.8, 1.2, 0.4, and $0.8p.u.$ in the interval [0 s, 20 s], [20 s, 40 s], [40 s, 60 s], [60 s, 80 s], [80 s, 100 s], respectively. The experimental waveforms of voltage and current are shown in Fig. 9.15, in which the voltage phase angles of DG1 and DG2 are always in phase. Therefore, the scheme can ensure the synchronous operation of DGs and maintain the stability of system.

The active power requirement curve of the load is shown in Fig. 9.16a. The frequency curve is shown in Fig. 9.16b. The active power allocations among DG1 and DG2 are depicted in Fig. 9.16c. The OEOF curves are depicted in Fig. 9.16d in the case of $P_L \in [0.1p.u., 1.4p.u.]$. Compared to Fig. 9.16d, the experimental results in Fig. 9.16c match the theoretical value approximately. The deviation of output power between Fig. 9.16c and d is because the line losses. As seen, the scheme can obtain the optimal economical operation without communications.

Table 9.2 Parameters for experiments

Parameters	Values	Parameters	Values
f^*(Hz)	50	$R_d(\Omega)$	3.3
m	03	L_{Line1}(H)	0.3e−3
V^*_{PCC}(V)	100	L_{Line2}(H)	1.6e−3
L_f(H) 0.6e−3	P_{max}(W)	200	
$R_f(\Omega)$	0.5	Q_{max}(Var)	200
$C_f(\mu F)$	20	$P_i(p.u.)$	[0, 1]

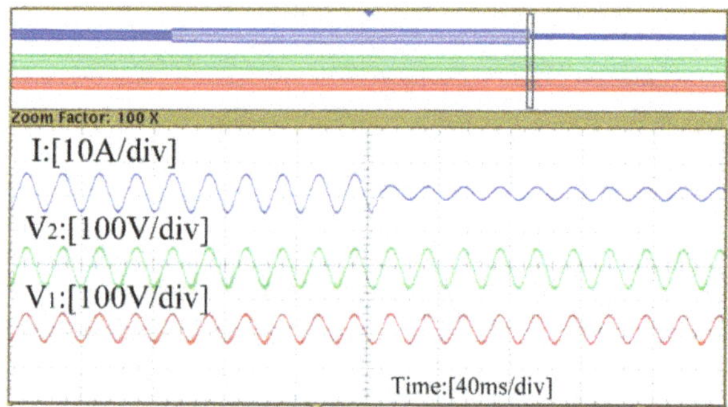

Fig. 9.15 Experimental waveforms with load stepping

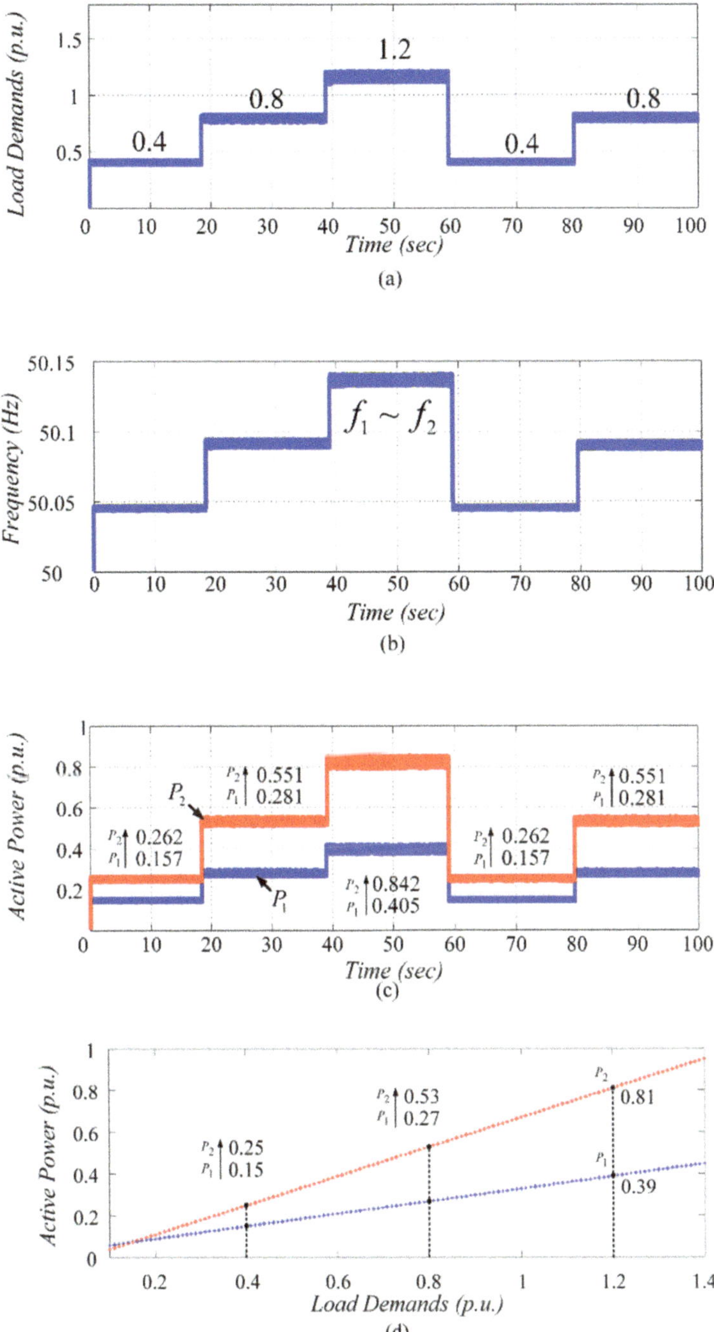

Fig. 9.16 Experimental results: (**a**) the active power load demands, (**b**) frequency, (**c**) active power allocations, and (**d**) OEOF

9.6 Conclusion

This chapter studied the optimal economical operation problem of the series-type microgrids. A communication-free control scheme ($\cos\varphi - f/P - V$) is introduced, which could achieve the global optimal economical operation easily. With this method, the excellent load voltage quality is guaranteed, and its implementation only needs the local information. Meanwhile, the synchronization of all DGs has been achieved autonomously under both the inductive and capacitive loads. Besides, the large-signal stability for the system suffering large disturbance like short-circuit faults will be investigated in the future.

References

1. L. Li, H. Ye, Y. Sun, et al., A communication-free economical-sharing scheme for cascaded-type microgrids. Int. J. Electr. Power Energy Syst. **104**, 1–9 (2019)
2. N. Pogaku, M. Prodanovic, T. Green, Modeling, analysis and testing of autonomous operation of an inverter-based microgrid. IEEE Trans. Power Electron. **22**,(2), 613–625 (2007)
3. Y.A.-R.I. Mohamed, E.F. El-Saadany, Adaptive decentralized droop controller to pre-serve power sharing stability of paralleled inverters in distributed generation microgrid. IEEE Trans. Power Electron. **23**(6), 2806–2816 (2008)
4. X.Q. Guo, Z.G. Lu, B.C. Wang, et al., Dynamic phasors-based modeling and stability analysis of droop-controlled inverters for microgrid applications. IEEE Trans. Smart Grid **5**(6), 2980–2987 (2014)
5. J.W. Simpson-Porco, F. Dörfler, F. Bullo, Synchronization and power sharing for droop-controlled inverters in islanded microgrids. Automatica **49**(9), 2603–2611 (2013)
6. I.U. Nutkani, P.C. Loh, P. Wang, et al., Linear decentralized power sharing schemes for economic operation of AC microgrids. IEEE Trans. Ind. Electron. **63**(1), 225–234 (2016)
7. H. Han, L. Li, L. Wang, et al., A novel decentralized economic operation in islanded AC microgrids. Energies **10**(6), 1–18 (2017)
8. C.L. Byrne, Iterative optimization, in *A First Course in Optimization* (CRC Press, Boca Raton, 2014)
9. L. Li, Y. Sun , H. Han, X. Hou, X. Yuan, W. Xiong, M. Su, Communication-free optimal economical dispatch scheme for cascaded-type microgrids with capacity constraints. IET Power Electron. **13**(3), 2866–2873 (2020)
10. L. Li, Y. sun, H. Han, X. Hou, M. Su, Z. Liu, Power factor angle consistency control for decentralized power sharing in cascaded-type microgrid. IET Gener. Trans. Distrib. **13**(6), 850–857 (2019)

Chapter 10
Decentralized SOC Balancing Control for Series-Type Storages

10.1 Decentralized SOC Balancing Control

10.1.1 Equivalent Model of Series Energy Storage System

Figure 10.1 illustrates an islanded CESS consisting of n ESUs, interfacing converters and local loads. In this configuration, the interfacing converters are single-phase DC/AC converters, all of which are connected in series to energize the loads. The by-pass switch is used to protect the ESU if the SOC of one ESU is not in its safe range or the ESU has failed.

For simplicity, each DC/AC converter is regarded as a controlled voltage source (CVS). Thus, the output voltage of the i-th converter is denoted as $V_i e^{j\theta_i}$ and the coupling point voltage is represented by $V_{pcc}e^{j\theta_{pcc}}$. The relationship between the coupling point voltage and the voltages of all controlled voltage sources is expressed as follows

$$V_{pcc}e^{j\theta_{pcc}} = \sum_{i=1}^{n} V_i e^{j\theta_i} \tag{10.1}$$

From Fig. 10.1, the output active power P_i and the reactive power Q_i of the i-th controlled voltage source are derived as follows

$$P_i + jQ_i = V_i e^{j\theta_i} \cdot \left(V_{pcc}e^{j\theta_{pcc}} / \left| Z_{load} \right| e^{j\theta_{load}} \right)^* \tag{10.2}$$

where the Z_{load} and θ_{load} are the load impedance magnitude and angle.

Y. Sun et al., *Series-Parallel Converter-Based Microgrids*, Power Systems,
https://doi.org/10.1007/978-3-030-91511-7_10

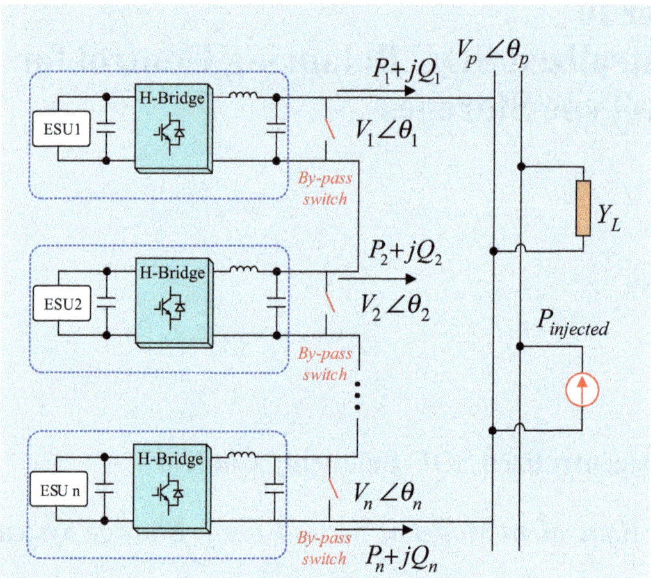

Fig. 10.1 The simplified structure diagram of Islanded CESS

The power transmission characteristics of the i-th controlled voltage source are obtained as

$$P_i = \frac{V_i}{|Z_{load}|}\sum_{j=1}^{n} V_j \cos\left(\theta_i + \theta_{load} - \theta_j\right) \tag{10.3}$$

$$Q_i = \frac{V_i}{|Z_{load}|}\sum_{j=1}^{n} V_j \sin\left(\theta_i + \theta_{load} - \theta_j\right) \tag{10.4}$$

10.1.2 Approximate Relationship Between SOC and Output Power

According to the definition of SOC, the SOC of the i-th ESU is expressed as follows

$$SoC_i = SoC_{0i} - \frac{1}{C_i}\int i_i^{in} dt \tag{10.5}$$

where SoC_{0i} is the initial SoC_i value. C_i and i_i^{in} denote the nominal capacity and terminal current of the i-th ESU, respectively. Neglecting the power losses in the inverter, the output power of inverter P_i yields,

$$P_i \approx P_i^{in} = v_i^{in} \cdot i_i^{in} \tag{10.6}$$

where P_i^{in} is the output power of the i-th ESU. Combining (10.5) and (10.6), the relationship between SOC and output power of i-th ESU is obtained, which plays a significant role in the following analysis.

$$SOC_i = SOC_{0i} - \frac{\int i_i^{in} dt}{C_i} \approx SOC_{0i} - \frac{\int P_i dt}{E_{\max\_i}} \tag{10.7}$$

where $E_{\max\_i} = v_i^{in} \cdot C_i$

10.1.3 SOC Balancing Control Method

According to reference [1], the mechanisms of autonomous synchronization in islanded series-type system are different due to the load impedance characteristic. Thus, in this chapter, four quadrant operation for storages, as illustrated in Fig. 10.2, should be taken into account. It includes: discharging with resistive-inductive load (1), charging with resistive-inductive load (2), charging with resistive-capacitive load (3), and discharging with resistive-capacitive load (4).

First of all, the rated power of the power sources should be pre-designed to avoid overload with considering the load demand. This is the precondition for ensuring the safe operation of the system. Then in order to achieve SOC balancing, the control scheme is designed as follows,

$$\begin{cases} \omega_i = \omega_0 + \operatorname{sgn}(Q_i)(m_i P_i - n_i SOC_i) \\ V_i = \varphi(E_{\max\_i}) V^* \end{cases} \tag{10.8}$$

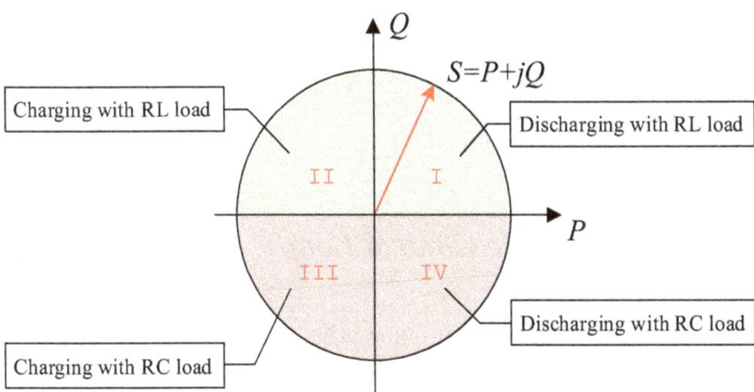

Fig. 10.2 Four quadrant working cases in CESS

where the sgn(.) denotes sign function. ω_i and V_i are the angular frequency reference and voltage amplitude reference of the i-th controlled voltage source, respectively. ω_0 represents the value of ω with no load. V^* is the nominal voltage amplitude value of the bus voltage. m_i is the adjustment coefficient of the controller which is designed to be inversely proportional to their control gains, i.e., $m_i \cdot E_{\max_i} = K$ where K is a constant. n_i is a SOC weighted coefficient to modify the droop control adaptively. For achieving SOC balancing, we choose $n_1 = \cdots = n_n = N$. The $\varphi\left(E_{\max\_i}\right)$ is designed as follows

$$\varphi\left(E_{\max\_i}\right) = E_{\max\_i} \Big/ \sum_{j=1}^{n} E_{\max\_j} \tag{10.9}$$

When the CESS has reached the steady state, the frequency synchronization and SOC balancing can be achieved. Then we can obtain

$$m_i \operatorname{sgn}(Q_i) P_i = m_j \operatorname{sgn}\left(Q_j\right) P_j \tag{10.10}$$

Define that $\varphi_i(\varphi_j)$ is the phase angle difference between the output voltage $V_i(V_j)$ and the common current I of the i-th (j-th) CVS. Combine (10.8), (10.9), and (10.10), then

$$\frac{KV^*I \operatorname{sgn}(Q_i)\cos\varphi_i}{\sum\limits_{k=1}^{n} E_{\max\_k}} = \frac{KV^*I \operatorname{sgn}(Q_i)\cos\varphi_j}{\sum\limits_{k=1}^{n} E_{\max\_k}} \tag{10.11}$$

From (10.11), it is obvious that $\operatorname{sgn}(\sin\varphi_i)\cos\varphi_i = \operatorname{sgn}\left(\sin\varphi_j\right)\cos\varphi_j$. Then $\varphi_i = \varphi_j$ or $\varphi_i =- \varphi_j$ can be obtained. As the $\varphi_i =- \varphi_j$ is the unfeasible solution, the power angle can keep the same for all CVS in steady state. Thus the bus voltage amplitude equals the sum of voltage amplitude of all the CVS, i.e., the nominal voltage amplitude reference of the bus voltage.

$$V_p = \sum_{i=1}^{n} V_i = V^* \tag{10.12}$$

10.1.4 Design of Double Control Loop

To obtain a better tracking performance, the double control loop is applied in this chapter, which comprises voltage control and current control. As the voltage reference calculated by this method is an AC variable, the proportional-resonant (PR) is more applicable than PI control to track the AC variables [2]. The voltage and current control are respectively designed as follows:

$$G_v(s) = k_{pV} + \frac{2k_{rV} \cdot \omega_{cV} \cdot s}{s^2 + 2\omega_{cV} \cdot s + \omega_r^2} \tag{10.13}$$

$$G_i(s) = k_{pI} + \frac{2k_{rI} \cdot \omega_{cI} \cdot s}{s^2 + 2\omega_{cI} \cdot s + \omega_r^2} \tag{10.14}$$

where k_{pV} and k_{rV} are the proportional and resonant parameters of the voltage control, respectively. The k_{pI} and k_{rI} are the proportional and resonant parameters of the current control, respectively. In addition, the ω_r is the resonant frequency and the ω_{cV} (ω_{cI}) represents the cutoff frequency of the voltage control (current control).

10.2 Stability Analysis of the Decentralized SOC Balancing Control

10.2.1 Singular Perturbation Theory

As the studied system is a two-time scale system, it can be expressed into the following singular perturbation system [3, 4].

$$\begin{cases} \varepsilon \dfrac{dx}{d\tau} = f(x, y) \\ \dfrac{dy}{d\tau} = g(x, y) \end{cases} \tag{10.15}$$

where x, y, and ε are the fast variable, the slow variable, and the timescale parameter, respectively.

Rewrite the model in the stretched timescale $t = \tau/\varepsilon$, as follows

$$\begin{cases} \dfrac{dx}{dt} = f(x, y) \\ \dfrac{dy}{dt} = \varepsilon g(x, y) \end{cases} \tag{10.16}$$

where τ and t are referred to as the slow and the fast timescales, respectively. The models in (10.15) and (10.16) represent the corresponding slow and fast system respectively.

In the slow system, it is usually considered that the fast variable achieves quasi-steady state while the slow variable is still varying [4]. Setting $\varepsilon = 0$ in (10.15), the outer system is expressed as

$$\begin{cases} 0 = f\left(x_s, y\right) \\ \dfrac{dy}{d\tau} = g\left(x_s, y\right) \end{cases} \tag{10.17}$$

where x_s is the quasi-steady-state value of x.

While in the fast system, the slow variables act several ten times slower than the fast variables. The variable y can be considered as a constant parameter. Define $\tilde{x} = x - x_s$, then the boundary layer system is expressed as

$$\begin{cases} \dfrac{d\tilde{x}}{dt} = f\left(\tilde{x}, y\right) \\ \dfrac{dy}{dt} = 0 \end{cases} \tag{10.18}$$

10.2.2 System Model

From (10.7) and (10.8), the dynamic equations (10.19) of the i-th controlled voltage source can be written

$$\begin{cases} \dfrac{d\theta_i}{dt} = \omega_0 + \text{sgn}\left(Q_i\right)\left(m_i \cdot P_i - N \cdot SOC_i\right) \\ \dfrac{dSOC_i}{dt} = -\dfrac{m_i P_i}{K} \end{cases} \tag{10.19}$$

Define $\tilde{\theta}_i = \theta_i - \theta_s$, $\varepsilon = {}^1/_K$, then

$$\begin{cases} \dfrac{d\tilde{\theta}_i}{dt} = \omega_0 + \text{sgn}\left(Q_i\right)\left(m_i \cdot P_i - N \cdot SOC_i\right) - \omega_s \\ \dfrac{dSOC_i}{dt} = -\varepsilon m_i P_i \end{cases} \tag{10.20}$$

where $\theta_s = \omega_s t + \theta_0$. ω_s is a quasi-steady state value of ω_i, and θ_0 depends on the selection of reference phase angle and ε is very small.

Rewrite the model (10.20) in the timescale: $\tau = \varepsilon t$, then

$$\begin{cases} \varepsilon\dfrac{d\tilde{\theta}_i}{d\tau} = \omega_0 + \text{sgn}\left(Q_i\right)\left(m_i \cdot P_i - N \cdot SOC_i\right) - \omega_s \\ \dfrac{dSOC_i}{d\tau} = -m_i P_i \end{cases} \tag{10.21}$$

10.2.3 Analysis on the Outer System

Setting $\varepsilon = 0$ in (10.21), the outer system of the i-th controlled voltage source is expressed as

$$\begin{cases} \omega_s = \omega_0 + \text{sgn}\left(Q_{is}\right)\left(m_i \cdot P_{is} - N \cdot SOC_i\right) \\ \dfrac{dSOC_i}{dt} = -m_i P_{is} \end{cases} \tag{10.22}$$

where P_{is} and Q_{is} are the quasi-steady state value of P_i and Q_i, respectively.

In the quasi-steady state, by combining the outer systems of both i-th and k-th controlled voltage sources we have

$$\frac{dSOC_i}{dt} + N \cdot SOC_i = \frac{dSOC_k}{dt} + N \cdot SOC_k \tag{10.23}$$

Define $z = SOC_i - SOC_k$ and (10.23) can be rewritten as

$$\gamma \frac{dz}{dt} + z = 0 \tag{10.24}$$

where $\gamma = 1/N$

The solution of (10.24) is derived as

$$z\left(t\right) = z\left(0\right) e^{-\frac{t}{\gamma}} \tag{10.25}$$

Clearly, z converges to zero as time goes to infinity. The convergence rate depends on N. The larger the N, the faster the convergence rate is. Thus, in the steady state we have

$$SOC_i = SOC_k \tag{10.26}$$

From the analysis above, the SOC balancing of all controlled voltage sources can be achieved in steady state under this control.

10.2.4 Analysis on the Boundary Layer System

For simplicity, we assume that $\varphi\left(E_{\max\_i}\right) = \varphi\left(E_{\max\_j}\right) = \frac{1}{n}$, where n is the number of CVS in CESS. Setting $\varepsilon = 0$ in (10.20), we can regard the ω_s and SOC_i

as invariants. Substitute (10.3) into (10.20), then

$$\frac{d\tilde{\theta}_i}{d\tau} = \omega_i' + \mathrm{sgn}\left(Q_i\right) \frac{m_i V^{*2}}{n^2 \left|Z_{load}\right|} \left(\sum_{j=1}^{n} \cos\left(\tilde{\theta}_i - \tilde{\theta}_j + \theta_{load}\right) \right) \quad (10.27)$$

where $\omega_i' = \omega_0 - \omega_s - N \cdot SOC_i$

To prove the stability of the boundary layer system (10.27), the small-signal analysis is used [5].

It is reasonable to assume the state variable in equilibrium point is $\left[\tilde{\theta}_{1s}\ \tilde{\theta}_{2s} \cdots \tilde{\theta}_{ns}\right]$, where the $\tilde{\theta}_{is}$ is the quasi-steady state value of $\tilde{\theta}_i$. By linearizing (10.27), the small-signal equations are derived as:

$$\frac{d\Delta\tilde{\theta}_i}{d\tau} = -\,\mathrm{sgn}\left(Q_{is}\right) \frac{m_i V^{*2}}{n^2 \left|Z_{load}\right|} \sum_{j=1, j\neq i}^{n} \sin\left(\theta_{load} + \tilde{\theta}_{is} - \tilde{\theta}_{is}\right)\left(\Delta\tilde{\theta}_i - \Delta\tilde{\theta}_j\right) \quad (10.28)$$

where $Q_{is} = \frac{V^{*2}}{n^2|Z_{load}|} \sum\limits_{j=1}^{n} \sin\left(\tilde{\theta}_{is} - \tilde{\theta}_{js} + \theta_{load}\right)$

Rewrite (10.28) in the matrix form

$$\mathbf{\Delta}\dot{\tilde{\boldsymbol{\theta}}} = \mathbf{A} \cdot \mathbf{\Delta}\tilde{\boldsymbol{\theta}} \quad (10.29)$$

where **A** is the coefficient matrix.

$$\mathbf{A} = -\mathbf{B} \cdot \mathbf{L} \quad (10.30)$$

where $\mathbf{B} = diag\left[b_1\ b_2 \cdots b_n\right]$, $b_i = \mathrm{sgn}\left(Q_{is}\right) \frac{m_i V^{*2}}{n^2|Z_{load}|}$,

$$\mathbf{L} = \begin{bmatrix} \sum\limits_{j=2}^{n} l_{1j} & -l_{12} & \cdots & -l_{1n} \\ -l_{21} & \sum\limits_{j=1, j\neq 2}^{n} l_{2j} & \cdots & -l_{2n} \\ \vdots & \vdots & \ddots & \vdots \\ -l_{n1} & -l_{n2} & \cdots & \sum\limits_{j=1}^{n-1} l_{nj} \end{bmatrix}_{n\times n}, \text{ and } l_{ij} = \sin\left(\theta_{load} + \tilde{\theta}_{is} - \tilde{\theta}_{js}\right)$$

According to Gerschgorin circle theorem, the eigenvalues of the system coefficient matrix A fall within the Gerschgorin regions.

$$\lambda_i\left(\mathbf{A}\right) \in G = \bigcup_{j=1}^{n} G_j \quad (10.31)$$

$$G_i(\mathbf{A}) = \left\{ z \in \mathbf{C}^{n\times n} \left\| z + \sum_{j=1, j\neq i}^{n} b_i l_{ij} \right| \leqslant R_i \right\}, \qquad i = 1, 2, \cdots n \tag{10.32}$$

where R_i is the i-th Gerschgorin circle radius, $R_i = \sum\limits_{j=1, j\neq i}^{n} \left|b_i l_{ij}\right|$.

To ensure stability, the eigenvalues of the coefficient matrix $\mathbf{A}$ should be on the left half plane, i.e., all Gerschgorin circles should be on the left half plane. From (10.28)–(10.32), if all $b_i l_{ij} > 0$, the system will be stable.

As $m_i V^{*2} / |Z_{load}|$ is positive, the stability condition is simplified as

$$\text{sgn}\,(Q_{is}) \sin\left(\theta_{load} + \tilde{\theta}_{is} - \tilde{\theta}_{js}\right) > 0 \tag{10.33}$$

From (10.33), the system is stable, if the following condition holds.

$$\max_{i,j\in\{1,2\ldots n\}} \left|\tilde{\theta}_{is} - \tilde{\theta}_{js}\right| < |\theta_{load}| \tag{10.34}$$

From (10.3), the quasi-steady state active powers of i-th and k-th can be expressed as

$$P_{is} = \frac{V^{*2}}{n^2 |Z_{load}|} \sum_{j=1}^{n} \cos\left(\tilde{\theta}_{is} - \tilde{\theta}_{js} + \theta_{load}\right) \tag{10.35}$$

$$P_{ks} = \frac{V^{*2}}{n^2 |Z_{load}|} \sum_{j=1}^{n} \cos\left(\tilde{\theta}_{ks} - \tilde{\theta}_{js} + \theta_{load}\right) \tag{10.36}$$

Combining (10.35) and (10.36), we have

$$\sin\left(\frac{\tilde{\theta}_{iks}}{2}\right) = \frac{-(P_{is} - P_{ks})\, n^2 |Z_{load}|}{2V^{*2} \sum\limits_{j=1}^{n} sin\left(\frac{\tilde{\theta}_{ijs} + \tilde{\theta}_{kjs} + 2\theta_{load}}{2}\right)} \tag{10.37}$$

where $\tilde{\theta}_{iks} = \tilde{\theta}_{is} - \tilde{\theta}_{ks}$.

In the quasi-steady state, all controlled voltage sources share the same frequency. From (10.8), we can obtain

$$P_{is} - P_{ks} = N\,(SOC_i - SOC_k) / m_i \tag{10.38}$$

Substituting (10.38) in (10.37) results in

$$\sin\left(\frac{\tilde{\theta}_{iks}}{2}\right) = \frac{-N\left(SOC_i - SOC_k\right)n^2\left|Z_{load}\right|}{2m_i V^{*2}\sum_{j=1}^{n} sin\left(\frac{\tilde{\theta}_{ijs}+\tilde{\theta}_{kjs}+2\theta_{load}}{2}\right)} \tag{10.39}$$

From (10.39), if $SOC_i = SOC_k$, then $\tilde{\theta}_{iks} = 0$, i.e., all phase angles are the same. If $SOC_i \neq SOC_k$, it is suggested to choose appropriate mi and N to satisfy (10.34). So the feasible range of m and n should be discussed. Taking the stability condition and allowable frequency deviation into condition, the boundary condition can be derived as follows

$$\sin\left(\max_{i,j\in\{1,2\ldots n\}}\left|\tilde{\theta}_{is} - \tilde{\theta}_{js}\right|/2\right) < \sin\left|\theta_{load}/2\right| \tag{10.40}$$

$$\left|\Delta\omega_i\right| = \left|m_i P_i - N \cdot SOC_i\right| \leqslant \Delta\omega_{\max} \tag{10.41}$$

where the $\Delta\omega_{\max}$is the maximum allowable frequency deviation which is often set to be 1 in the microgrid system [6]. The power P_i has been limited in $[-P^*_{\max}, P^*_{\max}]$, and the $P^*_{\max}$ is the nominal rated power, which is set to be 1500 W. The safe range of SOC is set to be [0.1, 0.9].

To facilitate visualization of the feasible range of m and N, the results of Fig. 10.4 are based on small-signal model around one quasi-steady state point ($t = 10\,s$ and $Z_{load} = 4 + j8$) and the assumption that all the CVS have the same capacities.

From Fig. 10.3, the N and m are specified in the N-axis and m-axis and the boundaries corresponding to stability and frequency deviation. It is easy to see that

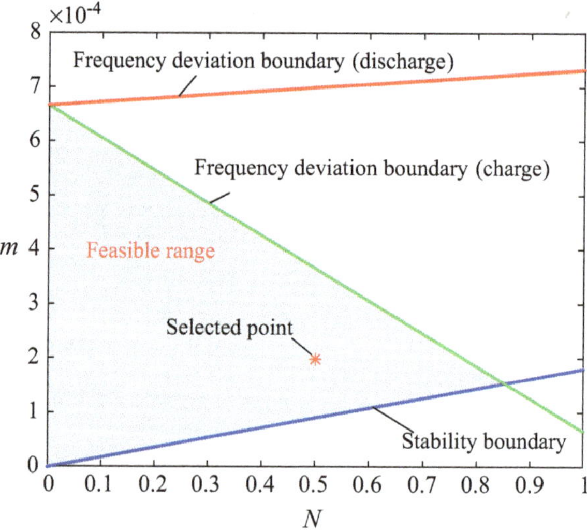

Fig. 10.3 Feasible range of m and N

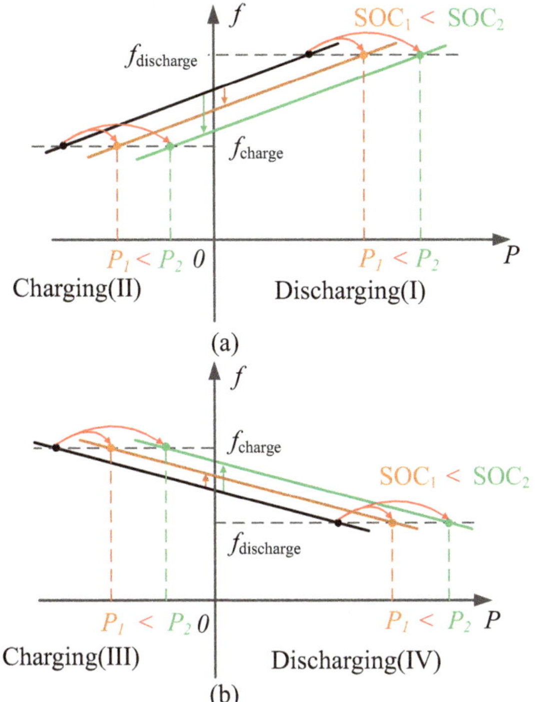

Fig. 10.4 Principles of the SOC balancing control method: (**a**) resistive-inductive load, (**b**) resistive-capacitive load

a large m and small N can extend the stable range of the system. But it is worth mentioning that a very large m may lead to unallowable frequency deviation and a very small N may decrease the SOC convergence rate. So selecting a point further away from the boundary line and a lager N for faster SOC convergence rate can make the system achieve better performance. Therefore, the selection of m_i and N is a trade-off among the stability, frequency deviation, and the SOC convergence rate.

As in whole, according to the control function (10.8), the intercept of $P - f$ curve is modified by the local SOC information, which is to adjust the output power of ESUs by the different output voltage angles. The basic principle is described in Fig. 10.4. In the discharging mode, higher SOC energy storage units result in greater power and lower SOC ones result in smaller power. Likewise, in the charging mode, units with lower SOC are injected with more power, and ones with larger SOC are injected with less power. Based on this principle, the SOC difference will be decreased and the SOC balancing will be achieved ultimately.

Remark The distinctions of this method in this chapter from the existing decentralized methods [7–10] are demonstrated in the following: (1) Previous works focus on the parallel-type ESS, while this chapter researched on the series-type ESS. (2) The control algorithms in [7–10] are based on droop control. In contrast, our algorithm is derived from inverse droop control. (3) The SOC balancing is proved based on multi-time scale analysis. Theoretical proof about SOC balancing has not been reported in the literatures.

10.3 Simulation Results

To validate the SOC balancing control method, the simulation model based on three controlled voltage sources is developed in the MATLAB/Simulink environment. Fig. 10.5 shows the detailed control diagram of the method, which contains power calculation loop, the control block, and the control block of voltage and current control loops. The control is to provide a voltage reference based on the local measured active and reactive power information. Then the voltage and current controllers use PR control to track the voltage reference. The system parameters are listed in Table 10.1 and the parameters of double control loop are listed in Table 10.1.

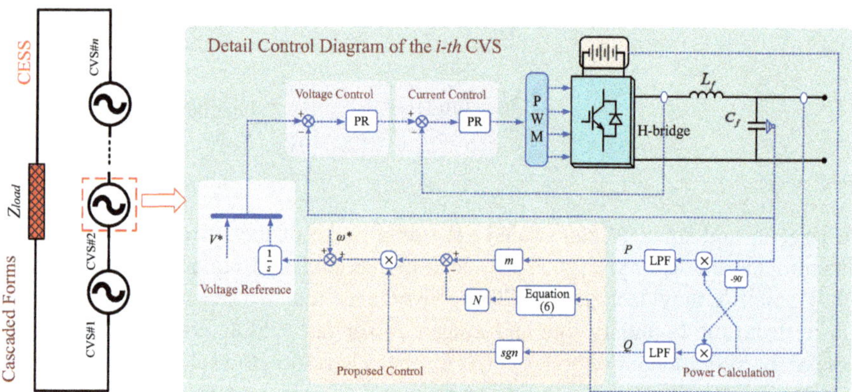

Fig. 10.5 Detailed control diagram of the method

Table 10.1 Parameters for experiments

Type	Symbol	Value
Voltage control	k_{pV}	50
	k_{rV}	03
	ω_{cV}	100
	ω_0	0.6e−3
Current control	k_{pI}	0.5
	k_{rI}	20
	ω_{cI}	100
	ω_0	0.6e−3

10.3.1 *Case 1: SOC Balancing in Four Quadrant Operations*

In this case, the control gains $m_1 = m_2 = m_3 = 0.0002$, and $N = 0.5$ are selected for the three energy storage units. Different initial SOC values of each energy storage units are selected as follows. SOC_{01}, SOC_{02}, and SOC_{03} are set to be 90, 80, 70% for the discharging mode and 10, 20, 30% for the charging mode, respectively. The resistive-inductive load $Z_l = 4 + j8\,\Omega$, and resistive-capacitive load $Z_l = 4 - j8\,\Omega$ are used in this case.

Figure 10.6a–d show the SOC and the output power of the energy storage units operating in the first, second, third, and fourth quadrant, respectively. As shown in Fig. 10.6, the deviation ΔSOC, ΔP, and ΔQ gradually decrease to zero in the steady state ($\Delta SOC = SOC_{\max} - SOC_{\min}$,and $\Delta P = P_{\max} - P_{min}$). From the results in Fig. 10.6, the SOC balancing controller performs well and the SOC balancing as well as power sharing are achieved eventually.

10.3.2 *Case 2: Mode Switching Between Discharging and Charging*

The mode switching between the charging and discharging processes is tested in this section. The parameters of the simulated system are the same with those in Case 1. The mode switching is enabled at 1000s.

Figure 10.7 presents the waveforms of energy storage units operating from the discharging mode to the charging mode under the resistive-inductive load. In the discharging mode, the SOC values of the three units are convergent, and the deviation ΔSOC is reduced from 20 to 1.9%. After switching to the charging mode, the convergence of SOC is still guaranteed, and the deviation ΔSOC gradually reduces to zero in the steady state. Moreover, the deviation ΔP has little fluctuations during the mode switching process.

Figure 10.8 shows the test results from charging mode to discharging mode. As shown in Fig. 10.8, the variation process of SOC and power are similar to the mode switching from the discharging to the charging. The SOC balancing and the power sharing are gradually achieved throughout the whole process and the fluctuations of mode switching are acceptable. From the simulation results above, the SOC balancing control has good performance during mode switching.

10.3.3 *Case 3: Simulation Tests Under Discharging Mode with Load Characteristics Changing*

The dynamic response during the load changing from inductive-resistive to capacitive-resistive one is shown in Fig. 10.9. In this case, all of the three energy

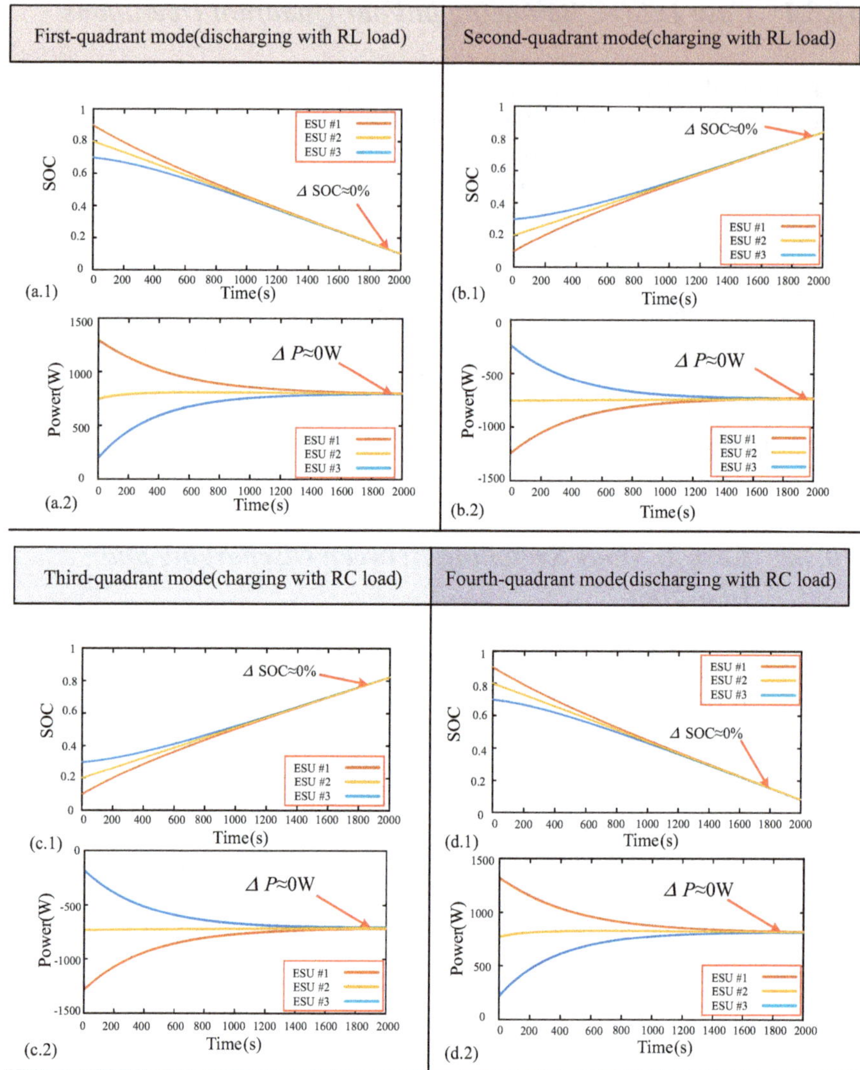

Fig. 10.6 Simulation results in case 1: (**a**) performance in first-quadrant mode; (**b**) performance in second-quadrant mode; (**c**) performance in third-quadrant mode; (**d**) performance in fourth-quadrant model

storage units operate in the discharging mode. At $t = 1000$ s, the load is changed from the inductive-resistive to the capacitive-resistive one. It shows that the convergences of SOC and power sharing are not affected by the changing of the load characteristics.

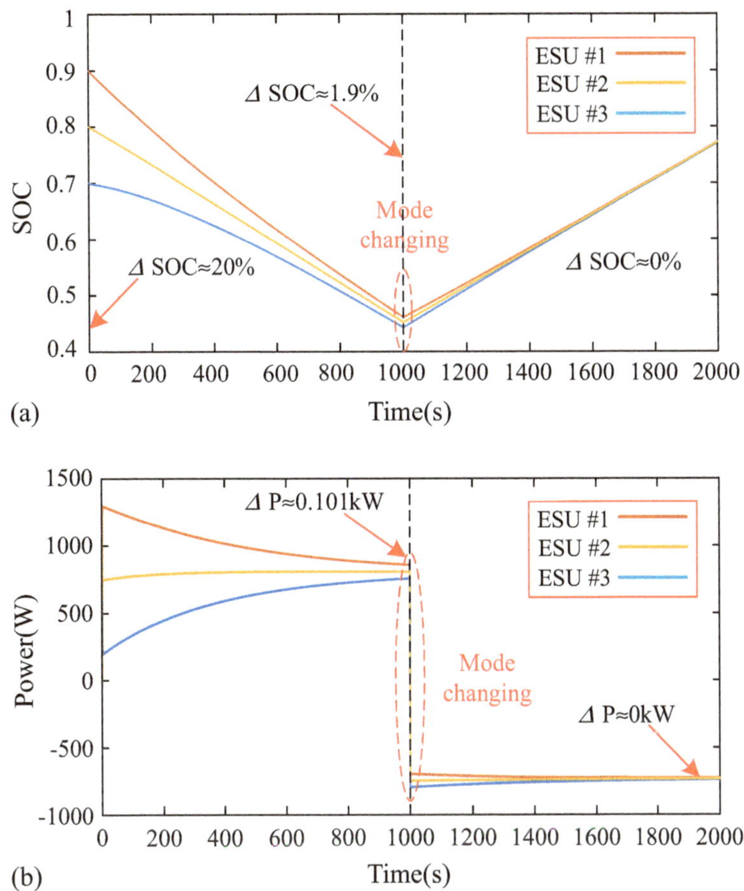

Fig. 10.7 Performance of transferring from discharging to charging process with the method: (**a**) SOC; (**b**) Active power

10.3.4 Case 4: Different Capacities of ESU

In order to show the effectiveness of the method with different capacities of ESU, the capacities of ESU#1, ESU#2, and ESU#3 are set to 6, 4, and $2A \cdot h$, respectively. Meanwhile, the droop coefficients are changed to be $4/3 \times 10^{-4}$, 2×10^{-4}, and 4×10^{-4}, respectively. Figure 10.10 shows the test results of SOC and output active power when the capacities of ESUs in CESS are different. In steady state, the SOC balancing has been already achieved and the output power values of ESUs are measured as 1.11, 0.738, and 0.364 kW, respectively. It is clearly shown that the power allocation can be described as an equation: $P_1 : P_2 : P_3 \approx 3 : 2 : 1$. As that the DC voltage are set to the same value, the output power is in proportion to its capacities, which can be derived as follows: $P_1 : P_2 : P_3 \approx C_1 : C_2 : C_3$. From the

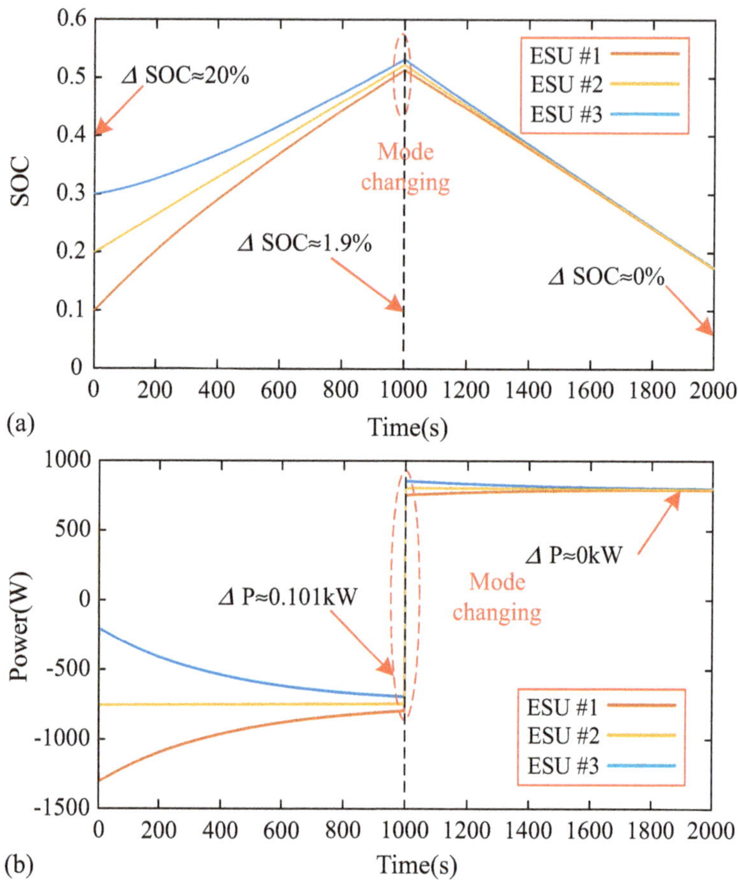

Fig. 10.8 Performance of transferring from charging to discharging mode with the method: (**a**) SOC; (**b**) Active power

simulation result, the SOC balancing and proportional power sharing are proved to be achieved with different capacities of ESU.

10.3.5 Case 5: Comparison of ESS with and Without SOC Balancing Control

The section provides the simulation results of comparison of ESS with and without SOC balancing control. The initial SOC values SOC_{01}, SOC_{02}, and SOC_{03} are set to be 50, 45, and 40%. Generally, the safe SOC range of Li-FePO4 battery is between 10 and 90%. From Fig. 10.11, the ESS without SOC balancing control works in unsafe range at 868 s, while the one with SOC balancing control still works

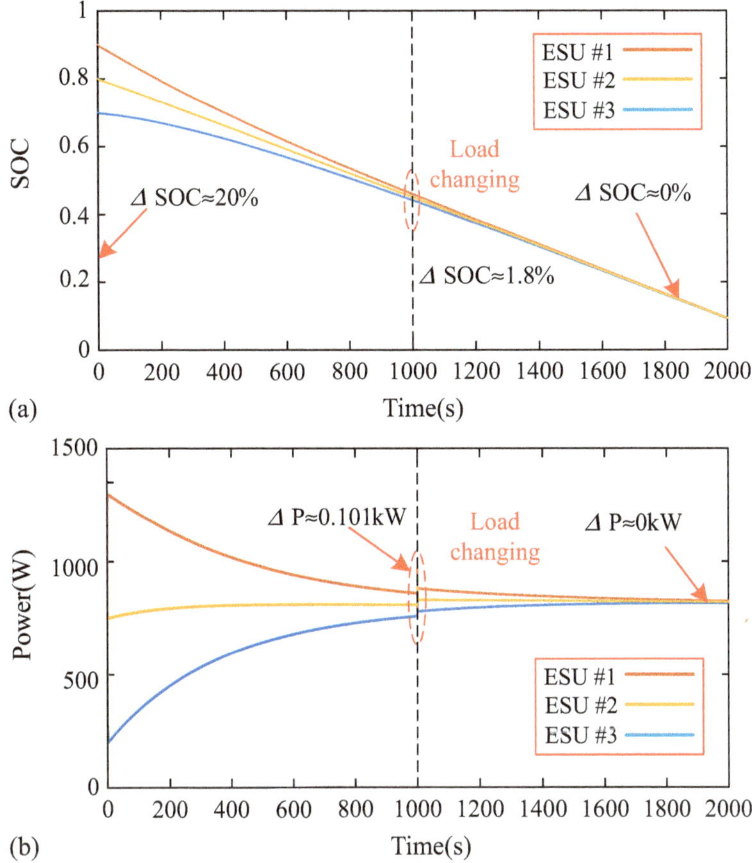

Fig. 10.9 Performance of load characteristic changing from inductive-resistive to capacitive-resistive in discharging mode: (**a**) SOC; (**b**) Active power

in safe range. To some extent, the depth of charge-discharge can be decreased with the control, i.e., the life of ESS can be increased [11].

10.4 Experimental Results

In this section, a series-type islanded system with two storage units is employed to validate the control method, and the main circuits of the experiment system are shown in Fig. 10.12. Each storage unit has 10 battery cells connected in series. The nominal voltage, capacity, impedance, max charge current, and max discharge current of battery cell are 3.2 V, 2 Ah, $\leq 1.3\,\mathrm{m}\Omega$, 10 C, and 30 C, respectively. The inverters are controlled by DSP program, to implement the SOC balancing control

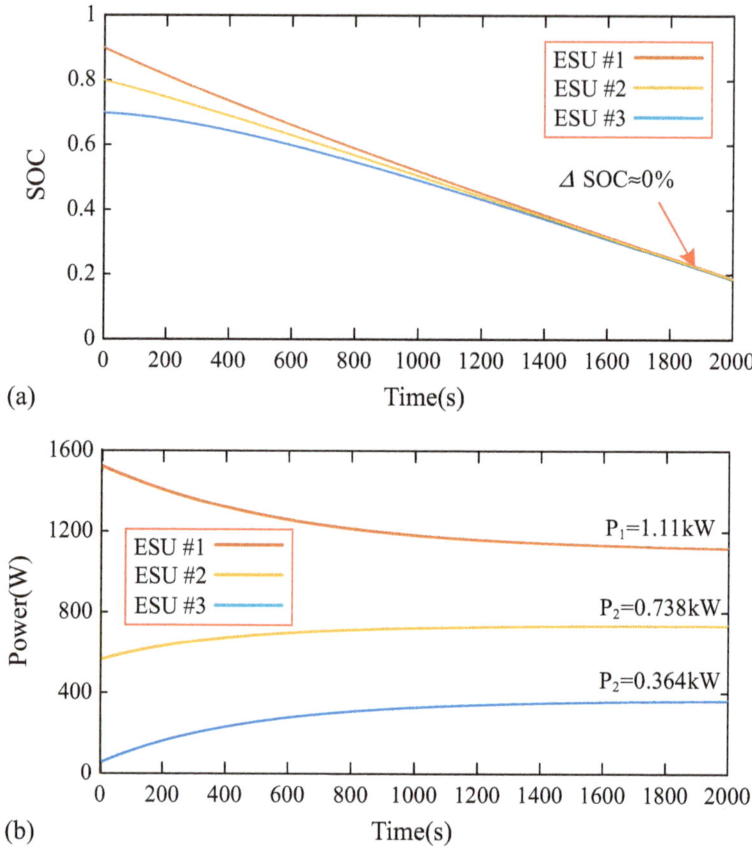

Fig. 10.10 Performance of load characteristic changing from inductive-resistive to capacitive-resistive in discharging mode: (**a**) SOC; (**b**) Active power

algorithm. The variables of active power and SOC are noted with 10s sampling period. The parameters of experiments are listed in Table 10.2. It is noted that the different SOC initial values under two different type loads are set to verify the validation of the SOC balancing control under various cases [12].

Figures 10.13 and 10.14 show the experimental results of the SOC balancing control under RL loads in discharging mode. In Fig. 10.13, the four measured waveforms represent the load voltage (V_{load}), the output voltage (V_1) of inverter 1, the output voltage (V_2) of inverter 2, and the output current of CESS (I), respectively. Figure 10.15 shows the measured SOC and the output power curves. In this case, the controller is enabled at $t = 7$ s and the initial SOC values of the two storage units are 0.82 and 0.78. In the beginning (i.e., $t < 7$ s), the corresponding real power outputs of the ESU #1 and ESU #2 are remaining the same. With the control strategies, the output power of ESUs has been changed. The ESU with larger SOC has greater power output and the one with smaller SOC has lower power

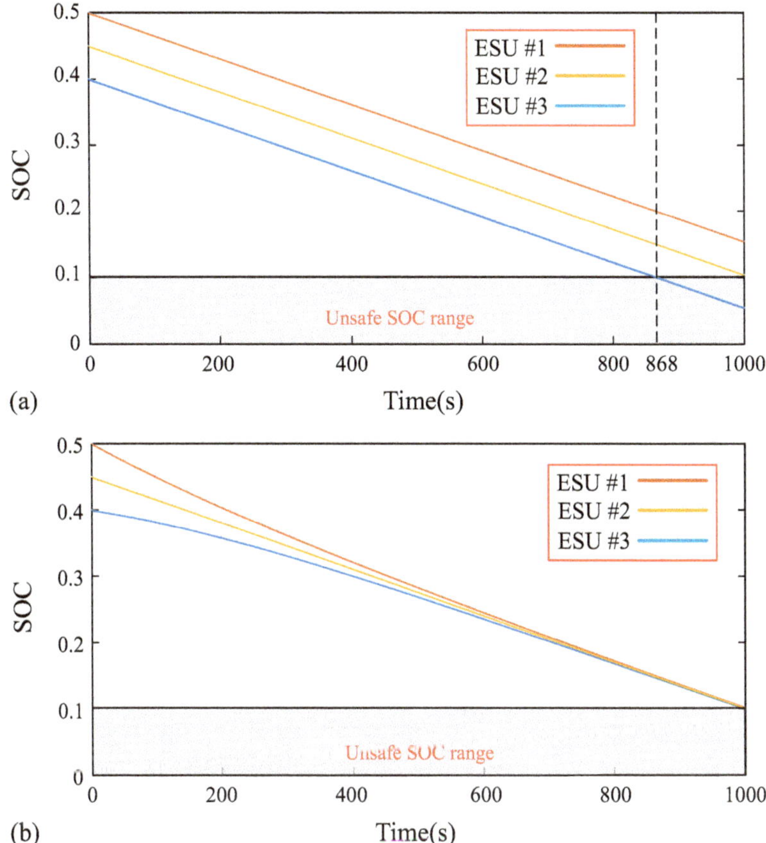

Fig. 10.11 Performance of the method with different capacities of ESU: (**a**) SOC; (**b**) Active power

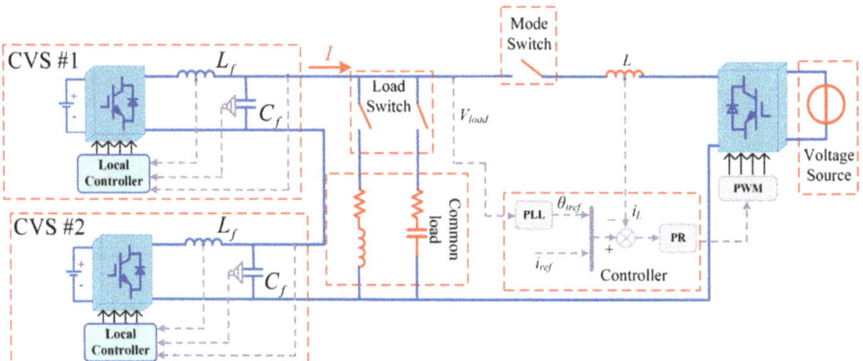

Fig. 10.12 Main circuits of the experiment system

Table 10.2 Experiment parameters

Symbol	Item	Value
V^*/f_{ref}	Voltage reference	25 V/50 Hz
$C_{e1} = C_{e2}$	Unit capacities	$2A \cdot h$
v^{in}	Output voltages of unit	32V
$m_1 = m_2$	Control coefficients	3.2e−4
N	SOC weighted coefficient	0.15
Z_{load}	Inductive-resistive (RL) load Capacitive-resistive (RC) load	$9.39 + j4.05\ \Omega$ $12.53 - j5.99\ \Omega$

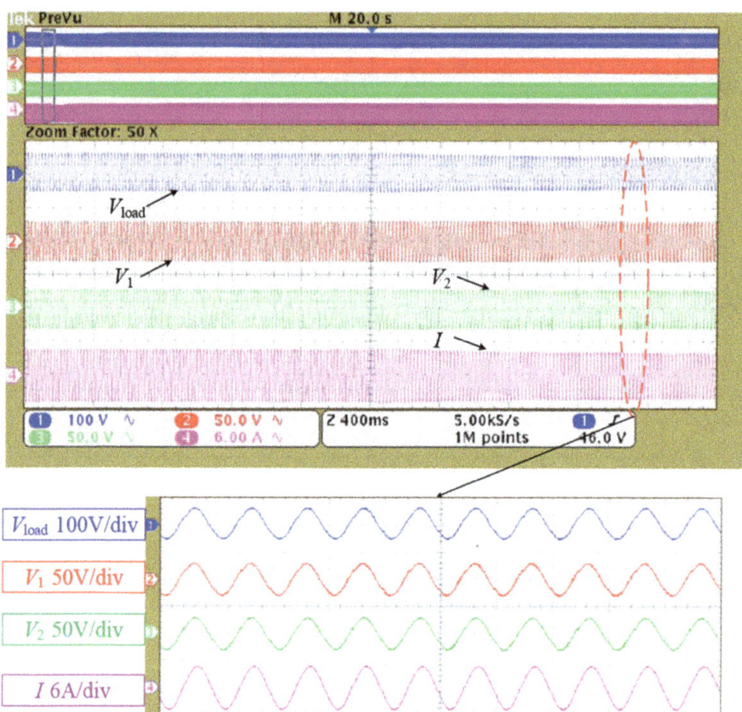

Fig. 10.13 Experimental waveforms of the control starting with RL loads in discharging mode

output, which means that the SOC_1 decreases faster than SOC_2. So the deviation of SOC and power between ESU #1 and ESU #2 decrease. Until $t = 1600s$, both the SOC balancing and the power sharing are realized.

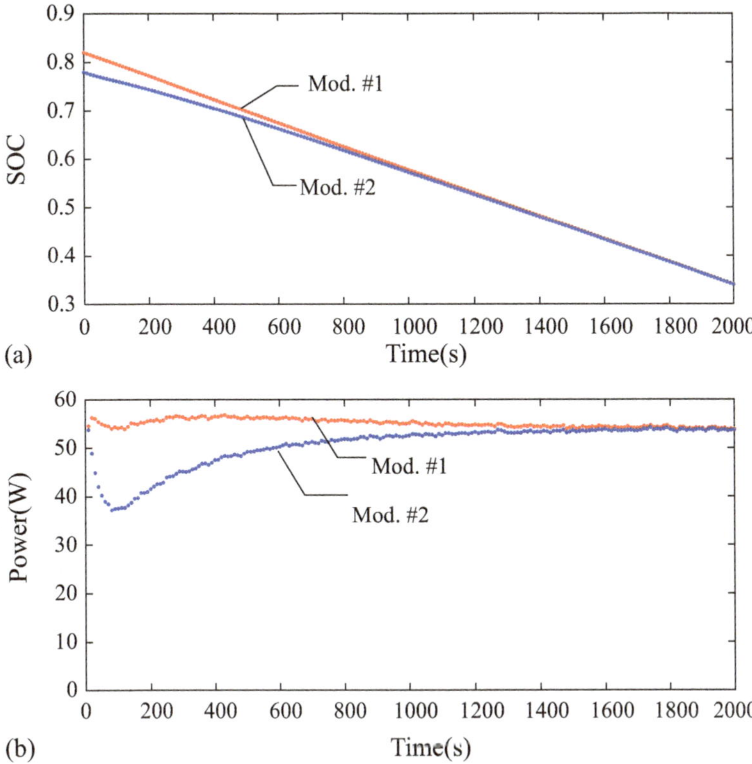

Fig. 10.14 Experimental results of the control method with RL loads in discharging mode: (**a**) SOC; (**b**) Power

Figures 10.15 and 10.16 show the measured waveforms under RC loads in discharging mode. The SOC balancing control is enabled at $t = 9\,\text{s}$, and the initial SOC values are: $SOC_{01} = 0.77$ and $SOC_{02} = 0.74$. As shown from Fig. 10.16, under the control, the output power of ESUs has been adjusted by following the SOC balancing principle. In the steady state, the purposes of SOC balancing and the power sharing are achieved. It is noted that the different SOC initial values under two different type loads are set to verify the validation of the SOC balancing control under various cases, which has no effect on whether we can obtain our conclusion or not.

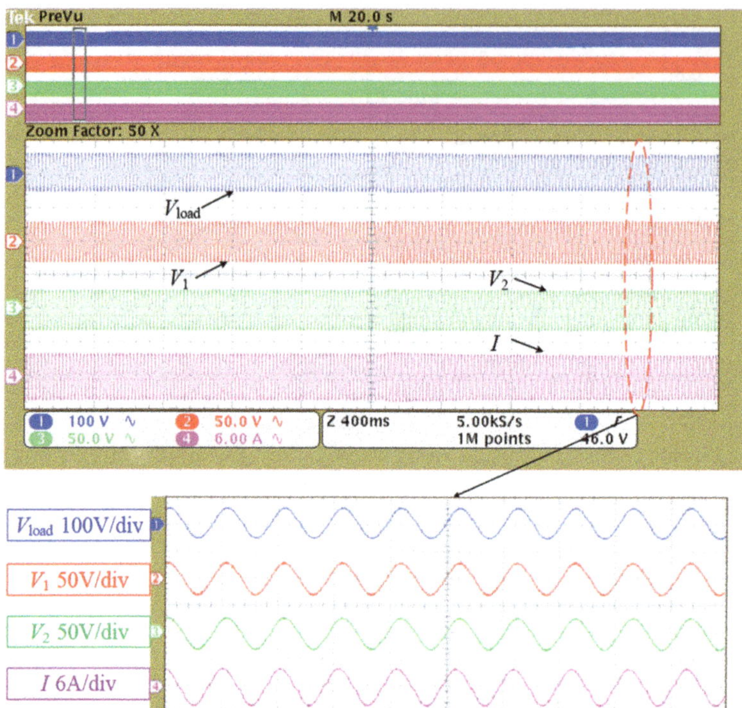

Fig. 10.15 Experimental waveforms of the control starting with RC loads in discharging mode

In charging mode, the DC voltage source is under the constant current control, which charges ESS and energizes loads. Figures 10.17 and 10.18 show the measured waveforms under RL loads in charging mode. The SOC balancing control is enabled at $t = 11$ s, and the initial SOC values are: $SOC_{01} = 0.23$ and $SOC_{02} = 0.27$. As shown from Fig. 10.18, under the control, the injected power and the SOC are balancing in the steady state, which validate the feasibility of the control in charging mode.

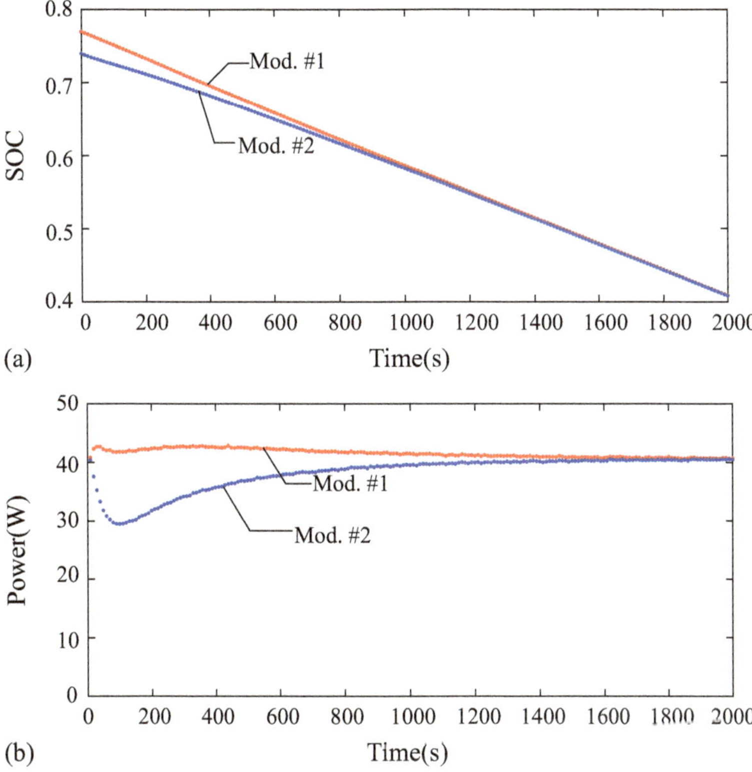

Fig. 10.16 Experimental results of the control method with RC loads in discharging mode: (**a**) SOC; (**b**) Power

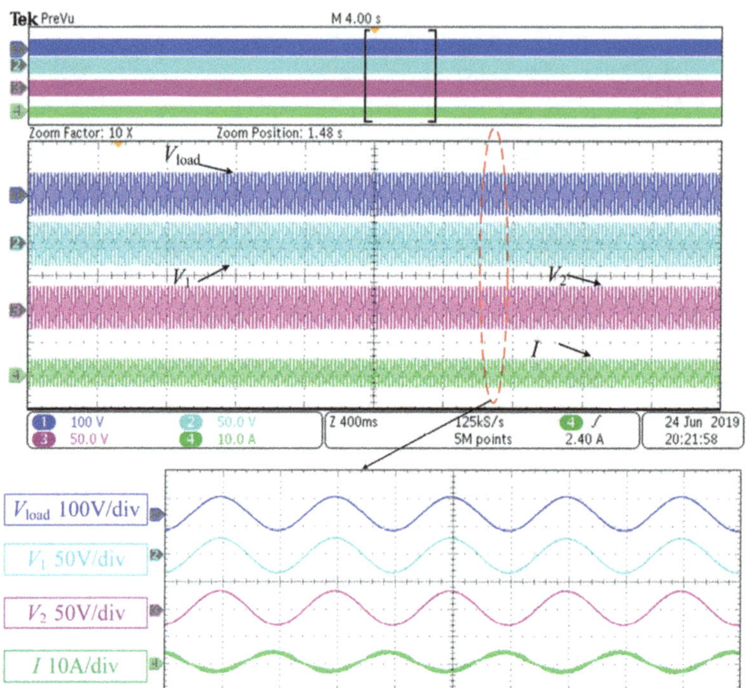

Fig. 10.17 Experimental waveforms of the control starting with RL loads in charging mode

10.5 Conclusion

In this chapter, we introduce a decentralized SOC balancing scheme for series-type energy storage system. This scheme achieves SOC balancing of energy storage units through adjusting $P - f$ curves. Without communication dependence, the decentralized control obtains higher reliability and lower construction cost. Meanwhile, the control has good extensibility, and its application is not limited by the number of ESUs. Moreover, based on the singular perturbation theory, the SOC steady-state values are proved to be equal. The stability of the system is proved and a sufficient condition for the stable operation is provided. Both the simulation and experimental results have validated the method.

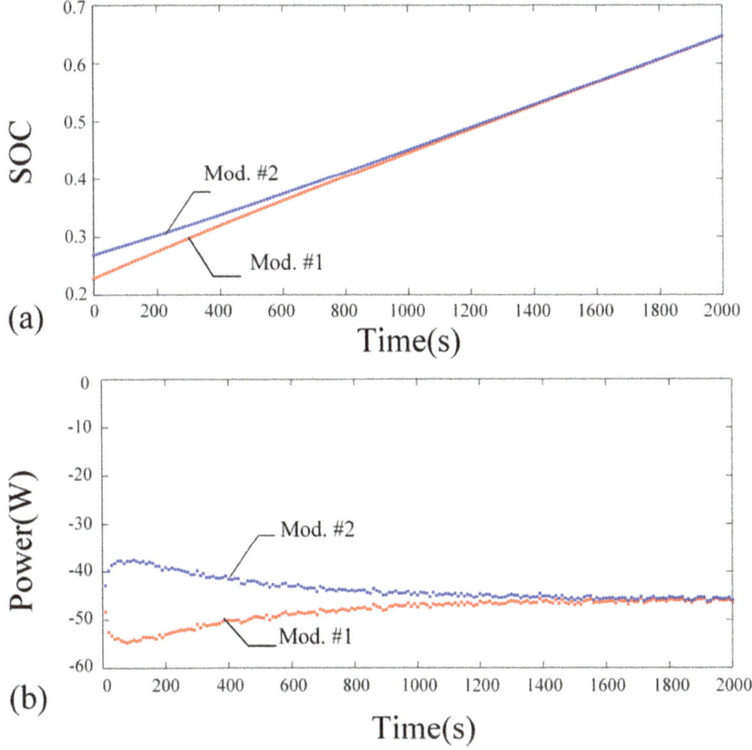

Fig. 10.18 Experimental results of the control method with RL loads in charging mode: (**a**) SOC; (**b**) Power

References

1. Y. Sun, G. Shi, X. Li, W. Yuan, M. Su, H. Han, X. Hou, An f-P/Q droop control in cascaded-type microgrid. IEEE Trans. Power Syst. **33**(1), 1136–1138 (2018)
2. M. Savaghebi, A. Jalilian, J.C. Vasquez, J.M. Guerrero, Secondary control scheme for voltage unbalance compensation in an islanded droop-controlled microgrid. IEEE Trans. Smart Grid **3**(2), 797–807 (2012)
3. R.E. O'Malley, *Introduction to Singular Perturbations* (Academic Press, New York, 1974)
4. A. Narang-Siddarth, J. Valasek, *Nonlinear Time Scale System in Standard and Non-standard Forms: Analysis and Control* (Society for Industrial and Applied Mathematics, Philadelphia, 2014)
5. E.A.A. Coelho, P.C. Cortizo, P.F.D. Garcia, Small-signal stability for parallel-connected inverters in stand-alone AC supply systems. IEEE Trans. Ind. Appl. **38**(2), 533–542 (2002)
6. I.U. Nutkani, P.C. Loh, P. Wang, F. Blaabjerg, Decentralized economic dispatch scheme with online power reserve for microgrids. IEEE Trans. Smart Grid **8**(1), 139–148 (2017)
7. M.S. Whittingham, History, evolution, and future status of energy storage. Proc. IEEE **100**, 1518–1534 (2012)
8. J.J. Justo, F. Mwasilu, J. Lee, J.-W. Jung, AC-microgrids versus DC-microgrids with distributed energy resources: a review. Renew. Sustain. Energy Re-views **24**, 387–405 (2013)

9. A. Mohd, E. Ortjohann, A. Schmelter, N. Hamsic, D. Morton, Challenges in integrating distributed energy storage systems into future smart grid, in *2008 IEEE International Symposium on Industrial Electronics* (2008), 1627–1632
10. C. Restrepo, A. Salazar, H. Schweizer, A. Ginart, Residential battery storage: is the timing right? IEEE Electr. Mag. **3**(3), 14–21 (2015)
11. T. Guena, P. Leblanc, How depth of discharge affects the cycle life of lithium-metal-polymer batteries, in *INTELEC06 - Twenty-Eighth International Telecommunications Energy Conference, Providence* (2006), pp. 1–8
12. G. Shi, H. Han, Y. Sun, Z. Liu, M. Zheng, X. Hou, A decentralized SOC balancing method for cascaded-type energy storage system. IEEE Trans. Ind. Electron. **68**(3), 2321–2333 (2021)

Chapter 11
Decentralized Control Strategies in Grid-Connected Mode

11.1 Decentralized Control for Grid-Connected Series-Connected Inverters

11.1.1 Equivalent Models of Grid-Connected Series-Connected Inverters

Figure 11.1 illustrates the schematic diagram of grid-connected series-type inverters. This configuration is beneficial to integrate low-voltage DC distributed generations (DGs) into medium voltage system. It is very common in PV grid-connected applications [1–4].

From Fig. 11.1, the output real power P_i and reactive power Q_i of i-th module are derived as follows

$$P_i + jQ_i = V_i e^{j\delta_i} \cdot \left((V_p e^{j\delta_p} - V_g e^{j\delta_g})/(|Z_{line}|\, e^{j\theta_{line}})\right)^* \tag{11.1}$$

where V_i and δ_i represent the output voltage amplitude and phase angle of i-th module. V_g and δ_g are the voltage amplitude and phase angle of utility grid. $|Z_{line}|$ and θ_{line} are the grid impedance amplitude and angle. Usually, the grid impedance is mainly inductive ($\theta_{line} \approx \pi/2$). The voltage $V_p e^{j\delta_p}$ at point of common coupling (PCC) is the sum of each module voltage.

$$V_p e^{j\delta_p} = \sum_{j=1}^{N} V_j e^{j\delta_j} \tag{11.2}$$

where N represents the total number of series modules.

From (11.1) and (11.2), the power transmission characteristic is given by

Y. Sun et al., *Series-Parallel Converter-Based Microgrids*, Power Systems,
https://doi.org/10.1007/978-3-030-91511-7_11

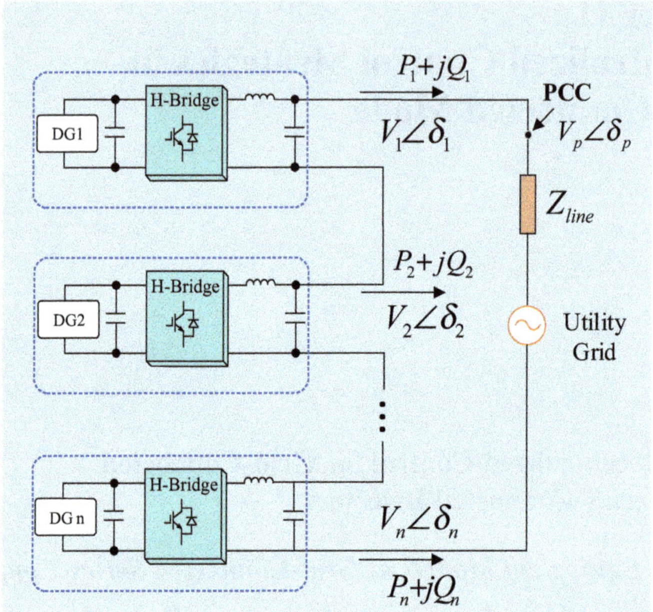

Fig. 11.1 Schematic diagram of grid-connected series-type inverters

$$P_i = \frac{V_i}{|Z_{line}|}\left(V_g \sin\left(\delta_i - \delta_g\right) - \sum_{j=1}^{N} V_j \sin\left(\delta_i - \delta_j\right)\right) \tag{11.3}$$

$$Q_i = \frac{V_i}{|Z_{line}|}\left(\sum_{j=1}^{N} V_j \cos\left(\delta_i - \delta_j\right) - V_g \cos\left(\delta_i - \delta_g\right)\right) \tag{11.4}$$

11.1.2 Decentralized $P - \omega$ Droop Control

To synchronize each module with the grid and realize power balance without communication, a decentralized control scheme is designed as

$$\omega_i = \omega^* - k \cdot (P_i - P^*) \tag{11.5}$$

$$V_i = V^* = V_g / M \tag{11.6}$$

where ω_i and V_i are the angular frequency and voltage amplitude references of i-th module, respectively. ω^* represents the nominal value of the grid angular frequency.

P^* represents the nominal rated power of each module. k is a positive coefficient of $P - \omega$ droop control. M is a critical parameter related with stability, which is designed later.

11.1.3 Steady State and Stability Analysis

In steady state, because the voltage amplitude reference V^* is same for all modules and each module shares the same grid current, the apparent power of each module is equal. Moreover, (11.7) is obtained due to the identical grid frequency from (11.5)

$$P_1 = P_2 = \cdots = P_N = P^* \tag{11.7}$$

Namely, the phase angles of all modules are equal in steady state ($\delta_i = \delta_j; i, j \in \{1, 2, \cdots, N\}$). Thus, the voltage amplitude and phase angle of PCC is derived from (11.2)

$$V_p = NV^* = (N/M) \cdot V_g;\ \delta_p = \delta_1 = \delta_2 = \cdots = \delta_N \tag{11.8}$$

Then, the steady-state power of each module is obtained from (11.3)–(11.6)

$$P^* = S_C \sin\bar{\delta};\ \bar{Q}_i = S_C\left(N/M - \cos\bar{\delta}\right) \tag{11.9}$$

where $S_C = V_g^2/(M \cdot |Z_{line}|)$ represents the power transfer capacity of a single module. $\bar{\delta} = \delta_p - \delta_g$ is referred to as the steady power angle. From (12.9), there are two steady points and they are $\bar{\delta} = \arcsin(\frac{P^*}{S_C})$ or $\bar{\delta} = \pi - \arcsin(\frac{P^*}{S_C})$.

Then, the small-signal analysis is carried out to test the system stability. Since $\dot{\delta}_i = \omega_i$, combining (11.3)–(11.6) and linearizing them around the steady-state points yield

$$\dot{\tilde{\delta}}_i = -k'\left(M(\cos\bar{\delta})\left(\tilde{\delta}_i - \tilde{\delta}_g\right) - \sum_{j=1, j\neq i}^{N}\left(\tilde{\delta}_i - \tilde{\delta}_j\right)\right) \tag{11.10}$$

where $k' = kV^{*2}/|Z_{line}|$, and $\tilde{\delta}_i$, $\tilde{\delta}_g$, $\tilde{\delta}_j$ denote small perturbations around the equilibrium point.

Rewrite (11.10) in matrix form as

$$\dot{\tilde{\delta}} = -k' \cdot L \cdot \tilde{\delta} \tag{11.11}$$

where $\tilde{\delta} = \left[\tilde{\delta}_1 \; \tilde{\delta}_2 \cdots \tilde{\delta}_N\right]^T$;

$$L = \begin{bmatrix} M\cos\bar{\delta} - N + 1 & 1 & \cdots & 1 \\ 1 & M\cos\bar{\delta} - N + 1 & \cdots & 1 \\ \vdots & \vdots & \ddots & \vdots \\ 1 & 1 & \cdots & M\cos\bar{\delta} - N + 1 \end{bmatrix}$$

$= (M\cos\bar{\delta} - N)I_{N\times N} + 1_N 1_N^T$

The eigenvalues of the system matrix $A = -k^{'}L$ are given by

$$\lambda_1(A) = -k^{'}M\cos\bar{\delta};\; \lambda_2(A) = \cdots = \lambda_N(A) = -k^{'}(M\cos\bar{\delta} - N) \quad (11.12)$$

Thus, the necessary and sufficient condition of system stability is obtained as follows

$$\Delta = M\cos\bar{\delta} - N > 0 \quad (11.13)$$

From (11.13), since both M and N are greater than zero, the power angle $\bar{\delta}$ should lie in $(-\pi/2, \pi/2)$ under the stability constraint. According to the above steady-state analysis, only $\bar{\delta} = \arcsin(P^*/S_C)$ is a stable equilibrium point in (11.13). From (11.9), to obtain a high power factor, Δ should be as small as possible. However, a too small is detrimental to stability from (11.12). Thus, a proper M should be designed by making a tradeoff between reactive power requirement and stability margin.

11.1.4 Simulation Results

The simulation tests are carried out to validate the ideas. The control block diagram of i-th module is presented in Fig. 11.1. The control algorithm in (11.5) and (11.6) is used to generate the voltage reference for the inner voltage current closed-loops. The simulation parameters of the system are listed in Table 11.1. Based on the steady-state analysis (11.9) and stability condition (11.13), three cases of different M are discussed (Fig. 11.2).

Table 11.1 Simulation parameters

Symbol	Value	Symbol	Value
V_g	311V	M	5.8(case1)
ω^*	$2\pi * 50$ rad/s	M	6.2(case2)
P^*	4 kW	M	7.0(case3)
k	1.2e−3	V^*	53.6 V(case1)
Z_{line}	$0.1 + j0.5\,\Omega$	V^*	50.1 V(case2)
N	6	V^*	44.4 V(case3)

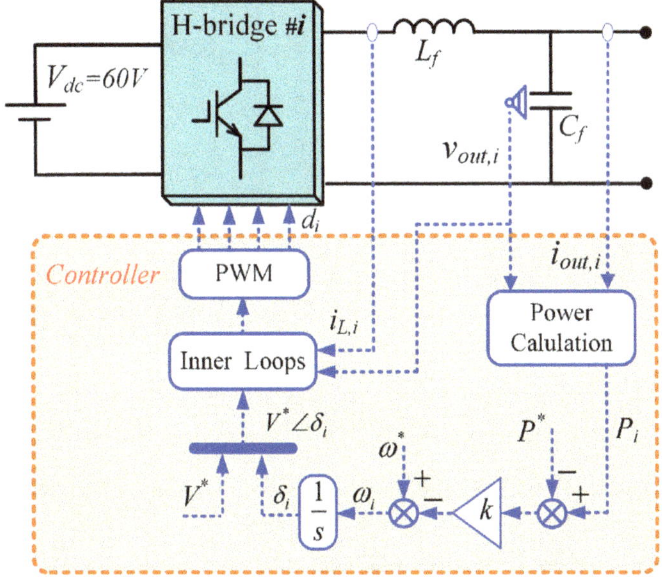

Fig. 11.2 Control diagram of i-th H-bridge inverter module

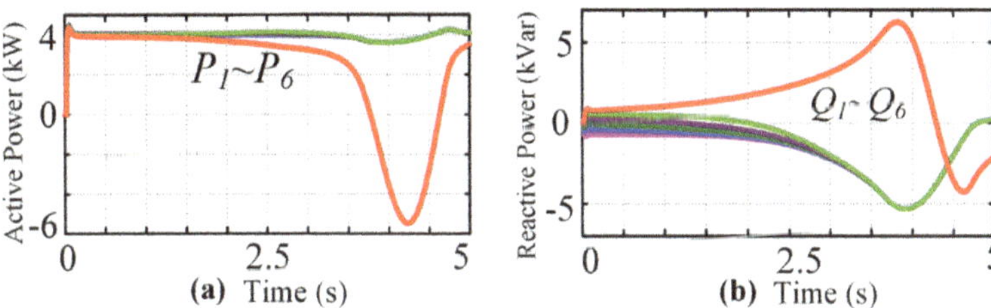

Fig. 11.3 Simulation results of case1. (**a**) Active power, (**b**) reactive power

Figure 11.3 shows the simulation results in case1. Since $M < N$, the stability condition of (11.13) is not met, thus the system is unstable and the output real power/reactive power of modules are unbalanced.

In case2, as the selected M satisfies the stability condition (11.13), the system works normally. The simulation results are illustrated in Fig. 11.4. As seen from Fig. 11.4a, b, real-power/reactive-power balance is achieved. Figure 11.4c indicates that the frequency synchronization is also obtained. Figure 11.4d shows the voltage/current waveforms at PCC.

Figures 11.5a–d show simulation results of case3. Compared with case2, the settling time becomes shorter, but the power factor becomes lower from 0.983 to 0.891 in Fig.11.6a. Therefore, in practice, M is designed by making a compromise between the power factor and dynamic response. In addition, Fig. 11.6b shows that

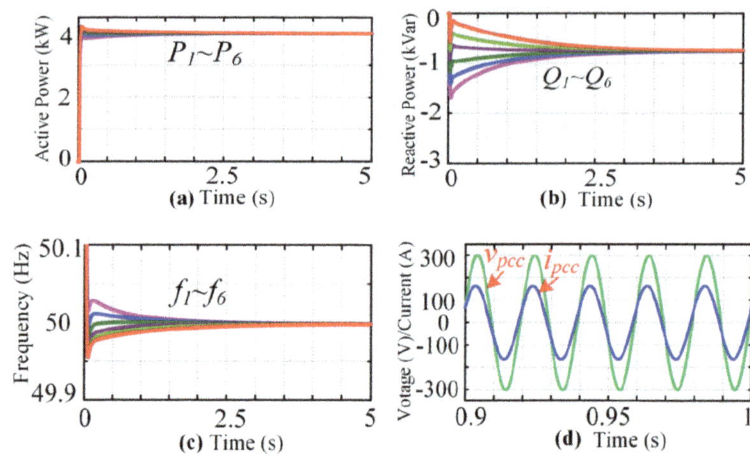

Fig. 11.4 Simulation results of case2. (**a**) Active power, (**b**) reactive power, (**c**) frequency, (**d**) voltage/current at PCC

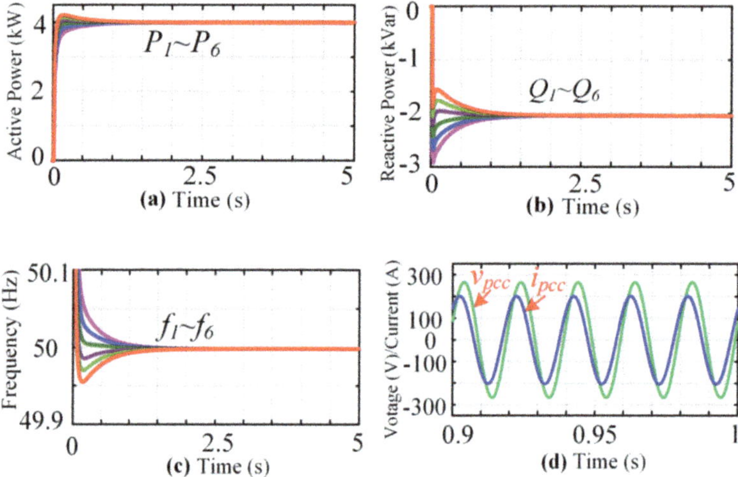

Fig. 11.5 Simulation results of case3. (**a**) Active power, (**b**) reactive power, (**c**) frequency, (**d**) voltage/current at PCC

the steady-state power angles of case2 and case3 lie in $(-\pi/2, \pi/2)$, which is in accordance with the former stability analysis.

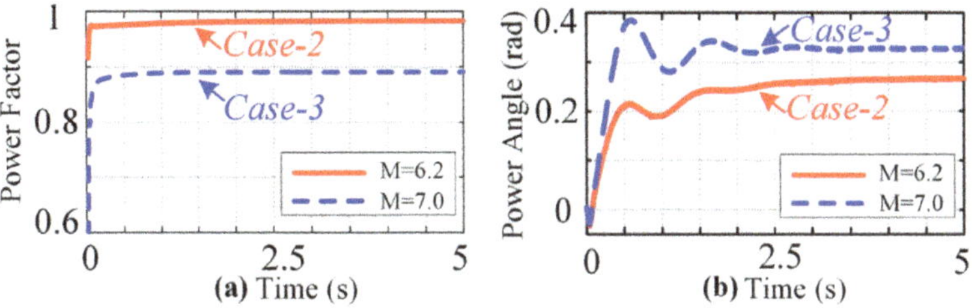

Fig. 11.6 Comparison of case2 and case3. (**a**) Power factor, (**b**) power angle

11.2 Decentralized Control for Series-Connected H-Bridge Rectifiers

11.2.1 Models of Series-Connected Rectifiers

Figure 11.7 shows the basic topology of series-connected rectifiers, which consists of *N* series-connected H-bridge modules. On the AC side, there is a filter inductance Z_{filt} between the point of common coupling (PCC) and rectifiers. On the DC side, each rectifier module has an independent DC-link capacitor and an isolated load. In this study, we focus on the series-connected rectifiers with the same power demand. In practice, it is very common in solid-state transformers and traction applications [4–7].

In Fig. 11.7, each rectifier module is controlled by an independent local controller, and the input characteristic of each module is equivalently treated as a voltage-controlled source. Then, the absorbed real power P_i and reactive power Q_i of i-th module from the utility grid (UG) are given by

$$P_i + jQ_i = V_i e^{j\delta_i} \cdot \left((V_g e^{j\delta_g} - V_p e^{j\delta_p}) / (|Z_f| e^{j\theta_f}) \right)^* \tag{11.14}$$

where V_i and δ_i represent the fundamental-frequency input voltage amplitude and phase angle of i-th module. V_g and δ_g are the voltage amplitude and phase angle of UG. $|Z_f|$ and θ_f are the amplitude and angle of the total grid impedance, including filter impedance Z_{filt} and line impedance Z_{line}.

The voltage at point of stacked input-voltage (PSIV) of modular rectifiers is the sum of all module input voltages.

$$V_p e^{j\delta_p} = \sum_{j=1}^{N} V_j e^{j\delta_j} \tag{11.15}$$

Generally, the total grid impedance is mainly inductive ($\theta_f \approx \pi/2$). In this case, the power transmission characteristic is given by

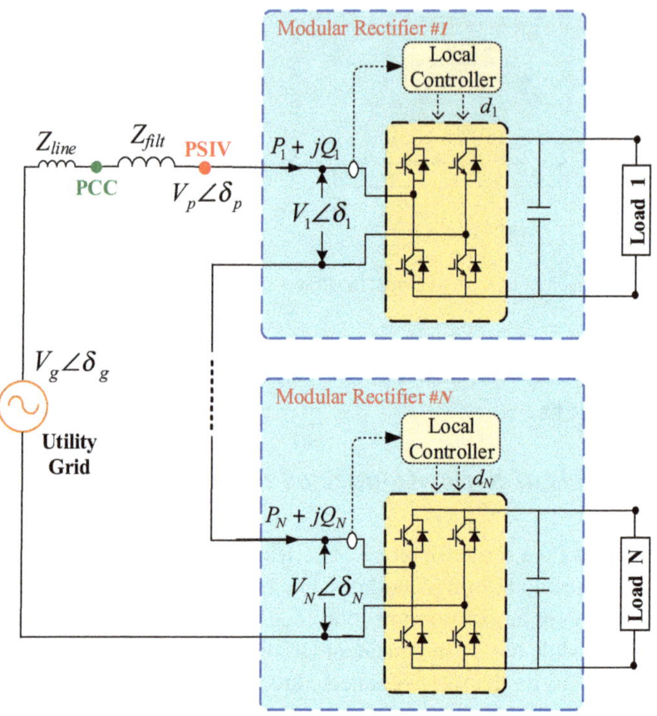

Fig. 11.7 Topology of series-connected rectifier modules

$$P_i = \frac{V_i}{|Z_f|}\left(\sum_{j=1}^{N} V_j \sin\left(\delta_i - \delta_j\right) - V_g \sin\left(\delta_i - \delta_g\right)\right) \tag{11.16}$$

$$Q_i = \frac{V_i}{|Z_f|}\left(V_g \cos\left(\delta_i - \delta_g\right) - \sum_{j=1}^{N} V_j \cos\left(\delta_i - \delta_j\right)\right) \tag{11.17}$$

11.2.2 Decentralized Control for Series-Connected H-Bridge Rectifiers

To synchronize each module with UG and flexibly control the absorbed active power, a decentralized control scheme is introduced

$$\begin{cases} \omega_i = \omega^* + k \cdot (P_i - P_i^*) \\ V_i = V^* \end{cases} \tag{11.18}$$

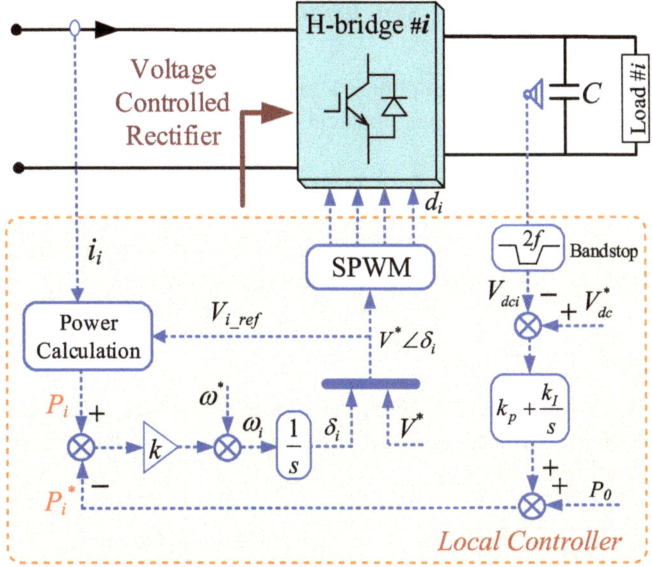

Fig. 11.8 The local controller of i-th voltage-controlled rectifier module

where ω_i and V_i are the angular frequency and voltage amplitude references of i-th module, respectively. ω^* represents the nominal value of grid angular frequency. k is a positive gain of $P - \omega$ control. V^* is a predesigned voltage parameter related to grid power factor and system stability, which will be assigned later. P_i^* represents the power reference absorbed from the grid, which is constructed as follows to balance the DC-link capacitor voltage V_{dci}.

$$P_i^* = P_0 + (k_p + \frac{k_I}{s})(V_{dc}^* - V_{dci}) \tag{11.19}$$

where k_p and k_I are the proportional-integral (PI) coefficients. V_d^*c is the dc-link voltage reference for each module. P_0 is a feed-forward power compensation term, which only impacts on the system startup dynamic. Generally, P_0 is equal to the rated load power. In steady state, the control objective of (11.19) is to balance the DC-link capacitor voltage which is a favorable indication of load demand.

According to (11.18) and (11.19), the voltage-controlled control diagram of i-th rectifier module is presented in Fig. 11.8. The voltage-generation reference is indicated by combining the voltage amplitude V^* and phase angle δ_i. The input active power is calculated based on the input current sampling and inner voltage-generation reference. It is noted that the notch filter is implemented to attenuate the twice line frequency ripple of DC-link voltage in the control loop [8].

11.2.3 Steady State and Synchronization Mechanism Analysis

In steady state, as the voltage amplitude V^* is same for all rectifiers and each module shares the same grid current, their apparent powers are equal.

$$S_1 = S_2 = \cdots = S_N \tag{11.20}$$

As the active powers of all DC-links feeding loads are the same, (11.21) is obtained from (11.18) due to the identical grid frequency ($\omega_1 = \cdots = \omega_N = \omega^*$)

$$P_1 = P_2 = \cdots = P_N = P^* \tag{11.21}$$

From (11.20) and (11.21), it can be deduced that the phase angles of all modules are equal in the steady state ($\delta_i = \delta_j; i, j \in \{1, 2, \cdots, N\}$). Further, the voltage at PSIV is derived from (11.14)

$$V_p = NV^*; \ \delta_p = \delta_1 = \delta_2 \cdots = \delta_N \tag{11.22}$$

From (11.16) and (11.17), the steady-state power of each module is given

$$\bar{P}_i = P^* = -S_C \sin\bar{\delta}; \ \bar{Q}_i = S_C \left(\cos\bar{\delta} - NV^*/V_g\right) \tag{11.23}$$

where $S_C = \left(V_g V^*/\left|Z_f\right|\right)$ represents the active power transfer capacity of a single module. $\bar{\delta} = (\delta_p - \delta_g)$ refers to the steady power angle. From (11.23), two steady points exist: $\bar{\delta} = -\arcsin(\frac{P^*}{S_C})$ or $\bar{\delta} = \pi + \arcsin(\frac{P^*}{S_C})$.

Then, the voltage amplitude reference V^* is designed according to the required grid power factor from (11.23) as $\tan\phi = \left(\bar{Q}_i/\bar{P}_i\right)$.

$$V^* = \frac{V_g}{N}(\tan\phi\sin\bar{\delta} + \cos\bar{\delta}) \tag{11.24}$$

where ϕ is the predesigned grid power factor angle.

For a given power factor in (11.24), V^* can be calculated if the total grid impedance is known. Normally, filter impedance is predesigned and easily known, and line impedance can be negligible when filter impedance is far greater than line impedance ($|Z_{filt}| \gg |Z_{line}|$). However, for a weak grid, line impedance profiles should be measured with the frequency scanning technique [9, 10].

For a better understanding of the control, the frequency synchronization mechanism of series modules is illustrated in Fig. 11.9. For simplicity, we assume two rectifier modules in series. The equivalent circuit and phasor diagrams are given in Fig. 11.9a, b, respectively.

In Fig. 11.9b, V_1 (green phasor) leads the steady-state voltage (red phasor); while V_2 (blue phasor) lags it. As two modules have the same voltage amplitude V^* and the same grid current I_g, their apparent powers are equal all the time ($S_1 = S_2$).

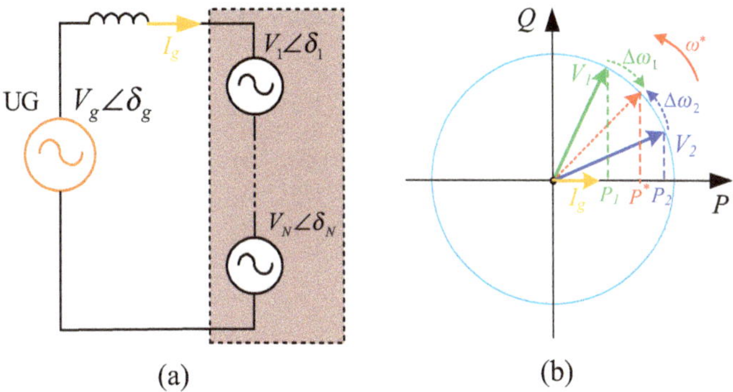

Fig. 11.9 A simplified series-connected rectifiers system for synchronization mechanism analysis. (**a**) Equivalent circuit, (**b**) Phasor diagram

Initially, $\delta_1 > \delta_2$, $P_1 < P^* < P_2$. Then, $\omega_1 < \omega^* < \omega_2$ is obtained from (11.18). As a result, δ_1 decreases ($\Delta\omega_1 = \omega_1 - \omega^* < 0$), while δ_2 increases ($\Delta\omega_2 = \omega_2 - \omega^* > 0$). The convergence process will continue until $\delta_1 = \delta_2$, $\omega_1 = \omega_2 = \omega^*$, and $P_1 = P_2 = P^*$.

It is worth noting that the synchronization mechanism of series modules is totally different from that of parallel converters. In the series modules, active power control is essentially a power factor control as the same flowing current and $P = V^* I \cos\phi$ [11]. While for the parallel modules, active power control is a power angle control as $P = \frac{V_1 V_2}{X}\sin\delta$, where δ denotes the power angle of power transmission .

Stability is essential for the normal operation of the series-connected rectifiers. Thus the small-signal stability analysis for the system is carried out in this section. First, the dynamic model of the DC capacitor is built.

$$C\frac{dV_{dci}}{dt} = \frac{P_i}{V_{dci}} - \frac{V_{dci}}{R_{load}} \tag{11.25}$$

where C is the value of DC capacitor. R_{load} implies an equivalent resistor of the load demand.

Then, linearizing (11.16), (11.18), (11.21), and (11.25) around the steady-state points yields

$$\begin{cases} \tilde{P}_i = \dfrac{V^*}{|Z_f|}\left(\displaystyle\sum_{j=1, j\neq i}^{N} V^*\left(\tilde{\delta}_i - \tilde{\delta}_j\right) - V_g(\cos\bar{\delta})\left(\tilde{\delta}_i - \tilde{\delta}_g\right)\right) \\ \dot{\tilde{\delta}}_i = \tilde{\omega}_i = k\cdot(\tilde{P}_i - \tilde{P}_i^*);\ \tilde{P}_i^* = -k_p\tilde{V}_{dci} - k_I\tilde{\Phi}_{dci} \\ \dot{\tilde{\Phi}}_{dci} = \tilde{V}_{dci};\ \dot{\tilde{V}}_{dci} = \dfrac{1}{V_{dc}^* C}\tilde{P}_i - \dfrac{P^*}{V_{dc}^{*2} C}\tilde{V}_{dci} - \dfrac{1}{R_{load} C}\tilde{V}_{dci} \end{cases} \tag{11.26}$$

where '' denotes small perturbation around equilibrium points.

Rewrite (11.26) in matrix form as

$$\begin{bmatrix} \dot{\tilde{\delta}} \\ \dot{\tilde{\Phi}}_{dc} \\ \dot{\tilde{V}}_{dc} \end{bmatrix} = [A_s]_{3N\times 3N} \begin{bmatrix} \tilde{\delta} \\ \tilde{\Phi}_{dc} \\ \tilde{V}_{dc} \end{bmatrix} = \begin{bmatrix} -kh_1A_1 & kk_I I & kk_p I \\ 0 & 0 & I \\ -h_3h_1A_1 & 0 & -h_4 I \end{bmatrix} \begin{bmatrix} \tilde{\delta} \\ \tilde{\Phi}_{dc} \\ \tilde{V}_{dc} \end{bmatrix} \tag{11.27}$$

where

$$\begin{cases} \tilde{\delta} = \left[\, \tilde{\delta}_1 \; \tilde{\delta}_2 \cdots \tilde{\delta}_N \,\right]^T; \; \tilde{\Phi}_{dc} = \left[\, \tilde{\Phi}_{dc1} \; \tilde{\Phi}_{dc2} \cdots \tilde{\Phi}_{dcN} \,\right]^T \\ \tilde{V}_{dc} = \left[\, \tilde{V}_{dc1} \; \tilde{V}_{dc2} \cdots \tilde{V}_{dcN} \,\right]^T; \; A_1 = h_2 I + 1_N 1_N^T \\ h_1 = \dfrac{V^{*2}}{|Z_f|}; \; h_2 = \dfrac{V_g \cos\bar{\delta}}{V^*} - N; \; h_3 = \dfrac{1}{V_{dc}^* C}; \; h_4 = \dfrac{P^*}{V_{dc}^{*2} C} + \dfrac{1}{R_{load} C} \end{cases} \tag{11.28}$$

The system stability is analyzed by inspecting the eigenvalues of matrix As, which are obtained by solving (11.29).

$$|\lambda I - A_s| = \left|\lambda^3 I + \lambda^2 (h_4 I + kh_1 A_1) + \lambda (h_4 + k_p h_3) kh_1 A_1 + kk_I h_1 h_3 A_1\right| = 0 \tag{11.29}$$

In (11.29), since the matrix $A1$ is symmetric and diagonalizable, then there exists a nonsingular matrix Z to meet

$$\begin{aligned} &\left|Z^{-1}(\lambda^3 I + \lambda^2 (h_4 I + kh_1 A_1) + \lambda (h_4 + k_p h_3) kh_1 A_1 + kk_I h_1 h_3 A_1) Z\right| \\ &= \left|\lambda^3 I + \lambda^2 (h_4 I + kh_1 \Lambda) + \lambda (h_4 + k_p h_3) kh_1 \Lambda + kk_I h_1 h_3 \Lambda\right| = 0 \end{aligned} \tag{11.30}$$

where $Z^{-1} A_1 Z = \Lambda = diag\{\mu_i\}; \quad i \in \{1, 2, \cdots, N\}$.

Then, the $3N$-order polynomial equation is decomposed into N cubic equations as

$$\lambda^3 + \lambda^2 (h_4 + kh_1 \mu_i) + \lambda (h_4 + k_p h_3) kh_1 \mu_i + kk_I h_1 h_3 \mu_i = 0; \quad i \in \{1, 2, \cdots, N\} \tag{11.31}$$

According to Routh-Hurwitz criterion, the necessary and sufficient conditions of system stability are obtained from (11.31)

$$\begin{cases} h_4 + kh_1 \mu_i > 0; \; (h_4 + k_p h_3) kh_1 \mu_i > 0 \\ kk_I h_1 h_3 \mu_i > 0 \\ (h_4 + kh_1 \mu_i)(h_4 + k_p h_3) kh_1 \mu_i > kk_I h_1 h_3 \mu_i \end{cases} ; \; i \in \{1, 2, \cdots, N\} \tag{11.32}$$

From (11.28), the eigenvalues of the matrix A1 are given by

$$\mu_1 = \frac{V_g \cos\bar{\delta}}{V*}; \ \mu_2 = \cdots = \mu_N = \frac{V_g \cos\bar{\delta}}{V*} - N. \tag{11.33}$$

As the constants h_1, h_3, h_4 in (11.28) are greater than zero, the stability conditions are simplified from (11.32) and (11.33) as follows

$$\begin{cases} \cos\bar{\delta} > 0 \\ k > 0 \\ V_g \cos\bar{\delta} - NV^* > 0 \\ k_p > -(h_4/h_3); \ (h_4 + kh_1h_2)(h_4/h_3 + k_p) > k_I > 0 \end{cases} \tag{11.34}$$

From the analysis above, we could conclude that:

1. As $\cos\bar{\delta} > 0$, the power angle $\bar{\delta}$ should lie in $(-\pi/2, \pi/2)$ in steady state. Thus, only $\bar{\delta} = -\arcsin(P^*/S_C)$ is the stable equilibrium point in (11.23), and $\bar{\delta}$ is normally around 0, which indicates .
2. From (11.34), the $P - \omega$ control gain k must be positive for grid-connected series rectifiers, which is different with the islanded series-connected inverters [12]. Moreover, a fast convergence rate of the frequency dynamic can be obtained by increasing k from (11.31). But a too large k will cause unacceptable frequency overshoots. Thus, a proper k could be designed by making a compromise between settling time and overshoot.
3. In (11.34), the predesigned voltage amplitude V^* should meet for system stability. From (11.24), a high power factor requires that V^* is close to the stability upper bound . However, a high value V^* is detrimental to stability from (11.34). Thus, a proper V^* should be designed by making a tradeoff between power factor requirement and stability margin.
4. For stability, the selection of PI coefficients k_p and k_I of DC-link capacitor control should meet (11.34).

11.2.4 Discussion and Comparisons Between the Introduced Control and Existing Methods

According to the stability condition $V_g > NV^*$ in (11.34), if control parameters are properly set, the system would be stable in normal-grid condition and the cases where there are grid voltage swell and small grid voltage dip. However, when a large grid dip occurs, and the condition is not further satisfied, the additional measures should be adopted.

Two potential measures can be taken. In the first one, a dynamic voltage regulator (DVR) can be series-connected with utility grid to guarantee a normal-grid voltage.

In the second one, we could modify the voltage reference V_i by adding a feed-forward compensation to (11.18):

$$V_i = V^* + \frac{V_g - V_g^*}{N} \tag{11.35}$$

where V_g and V_g^* are measured voltage amplitude and nominal amplitude of grid voltage. If each module samples the grid voltage, the cost will increase. Another method is to use communication to acquire V_g. As V_g is a slow variable, a low-bandwidth communication is needed, such as an optical fiber with a 5 Mb/s execution speed.

Table 11.2 presents a comparison between the centralized methods and the decentralized method. Compared to the centralized methods, the superior feature of the decentralized method is the self-synchronization, and it makes it possible for an arbitrary number of series modules. However, the method has lower performances of disturbance rejection ratio and resilient to grid dynamic. Moreover, as only the grid voltage amplitude is acquired via a distributed low-bandwidth communication, the transfer data latency and communication burden of the control strategy are significantly reduced (Table 11.3).

Table 11.2 Comparison with the existing methods

Features	Centralized methods [3–6]	The decentralized method
Synchronization method	Centralized communication	Self-synchronization
Communication bandwidth	High	Low
Cost	High	Low
Complexity	High	Low
Resilient to communication failure	Low	High
Resilient to grid dynamic	High	Low
Disturbance rejection ratio	High	Low
Scalability	Low	High
Application scale	Limited	Arbitrary

Table 11.3 Experiment parameters

Symbol	Value	Symbol	Value	Symbol	Value
V_g	311V	ω^*	$2\pi * 50$ rad/s	P_0	2 kW
C	3300 μF	Z_f	$0.08 + j1.0$	R_{load}	20 Ω
k	1.2e−4	k_p/k_I	8/8	V_d^*c	200 V
N	4	$\cos\phi$	0.995	V^*	75

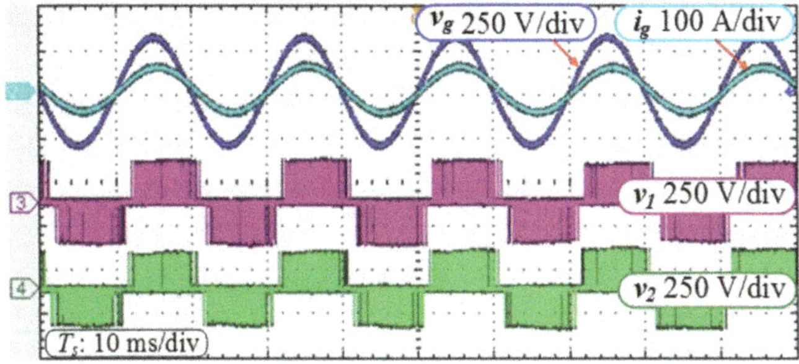

Fig. 11.10 Results under normal-grid condition. (From top to down: grid voltage v_g, grid current i_g, input voltage v_1 v_2 of module#1 and module#2)

11.2.5 Experimental Results

To verify the feasibility of the control scheme, a system comprised of four series-connected rectifier modules has been built and tested in the lab. The waveforms of the electrical quantities are recorded by the oscilloscope. The system parameters are designed according to steady-state analysis (11.10) and stability condition (11.21) as listed in Table 11.2. The switching frequency of rectifiers is 10 kHz. Moreover, beyond the research of this work, a power decoupling technology is used to suppress the twice line frequency ripple of DC capacitor in single-phase systems.

Performances under Normal-Grid Condition: Figs. 11.10 and 11.11 show the experimental results in normal-grid condition. From the voltage/current waveforms at utility grid in Fig. 11.10, a sinusoidal input current is guaranteed and a predesigned power factor 0.995 is enabled. Moreover, the system achieves real-power/reactive-power balances among four modules in Fig. 11.11a, b, autonomous grid frequency synchronization in Fig. 11.11c, equal DC capacitor voltage balances in Fig. 11.11d in steady state. Meanwhile, the DC-link capacitor voltage is maintained at the desired value 200 V and has a satisfactory response and good stability.

Performances under Load Change: This case study aims to evaluate the response of the control strategy under different load power levels. The experimental results are shown in Figs. 11.12 and 11.13. The four dc-link resistance loads of rectifiers are simultaneously changed from 2 to 1 kW at $t = 2$ s.

Figure 11.12 shows the grid voltage and current waveforms before and after the load change. Clearly, the grid current of $P^* = 2$ kW is twice as large as the grid current of $P^* = 1$ kW. From Fig. 11.13, the active power, reactive power, and dc-link capacitor voltage of four modules reach the new steady state after a regulation time of about two seconds. The dynamic response is smooth and satisfactory.

Performances under 2% Grid Voltage Dip: In this case, a small grid disturbance is considered to verify the effectiveness of the decentralized control scheme without adopting (11.34). Figure 11.14 shows the experimental results. According to the IEC 61000 standard, a 2% grid voltage dip is imposed at $t = 3$ s as shown in

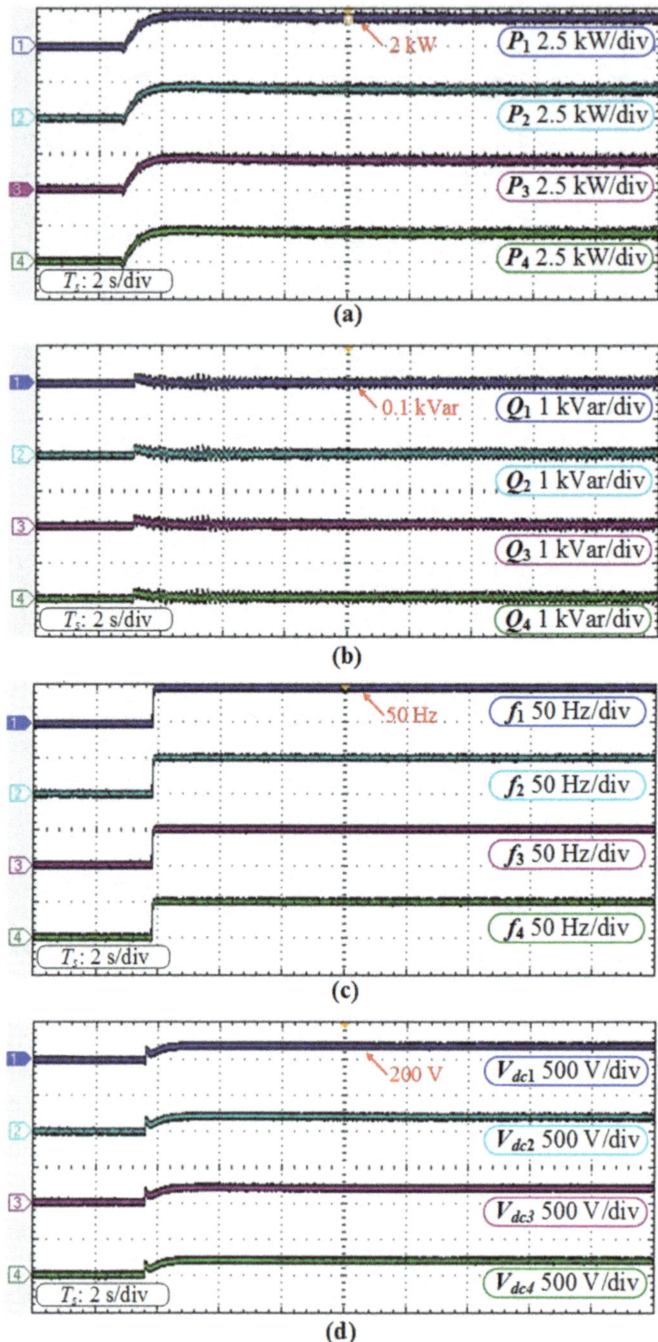

Fig. 11.11 Results of four modules under normal-grid condition. (**a**) Input active power P_1 P_4, (**b**) input reactive power Q_1 Q_4, (**c**) operation frequency f_1 f_4, and (**d**) DC-link capacitor voltage V_{dc1} V_{dc4}

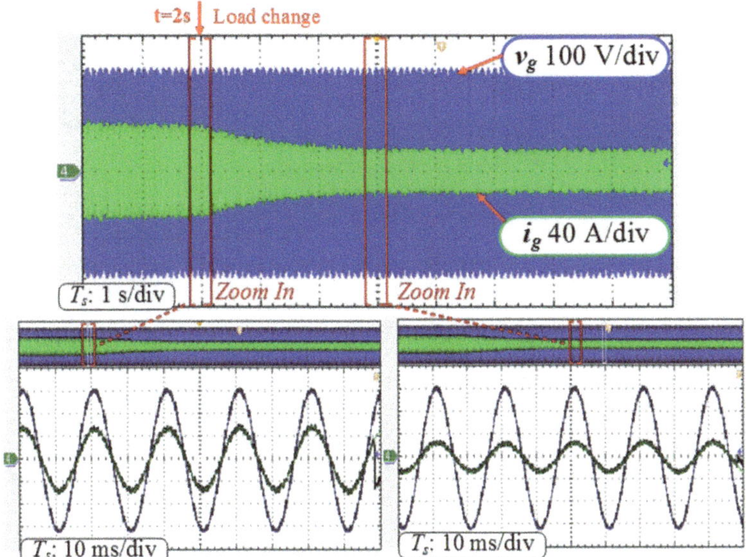

Fig. 11.12 Grid voltage v_g and grid current i_g waveforms under load change

Fig. 11.14a. For the dynamic response, the grid current and input active powers are almost unaffected. Only the reactive powers have a slight decrease. In the whole, the method of (11.18) is still feasible under a small grid disturbance.

Performances under 10% Grid Voltage Dip: For a large grid disturbance, the experimental test under abnormal-grid condition is conducted to verify the decentralized control of (11.18) and (11.34). Figure 11.15 shows the experimental results. A 10% grid voltage dip is imposed at $t = 4$ s as shown in Fig. 11.15a.

Initially, grid voltage is normal, and all modules work normally during t =0–4 s. When the grid voltage dip occurs, the system does not lose stability, and only active power has a small impulse. With the function of grid voltage feed-forward, the system returns to be normal after about five cycles in Fig. 11.15a. From this case, the control method can still work in the large grid disturbance condition.

11.3 Decentralized Control Scheme for Medium/High Voltage Series-Connected STATCOM

11.3.1 Models of Series-Connected STATCOM

Figure 11.16 illustrates the diagram of series STATCOM, which consists of N H-bridge modules [8, 9, 12].

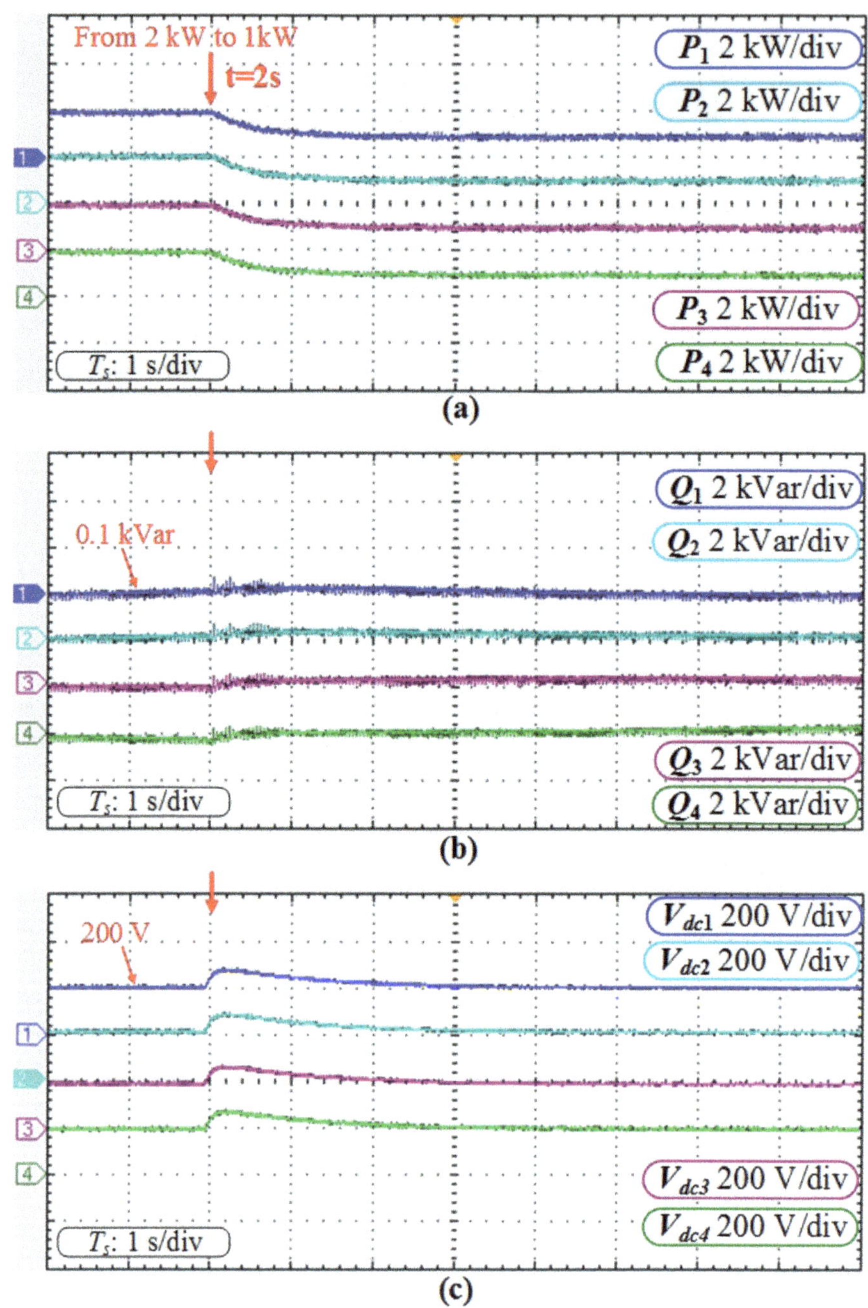

Fig. 11.13 Results of four modules under normal-grid condition. (**a**) Input active power P_1 P_4, (**b**) input reactive power Q_1 Q_4, and (**c**) DC-link capacitor voltage V_{dc1} V_{dc4}

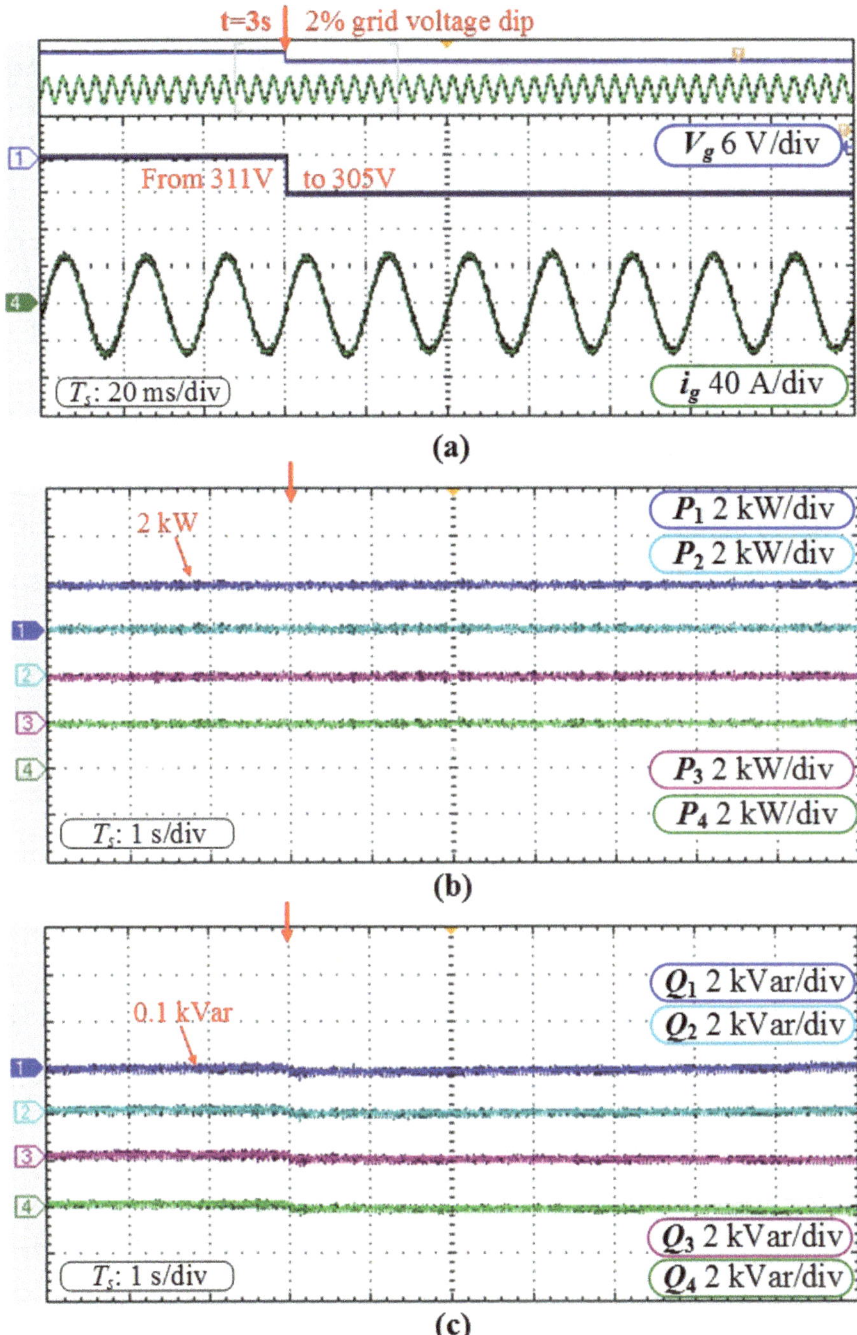

Fig. 11.14 Results under 2% grid voltage dip. (**a**) Grid voltage amplitude V_g, grid current i_g, (**b**) input active power P_1 P_4, and (**c**) input reactive power Q_1 Q_4

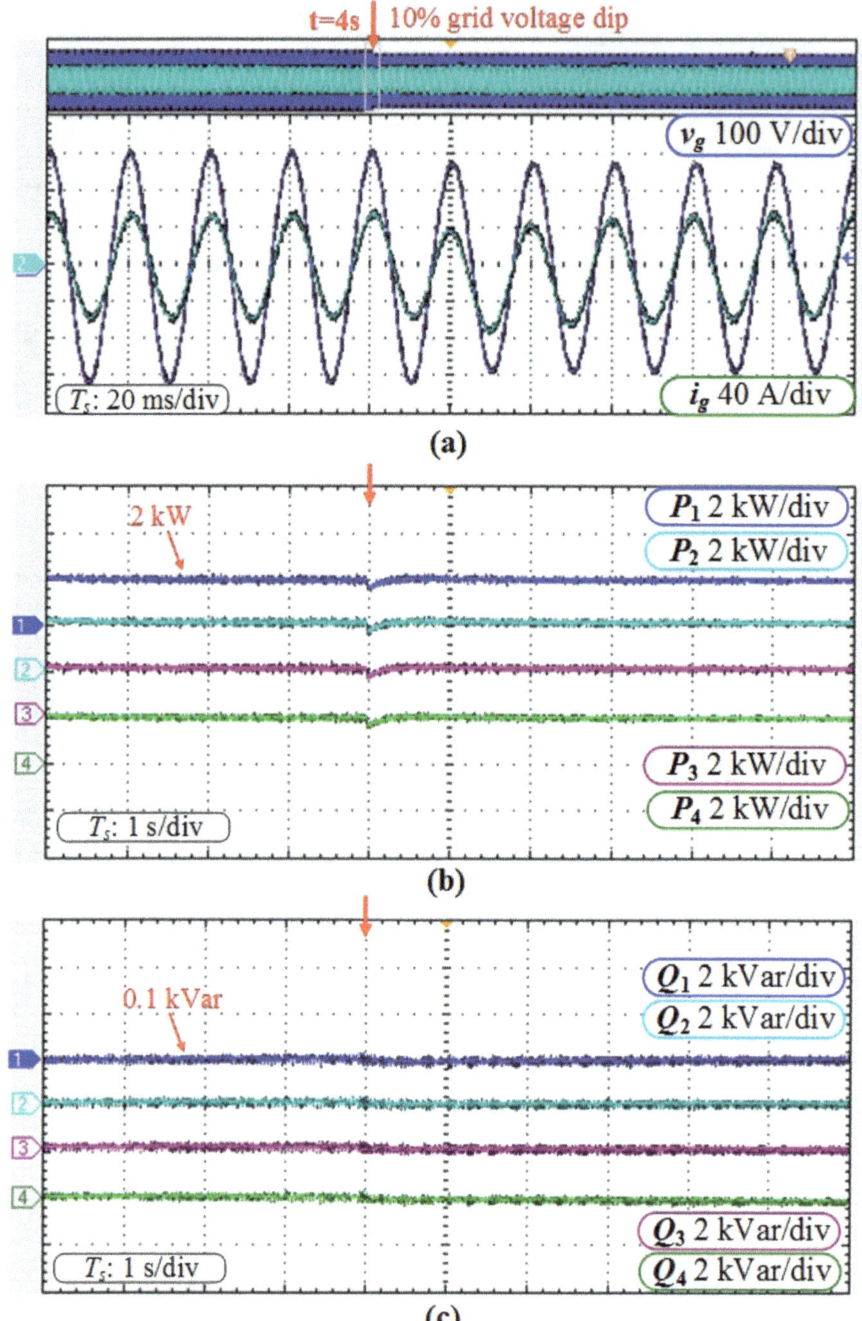

Fig. 11.15 Results under 10% grid voltage dip. (**a**) Grid voltage amplitude V_g, grid current i_g, (**b**) input active power P_1 P_4, and (**c**) input reactive power Q_1 Q_4

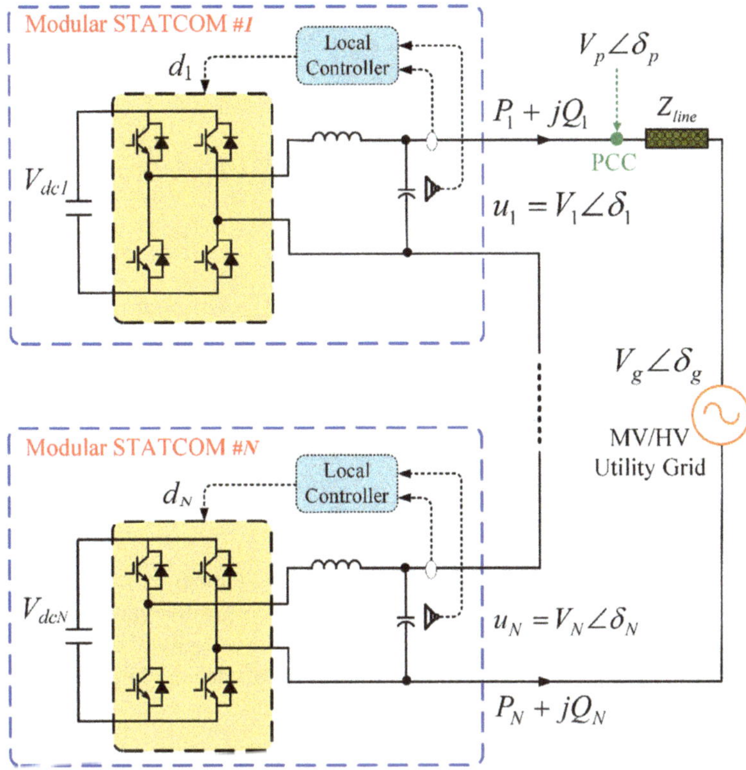

Fig. 11.16 Schematic diagram of MV/HV series STATCOM modules

From Fig. 11.16, the output active power P_i and reactive power Q_i of i-th module are expressed by

$$P_i + jQ_i = V_i e^{j\delta_i} \cdot \left((V_p e^{j\delta_p} - V_g e^{j\delta_g})/(|Z_{line}|\, e^{j\theta_{line}})\right)^* \tag{11.36}$$

where V_i and δ_i represent the output voltage amplitude and phase angle of i-th module. V_g and δ_g are the voltage amplitude and phase angle of utility grid. $|Z_{line}|$ and θ_{line} are the grid impedance amplitude and angle. Note that the grid impedance is shaped as a mainly resistive line ($\theta_{line} \approx 0$) to facilitate the $Q - \omega$ boost control [13]. The voltage $V_p e^{j\delta_p}$ at the point of common coupling (PCC) is the sum of all module voltages.

$$V_p e^{j\delta_p} = \sum_{j=1}^{N} V_j e^{j\delta_j} \tag{11.37}$$

From (11.36) and (11.37), the power transmission characteristic is given by

$$\begin{aligned} P_i &= \frac{V_i}{|Z_{line}|}\left(\sum_{j=1}^{N} V_j \cos\left(\delta_i - \delta_j\right) - V_g \cos\left(\delta_i - \delta_g\right)\right) \\ Q_i &= \frac{V_i}{|Z_{line}|}\left(\sum_{j=1}^{N} V_j \sin\left(\delta_i - \delta_j\right) - V_g \sin\left(\delta_i - \delta_g\right)\right) \end{aligned} \tag{11.38}$$

11.3.2 Decentralized Control for Series-Connected STATCOM

To synchronize each module with the grid and regulate the grid-injected reactive power, a decentralized $Q - \omega$ boost control scheme is introduced as

$$\omega_i = \omega^* + k_q(Q_i - Q^*) \tag{11.39}$$

where ω_i is the angular frequency reference of i-th module. ω^* represents the nominal value of the grid angular frequency. k_q is a positive gain of the $Q - \omega$ boost control. Q^* denotes the nominal reactive power capacity of each module.

In order to balance the DC capacitor voltage V_{dci} of each module, a decentralized $P - V$ control (11.40) is constructed as

$$V_i = V_0 + k_p(V_{dci} - V_{dc}^*) \tag{11.40}$$

where V_i is the voltage amplitude reference of i-th module. V_d^*c is the capacitor voltage reference of each module. V_0 represents the initial voltage. k_p is a gain of $P - V$ control, which will be designed in the following part.

11.3.3 Steady State and Stability Analysis

In steady state of normal-grid condition, reactive power sharing among modules is obtained due to the identical grid frequency ($\omega_1 = \cdots = \omega_N = \omega^*$) from (11.39)

$$Q_1 = Q_2 = \cdots = Q_N = Q^* \tag{11.41}$$

Meanwhile, because each module shares the same grid current, the active power losses of all modules are approximately equal in nominal state. Namely,

$$P_1 = P_2 = \cdots = P_N = -P^* \tag{11.42}$$

where P^* denotes a nominal power loss of a module.

From (11.41) and (11.42), the voltage amplitudes and phase angles of the modules are equal in steady state.

$$V_1 = V_2 \cdots = V_N = V^*;\ \delta_1 = \delta_2 \cdots = \delta_N \tag{11.43}$$

where V^* denotes the steady-state voltage amplitude. Thus, the voltage amplitude and phase angle of PCC derived from (11.37)

$$V_p = NV^*;\ \delta_p = \delta_1 = \delta_2 \cdots = \delta_N \tag{11.44}$$

Then, the steady-state power of each module is obtained by combing (11.37)–(11.38) and (11.41)–(11.44)

$$\bar{P}_i = -P^* = S_C\left(NV^*/V_g - \cos\bar{\delta}\right);\ \bar{Q}_i = Q^* = -S_C\sin\bar{\delta} \tag{11.45}$$

where $S_C = V^*V_g/|Z_{line}|$ represents the power transfer capacity of a single module. $\bar{\delta} = \delta_p - \delta_g$ refers to the steady power angle, which is equal to $\bar{\delta} = -\arcsin(\frac{Q^*}{S_C})$ in steady state from (11.45).

From (11.45), the delivered reactive power depends mostly on the power angle, and active power is predominately dependent on voltage amplitude difference. Thus, stability analysis of $Q-\omega$ and $P-V$ can be decoupled to facilitate their individual designs. In this section, to verify the stability of the methods, small-signal analysis is carried out.

1. Power Angle Stability

   Since $\dot{\delta}_i = \omega_i$, combining (11.38), (11.39), (11.43), and (11.44) and linearizing them around the steady-state points yield

$$\dot{\tilde{\delta}}_i = k\left(\sum_{j=1, j\neq i}^{N}\left(\tilde{\delta}_i - \tilde{\delta}_j\right) - M(\cos\bar{\delta})\left(\tilde{\delta}_i - \tilde{\delta}_g\right)\right) \tag{11.46}$$

   where $k = k_q V^{*2}/|Z_{line}|$ and $M = V_g/V^*$. $\tilde{\delta}_i$, $\tilde{\delta}_g$, $\tilde{\delta}_j$ denote small perturbations around the equilibrium point.

   From (11.46), the system matrix is given by

$$\dot{\tilde{\delta}} = -k \cdot L \cdot \tilde{\delta} \tag{11.47}$$

where $\tilde{\delta} = \left[\tilde{\delta}_1\ \tilde{\delta}_2 \cdots \tilde{\delta}_N\right]^T$;

$$L = \begin{bmatrix} M\cos\bar{\delta} - N + 1 & 1 & \cdots & 1 \\ 1 & M\cos\bar{\delta} - N + 1 & \cdots & 1 \\ \vdots & \vdots & \ddots & \vdots \\ 1 & 1 & \cdots & M\cos\bar{\delta} - N + 1 \end{bmatrix} =$$

$(M\cos\bar{\delta} - N)I_{N\times N} + 1_N 1_N^T$

The eigenvalues of the system matrix $A = -kL$ are given by

$$\lambda_1(A) = -kM\cos\bar{\delta};\ \lambda_2(A) = \cdots = \lambda_N(A) = -k(M\cos\bar{\delta} - N) \tag{11.48}$$

Thus, the necessary and sufficient condition of power angle stability is obtained as follows

$$M\cos\bar{\delta} - N > 0 \tag{11.49}$$

That is, the power angle stability condition is concluded by substituting $M = V_g/V^*$ into (11.49)

$$V_g\cos\bar{\delta} - NV^* > 0 \tag{11.50}$$

From (11.50), $\cos\bar{\delta}$ must be greater than zero, and the power angle should lie in $(-\pi/2, \pi/2)$ under the stability constraint. According to above steady-state analysis, $\bar{\delta} = -\arcsin(\frac{Q^*}{S_C})$ is a stable equilibrium point.

From (11.48), a fast convergence rate of the power angle dynamic can be obtained by flexibly regulating the $Q - \omega$ control coefficient k_q, but a too large k_q would cause unacceptable impulse frequency responses during the transient process. Thus, a proper k_q could be designed by making a compromise between settling time and overshoot.

2. DC-link Voltage Stability

   The dc-link voltage stability is essential for the normal operation of the series STATCOM. It is guaranteed by $P - V$ control (11.40). To analyze the voltage stability, the mathematical model of the fore-end DC-link capacitor is set up firstly.

$$C\frac{dV_{dci}}{dt} = -\frac{V_{dci}}{R_{loss}} - \frac{P_i}{V_{dci}} \tag{11.51}$$

   where C is the value of DC capacitor. R_{loss} implies an equivalent resistor of power losses, connected in parallel with the DC capacitor in the model.

Combining (11.37), (11.40), (11.42)–(11.44), (11.52) and linearizing them around the steady-state points yield

$$
\begin{cases}
\tilde{P}_i = \dfrac{V^*}{|Z_{line}|}\left[(N - M\cos\bar{\delta})\tilde{V}_i + \sum_{j=1}^{N}\tilde{V}_j\right] \\
\tilde{V}_i = k_p\tilde{V}_{dci};\ \dot{\tilde{V}}_{dci} = -\dfrac{1}{R_{loss}C}\tilde{V}_{dci} - \dfrac{P^*}{V_{dc}^{*2}C}\tilde{V}_{dci} - \dfrac{1}{V_{dc}^*C}\tilde{P}_i
\end{cases}
\tag{11.52}
$$

From (11.52), the system matrix of voltage stability is given by

$$
\dot{\tilde{V}}_{dc} = A_v\tilde{V}_{dc} \tag{11.53}
$$

where $\tilde{V}_{dc} = \left[\tilde{V}_{dc1}\ \tilde{V}_{dc2}\ \cdots\ \tilde{V}_{dcN}\right]^T;\ A_v = -\left[bI + k_pA_1\right]$

$$
\begin{cases}
A_1 = a\left[-dI_{N\times N} + 1_N1_N^T\right] \\
a = \dfrac{V^*}{|Z_{line}|\,V_{dc}^*C};\ b = \dfrac{1}{R_{loss}C} + \dfrac{P^*}{V_{dc}^{*2}C};\ d = M\cos\bar{\delta} - N
\end{cases}
\tag{11.54}
$$

The voltage stability can be analyzed by inspecting the eigenvalues of the matrix A_v, which are obtained as

$$
\lambda_1(A_v) = -(b - adk_p + aNk_p);\ \lambda_2(A_v) = \cdots = \lambda_N(A_v) = -(b - adk_p) \tag{11.55}
$$

From (11.54) and (11.55), as a, b, and d are greater than zero, the necessary and sufficient condition of voltage stability is concluded as follows

$$
-\frac{b}{a(N-d)} < k_p < \frac{b}{ad} \tag{11.56}
$$

11.3.4 Improved Decentralized Control for Abnormal-Grid Condition

When contingencies occur in power grid, an improved decentralized control is introduced by adding a feed-forward compensation to (11.40).

$$
V_i = V_0 + k_p(V_{dci} - V_{dc}^*) + \frac{V_g - V_g^*}{N_{few}} \tag{11.57}
$$

where V_g and V_g^* are real-time voltage amplitude and nominal voltage amplitude of grid. Note that the voltage compensation does not require all modules to take part in. That is, as long as N_{few} modules ($N_{few} \approx 10\%N\ 20\%N$) need to acquire the grid information, the adverse effect of the contingencies can be compensated. For these N_{few} modules, their reactive power references should be regulated according to their individual voltage amplitude.

$$\omega_i = \omega^* + k_q(Q_i - \frac{V_i}{V^*}Q^*) \tag{11.58}$$

11.3.5 Simulation Results

The simulations are carried out to verify the ideas of $Q-\omega$ control (11.39) and $P-V$ control (11.40). The local control diagram of i-th module is presented in Fig. 11.17. The system parameters of the medium voltage STATCOM are listed in Table 11.4. Typically, the physical parameters of each H-bridge module are firstly assumed: the rated capacity 100 kVA, rated AC voltage 600 V/50 Hz, dc-link voltage 900 V, and conversion efficiency 95%. Then, the detailed designs of control parameters are presented as follows: (1) According to the grid voltage class $V_g^* = 6\,\text{kV}$, we choose the number of series modules $N = 10$. (2) The nominal reactive power capacity of each STATCOM module $Q^* = 100\,\text{kVar}$ is set. Normally, the active power loss of each module is about 5% of the reactive power capacity ($P^* \approx 5\,\text{kW}$). As the dc-link voltage $V_d^*c = 900\,\text{V}$, an equivalent resistor of power losses $R_{loss} = 162\,\Omega$ is calculated. (3) The $Q-\omega$ control gain k_q is chosen according to (11.48)–(11.50). (4) The $P-V$ control gain k_p is chosen according to (11.56). (5) The initial voltage V_0 should reach the steady-state voltage V^* as close as possible, which is derived from (11.45).

Simulation Results in Normal-Grid Condition: To illustrate the effectiveness, the simulation in normal-grid condition ($V_g = V_g^*$) is firstly carried out. The different initial phase angles of modules are set. Figure 11.18 shows the simulation results of the decentralized control. As seen from Fig. 11.18a, the function of nominal reactive-power compensation ($Q_1 = \cdots = Q_10 = 100\,\text{kVar}$) is realized. From Fig. 11.18b, the absorbed active power is about 5 kW, which is equal to the power losses of each module. Figure 11.18c indicates that the autonomous frequency synchronization with the utility grid is obtained. Figure 11.18d shows the output voltage amplitudes of ten modules. Figure 11.18e reveals that the DC capacitor voltage balance among modules is achieved. Moreover, Fig. 11.18f shows that the steady-state power angle between PCC and utility grid is close to 0, which is in accordance with stability analysis where $\bar{\delta} \in (-\pi/2, \pi/2)$.

Table 11.4 Simulation parameters

Symbol	Value	Symbol	Value	Symbol	Value
V_g^*	6 kV	N	10	Z_{line}	$0.8 + j0.04\,\Omega$
ω^*	$2\pi * 50$ rad/s	V_0	599.2 V	C	5000 μF
k_q	5e−5	k_p	1e−3	V_d^*c	900 V
Q^*	100 kVar	R_{load}	162 Ω	P^*	$\approx 0.05Q^*$

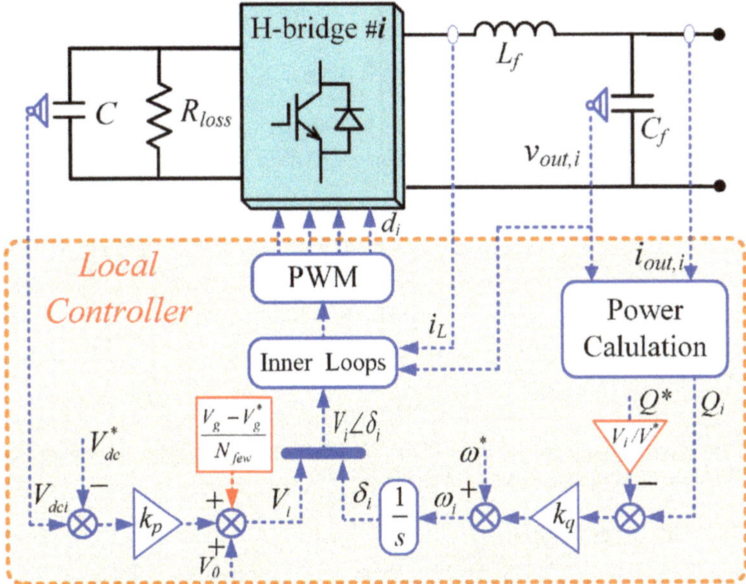

Fig. 11.17 The local control diagram of i-th STATCOM module

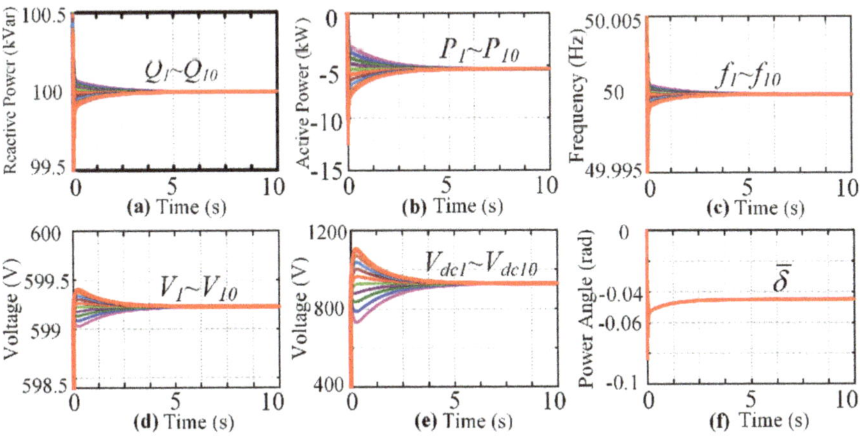

Fig. 11.18 Results in normal-grid condition. (**a**) Reactive power. (**b**) Active power. (**c**) Frequency. (**d**) Voltage amplitude. (**e**) DC-link voltage. (**f**) Power angle

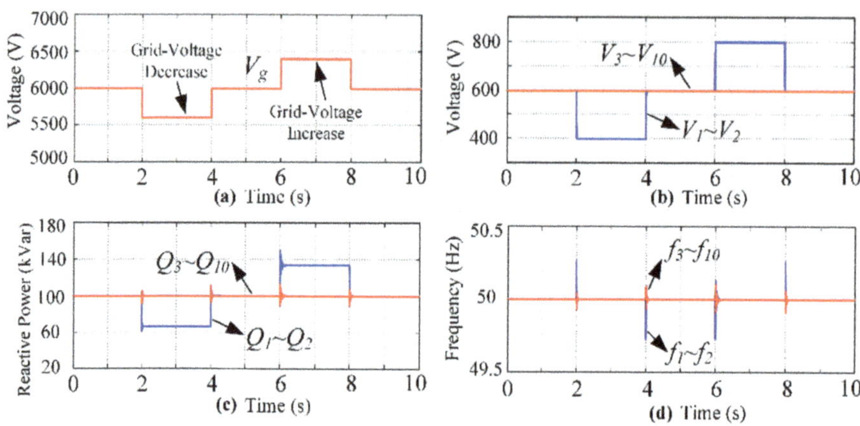

Fig. 11.19 Results in abnormal-grid condition. (**a**) Grid voltage amplitude. (**b**) Voltage amplitude of ten modules. (**c**) Reactive power. (**d**) Frequency

Table 11.5 HIL test parameters

Symbol	Value	Symbol	Value	Symbol	Value
V_g	1 kV	N	4	Z_{line}	$0.8 + j0.04\Omega$
ω^*	$2\pi * 50$ rad/s	V_0	247.8 V	C	1000 μF
k_q	5e−5	k_p	1e−3	V_d^*c	400 V
Q^*	200 kVar	R_{load}	160 Ω	P^*	1 kW

Simulation Results in Abnormal-Grid Condition: The simulation in abnormal-grid condition is conducted to verify the validity of the improved decentralized control (11.57) and (11.58). In this case, mudul#1 and mudule#2 are chosen as the voltage compensation modules ($N_{few} = 2$). Figure 11.19a shows the real-time grid voltage amplitude. Initially, grid voltage is normal, and all modules work normally during t =0–2 s. Then 400 V grid voltage sag and 400 V grid overvoltage occur at $t = 2$ s and $t = 6$ s, respectively. From Fig. 11.19b, V1 and V2 decrease by 200 V at $t = 2$ s and increase by 200 V at $t = 6s$ to adapt the grid dynamics, while V_3 V_{10} are identical and unchanged. Accordingly, Q_1 and Q_2 decrease at $t = 2$ s and increase at $t = 6$ s according to (11.58) in Fig. 11.19c, while Q_3 Q_{10} are equal and remain unchanged. Figure 11.19d reveals that the autonomous frequency synchronization with the grid in steady state is also achieved under grid dynamics (Table 11.5).

11.4 Conclusion

This study exploits decentralized control strategies for series-connected inverters, series-connected H-bridge rectifiers, and series-connected H-bridge inverter-based static compensators (STATCOM) in medium/high voltage (MV/HV) power network without any communication, and each module makes decisions based on its own local information. They can realize accurate power balance and frequency synchronization autonomously, and the transfer data latency and communication burden of the control strategy are significantly reduced. For readers, this chapter explores the possibilities of inspiring new methods in a power network with series-type inverters, series-type H-bridge rectifiers, and series-type H-bridge inverter-based STATCOM, respectively [14–17].

References

1. Y. Sun, X. Hou, J. Yang, H. Han, M. Su, J.M. Guerrero, New perspectives on droop control in AC microgrid. IEEE Trans. Ind. Electron. **64**(7), 5741–5745 (2017)
2. Y. Sun, G. Shi, X. Li, W. Yuan, M. Su, H. Han, X. Hou, An f-P/Q droop control in cascaded-type microgrid. IEEE Trans. Power Syst. **33**(1), 1136–1138 (2018)
3. M. Malinowski, K. Gopakumar, J. Rodriguez, M. Pérez, A survey on cascaded multi-level inverters. IEEE Trans. Ind. Electron. **57**(7), 2197–2206 (2010)
4. Y. Yu, G. Konstantinou, B. Hredzak, V.G. Agelidis, Power balance of cascaded H-bridge multilevel converters for large-scale photovoltaic grid integration. IEEE Trans. Power Electron. **31**(1), 292–303 (2016)
5. C. Cecati, A.D. Aquila, M. Liserre, V.G. Monopoli, Design of H-bridge multilevel active rectifier for traction systems. IEEE Trans. Ind. Appl. **39**(5), 1541–1550 (2003)
6. D. Sha, G. Xu, Y. Xu, Utility direct interfaced charger/discharger employing unified voltage balance control for cascaded H-bridge units and decentralized control for CF-DAB modules. IEEE Trans. Ind. Electron. **64**(10), 7831–7841 (2017)
7. L. Wang, D. Zhang, Y. Wang, B. Wu, H.S. Athab, Power and voltage balance control of a novel three-phase solid-state transformer using multilevel cascaded H-bridge inverters for microgrid applications. IEEE Trans. Power Electron. **31**(4), 3289–3301 (2016)
8. G. Farivar, C.D. Townsend, B. Hredzak, J. Pou, V.G. Agelidis, Passive reactor compensated cascaded H-Bridge multilevel LC-StatCom. IEEE Trans. Power Electron. **32**(11), 8338–8348 (2017)
9. Y. Liu, A.Q. Huang, W. Song, S. Bhattacharya, G. Tan, Small-signal model-based control strategy for balancing individual DC capacitor voltages in cascade multilevel invert-er-based StatCom. IEEE Trans. Ind. Electron. **56**(6), 2259–2269 (2009)
10. X. Hou, Y. Sun, H. Han, Z. Liu, W. Yuan, M. Su, A fully decentralized control of grid-connected cascaded inverters. IEEE Trans. Sustain. Energy **10**(1), 315–317 (2019)
11. X. Hou, Y. Sun, H. Han, Z. Liu, M. Su, B. Wang, X. Zhang, A general decentralized control scheme for medium/high voltage cascaded STATCOM. IEEE Trans. Power Syst. **33**(6), 7296–7300 (2018)
12. Y. Liang, C.O. Nwankpa, A new type of StatCom based on cascading voltage-source inverters with phase-shifted unipolar SPWM. IEEE Trans. Ind. Appl. **35**(5), 1118–1123 (1999)
13. W. Chen, X. Ruan, H. Yan, C.K. Tse, DC/DC conversion systems consisting of multiple converter modules: stability, control, and experimental verifications. IEEE Trans. Power Electron. **24**(6), 1463–1474 (2009)
14. K. Sun, X. Lin, Y. Li, Y. Gao, L. Zhang, Improved modulation mechanism of parallel-operated T-type three-level PWM rectifiers for neutral-point potential balancing and circulating current suppression. IEEE Trans. Power Electron. **33**(9), 7466–7479 (2018)
15. G. Xu, D. Sha, X. Liao, Decentralized inverse-droop control for input-series-output-parallel DC-DC converters. IEEE Trans. Power Electron. **30**(9), 4621–4625 (2015)

16. L. Shu, W. Chen, X. Jiang, Decentralized control for fully modular input-series-output-parallel (ISOP) inverter system based on the active power inverse-droop method. IEEE Trans. Power Electron. **33**(9), 7521–7530 (2018)
17. X. Hou, Y. Sun, X. Zhang, G. Zhang, J. Lu, F. Blaabjerg, A self-synchronized decentralized control for series-connected H-bridge rectifiers. IEEE Trans. Power Electron. **34**(8), 7136–7142 (2019)

Chapter 12
A Master–Slave Control in Grid-Connected Applications

12.1 Hybrid Voltage/Current Control

12.1.1 Control Configuration of Series Inverters

Figure 12.1 presents the overall system and control configuration of grid-connected series H-Bridge inverters. On the DC side, each inverter is connected with a primary source of distributed generation (DG). On the AC side, each inverter has an output LC filter to achieve a flexible real/reactive power control and a closed-loop voltage/current tracking [1, 2]. For the hybrid voltage/current control configuration of series inverters as shown in Fig. 12.2, some main features are:

1. Individual local controllers for inverters. In contrast to the existing multilevel-based methods[3–6], each inverter is controlled by a local controller without the need for a central controller or high-bandwidth global communication among the units.
2. Sharing a common grid current. The obvious feature of series inverters is that they share the common grid current. Thus, the current-phase component can be used as a natural synchronization carrier, and the synchronization of voltage phase can be realized by power factor angle consistency.
3. A few m current-controlled inverters. To directly regulate grid current and guarantee a desired grid PF, some inverters near to the grid are controlled as CCIs, which are connected in parallel to jointly adapt to the grid variations. The current-phase synchronization with voltage at point of common coupling (PCC) is required.
4. Most $n - m$ voltage-controlled inverters. The most series inverters are controlled as VCIs to achieve independent power regulation and voltage phase self-synchronization. Generally, $m \approx 10\%n$ by considering $\pm 10\%$ grid voltage variations according to the IEEE Standard 1547–2018 for interconnection and

Y. Sun et al., *Series-Parallel Converter-Based Microgrids*, Power Systems,
https://doi.org/10.1007/978-3-030-91511-7_12

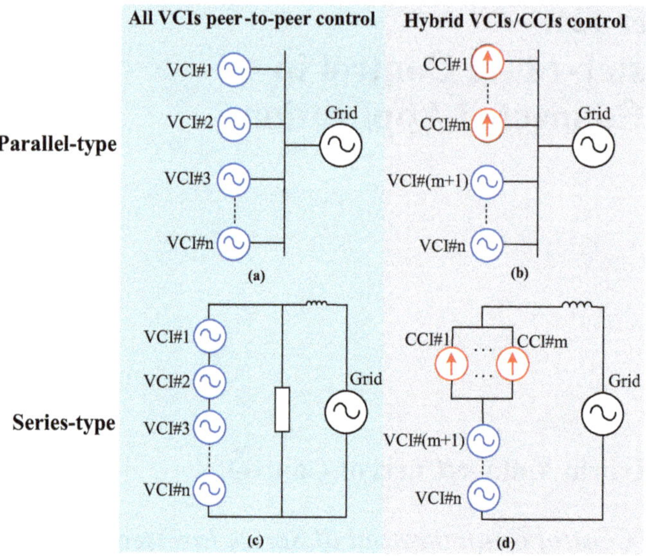

Fig. 12.1 Comparative control architecture of parallel and series inverters. (**a**) All VCIs peer-to-peer control of parallel-type, (**b**) hybrid VCIs/CCIs of parallel-type, (**c**) all VCIs control of series-type, (**d**) hybrid VCIs/CCIs control framework of series-type)

interoperability of distributed resources with associated electric power systems interfaces [7].

12.1.2 *Hybrid Voltage/Current Control in Decentralized Manner*

The hybrid voltage/current control configuration of all series inverters is shown in Fig. 12.2, which includes m CCIs and $n-m$ VCIs. For the point of common coupling (PCC), the grid current $i_g = I_g \angle\theta_{Ig}$ is the sum of all CCIs ($i_g = i_1 + \cdots + i_m$).

Figure 12.3a shows the control scheme of CCI#*i*. The output current reference $i_i^* = I_i \angle\theta_{Ii}$ of CCI#*i* is derived from

$$\begin{cases} I_i = \left(K_{Pi} + \dfrac{K_{Ii}}{s}\right)\left(P_i^* - P_i\right) \\ \theta_{Ii} = \theta_p - \varphi^* \end{cases} \quad (i = 1, \ldots, m) \tag{12.1}$$

where P_i^* implies the real-time available power from the primary DG-*i* source. P_i is the output real power. K_{Pi} and K_{Ii} are proportional-integral (PI) coefficients. θ_p is the acquired voltage phase at PCC. φ^* is the predesigned same PF angle for all

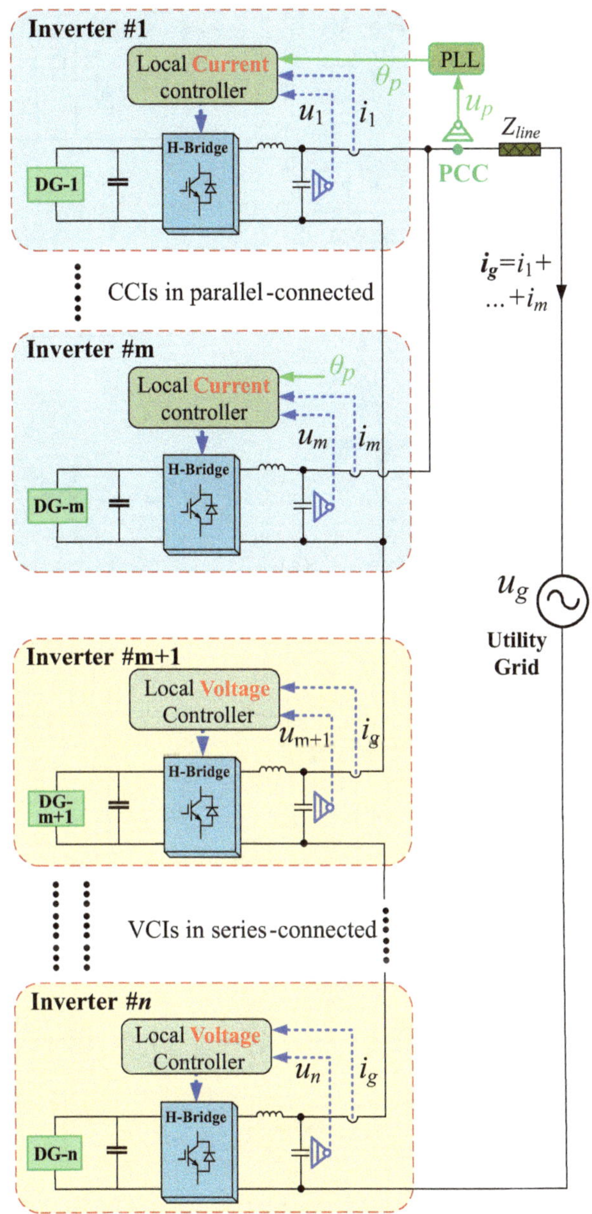

Fig. 12.2 The hybrid voltage/current control configuration of series inverters (Generally, $m \approx 10\% n$ by considering $\pm 10\%$ grid voltage variations)

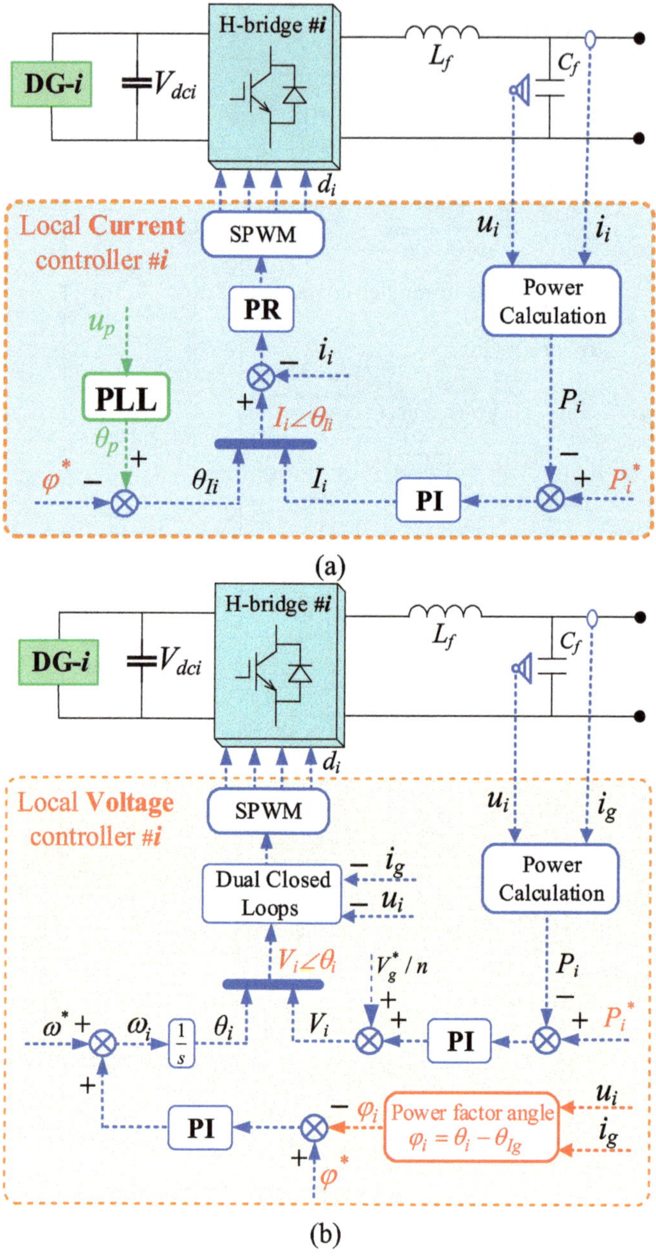

Fig. 12.3 Control scheme of inverters. (**a**) Current-controlled inverter#i ($i = 1, \ldots, m$), (**b**) voltage-controlled inverter#i ($i = m + 1, \ldots, n$)

modules, which is offline set by considering the practical grid-code requirement of IEEE Standard 1547–2018. Specially, for series PV micro-inverters, P_i^* is usually determined by maximum power point tracking (MPPT) algorithm to take full utilization of DGs, and unity PF is realized by setting $\varphi^* = 0$. And for series storage units, the power reference P_i^* and power factor angle φ^* are indicated by the upper supervisory controller to participate the grid ancillary services and provide the reactive power compensation for a weak remote voltage feeder [8, 9].

Figure 12.3b shows the control scheme of VCI#i. The voltage reference $u_i^* = V_i \angle_i$ of VCI#i is derived from a real power- amplitude (P-V) and PF angle-frequency $(\varphi - \omega)$ control.

$$\begin{cases} V_i = \dfrac{V_g^*}{n} + \left(K_{Pvi} + \dfrac{K_{Ivi}}{s}\right)(P_i^* - P_i) \\ \dot{\theta}_i = \omega_i = \omega^* + \left(K_{P\omega i} + \dfrac{K_{I\omega i}}{s}\right)(\varphi^* - \varphi_i) \end{cases} \quad (i = m+1, \ldots, n) \tag{12.2}$$

where P_i is output real power of VCI-i. K_{Pvi} and K_{Ivi} are PI coefficients of voltage amplitude control. V_g^* is the rated grid voltage amplitude. ω^* is the rated grid angular frequency. $K_{P\omega i}$ and $K_{I\omega i}$ are PI coefficients of voltage phase control. φ^* is the predesigned PF angle. φ_i is output PF angle of VCI-i, obtained by the phase-difference of the output voltage and the sharing grid current $(\varphi_i = \theta_i - \theta_{Ig})$.

Moreover, P_i^* implies the real-time available power from the primary DG-i. Normally, to balance the DC-link capacitor voltage V_{dci} of each converter, P_i^* is constructed as follows

$$P_i^* = (k_p + \frac{k_I}{s})(V_{dci} - V_{dci,ref}) \tag{12.3}$$

where k_p and k_I are the proportional-integral (PI) coefficients. $V_{dci,ref}$ is the dc-link voltage reference for each module. Specially, for single-stage series PV micro-inverters, $V_{dci,ref}$ is usually determined by maximum power point tracking (MPPT) algorithm [10]. And for two-stages series power conversion units, $V_{dci,ref}$ is assumed as a constant value to balance the DC-link capacitor voltage [11].

12.2 Performance Discussion and Comparison

12.2.1 Steady-State Analysis and Comparison with Existing Methods

According to the system physical topology of Fig. 12.2, m parallel CCIs have a same output voltage $(u_1 = u_2 = \cdots = u_m)$. Due to the current-phase control of CCIs in (12.1), the same current-phase θ_{Ii} of CCI#i is obtained. Some equations are given by

$$\begin{cases} V_1 = V_2 = \cdots = V_m;\ \theta_1 = \theta_2 = \cdots = \theta_m \\ I_{Ig} = I_{I1} + I_{I2} + \cdots + I_{Im}; \\ \theta_{I1} = \theta_{I2} = \cdots = \theta_{Im} = \theta_{Ig} = (\theta_p - \varphi^*) \end{cases} \tag{12.4}$$

Due to the zero steady-state error of PI control in (12.2), the PF angles of $n - m$ VCIs would be identical ($\varphi_{m+1} = \varphi_{m+2} = \cdots = \varphi_n = \varphi^*$). Then, by combining (12.1)–(12.3), same voltage phase θ_i of all inverters is obtained due to the sharing grid current and the same power factor angle.

$$\theta_p = \theta_1 = \cdots = \theta_m = \theta_{m+1} = \cdots = \theta_n = (\theta_{Ig} + \varphi^*) \tag{12.5}$$

Then, due to the same PF angle φ^* and same voltage amplitude V_i of m parallel CCIs, the real power flow $P_i^* = V_i I_i \cos\varphi^*(i = 1,\ldots,m)$ is proportional to the current amplitude I_i in the steady state.

$$P_1^* : P_2^* : \cdots : P_m^* = I_1 : I_2 : \cdots : I_m;\ For CCIs \tag{12.6}$$

Then, due to the same PF angle φ^* and same grid current I_g of $n - m$ VCIs, the real power $P_i^* = V_i I_g \cos\varphi^*(i = m + 1,\ldots,n)$ and reactive power $Q_i^* = V_i I_g \sin\varphi^*(i = m + 1,\ldots,n)$ are proportional to the voltage amplitude Vi in the steady state.

$$\begin{cases} P_{m+1}^* : P_{m+2}^* : \cdots : P_n^* = V_{m+1} : V_{m+2} : \cdots : V_n \\ Q_{m+1}^* : Q_{m+2}^* : \cdots : Q_n^* = V_{m+1} : V_{m+2} : \cdots : V_n \end{cases};\ For VCIs \tag{12.7}$$

By combing (12.4)–(12.7), the steady-state equations of all inverters are built as follows

$$\begin{cases} I_{Ig} = I_{I1} + I_{I2} + \cdots + I_{Im} \\ \theta_{Ig} = \theta_{I1} = \theta_{I2} = \cdots = \theta_{Im} \\ I_1 : I_2 : \cdots : I_m = P_1^* : P_2^* : \cdots : P_m^* \end{cases} \tag{12.8}$$

$$\begin{cases} V_1 = V_2 = \cdots = V_m \\ V_{1,2\ldots,m} : V_{m+1} : \cdots : V_n = \left(\sum_{i=1}^{m} P_i^*\right) : P_{m+1}^* : \cdots : P_n^* \\ \theta_1 = \cdots = \theta_m = \theta_{m+1} = \cdots = \theta_n = (\theta_{Ig} + \varphi^*) \end{cases} \tag{12.9}$$

From above analysis, the parallel CCIs can be treated as an equivalent large-capacity CCI unit in steady-state analysis. Thus, we assume $m = 1$ to clearly present

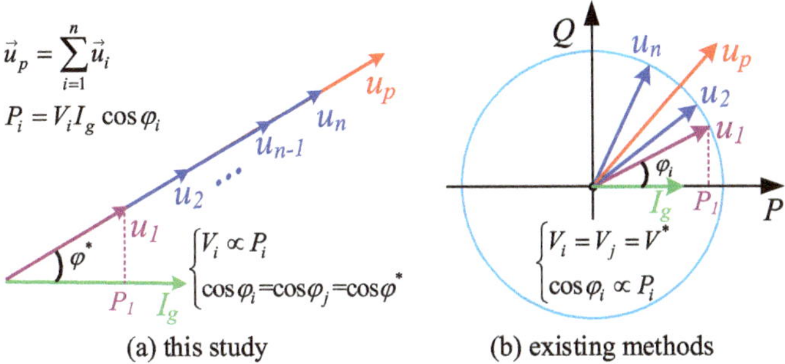

Fig. 12.4 Comparison of steady-state voltage phasors. (**a**) "Varied-amplitude & fixed-phase" voltage control in this study, (**b**) "fixed-amplitude & varied-phase" voltage control in [12]

the steady-state voltage-phasor diagram in Fig. 12.4a, which shows that all inverters have a same voltage phase, and the voltage amplitude V_i is proportional to the real power flow P_i.

It is worth noting that the voltage control of VCIs in this study is totally different from that in [12]. In the voltage control of [12], the voltage amplitude reference is fixed and the output power regulation is varied with voltage phase (herein called "fixed-amplitude & varied-phase" in Fig. 12.4b), which is just feasible for fully modular inverters with equal power capacity. In this study, the voltage amplitude is varied according to the source power of each inverter, and the voltage phase synchronization is automatically realized by power factor angle consistency (herein called 'varied-amplitude & fixed-phase'). Thus, the method not only can be used for fully modular inverters, but also series inverters with seriously unbalanced power sources are applicable.

12.2.2 Synchronization Mechanism

As the grid current is shared by all VCIs, the common current-phase component is used as a natural synchronization carrier. That is, the voltage phase synchronization of n-m VCIs can be achieved by power factor angle consistency. To help understanding the PF consistency control, a synchronization mechanism is analyzed in Fig. 12.5, where the equivalent circuit includes one CCI and two VCIs. For simplicity, unity PF operation is assumed by setting $\varphi^* = 0$.

In Fig. 12.5b, u_i (blue phasor) leads the steady-state grid current (green phasor); while u_j (red phasor) lags it. Initially, $\varphi_i > 0, \varphi_j < 0$. Then, $\omega_i < \omega^* < \omega_j$ is obtained from (12.2). As a result, φ_i decreases ($\Delta\omega_i = \omega_i - \omega^* < 0$), while φ_j increases ($\Delta\omega_i = \omega_j - \omega^* > 0$). The convergence process will continue until $\varphi_i = \varphi_j = \varphi^* = 0$, and $\theta_i = \theta_j = (\theta_{Ig} + \varphi^*)$.

It is worth noting that the synchronization mechanism of inverters is totally different from that in [2, 10, 13]. In the voltage phase synchronization of [2, 10, 13],

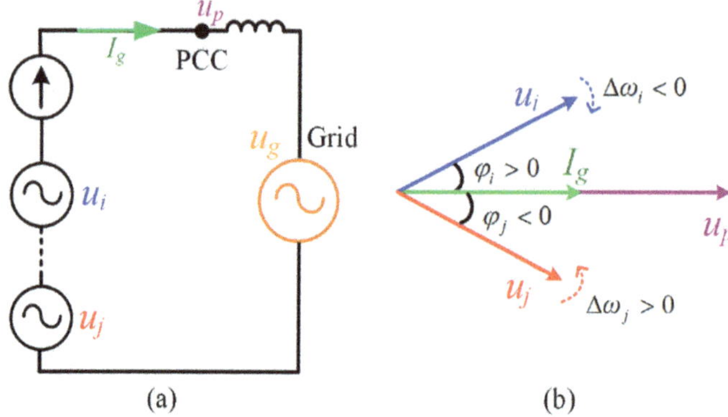

Fig. 12.5 A series inverter system for synchronization mechanism analysis. (**a**) Equivalent circuit and (**b**) phasor diagram

all inverters rely on the global voltage phase signal of utility grid. The acquired AC phase signal will be delayed and increase cost in long distributed generation, which limits its performances and application scope [1]. In this study, the current-phase is a local signal for each inverter, which is convenient to acquire. Thus, the method can realize the self-synchronization via the natural synchronization carrier of current-phase component.

Moreover, the hybrid voltage/current control scheme focuses on the grid-connected mode, whose dynamic model and operation mechanism are different from islanded mode of [1, 8, 14]. In islanded mode, the main dynamic coupling of system characteristic is the "source-load" power interaction [14]. While for grid-connected mode, the dynamic characteristic is the "source-grid" power interaction [15].

12.2.3 Discussion of One-to-All-Failure Redundancy

Figure 12.6 presents a hybrid VCIs and CCIs system configuration with one-fault redundancy, which includes m CCIs and $n - m$ VCIs. To overcome a single point of failure, multiple CCIs are connected in parallel to jointly adapt to the grid variations ($m \approx 10\%n$ by considering 10% grid voltage variations, and $m \geqslant 2$). It is worth noting that each CCI is independently controlled by local control as shown in Fig. 12.2 rather than a central controller. If one CCI unit fails, it should be disconnected, and the rest $m - 1$ CCIs operate well. Moreover, if a failure occurs in one VCI unit, the bypass switch should turn ON to separate it. In this way, the whole system has a redundancy to one-fault [16].

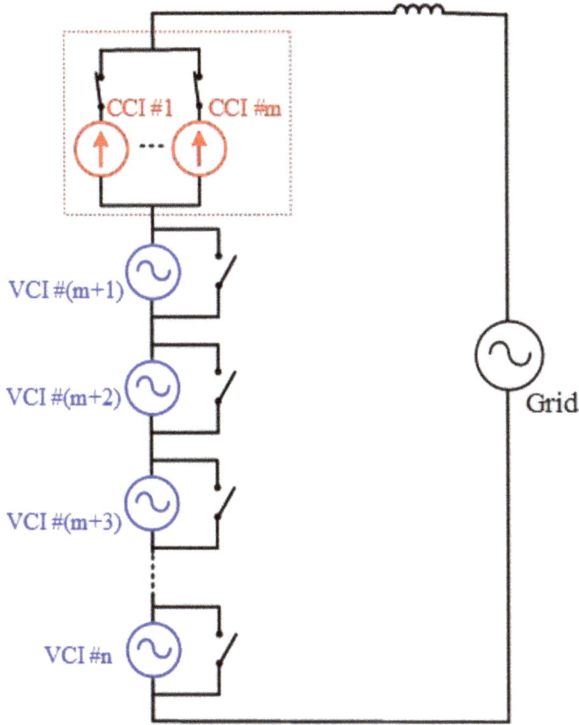

Fig. 12.6 Hybrid VCIs and CCIs system configuration with one-fault redundancy

12.3 Experimental Results

To test the effectiveness of the hybrid voltage/current control method, a grid-connected system has been built in the lab. The electrical waveforms are recorded by the oscilloscope. The system parameters are presented in Table 12.1.

In case 1–6, a system with three inverters is tested. Inverter#1 is chosen as a CCI. Inverter#2 and inverter#3 are VCIs. To further verify the one-fault redundancy, a system with four inverters is tested in case 7–8, where inverter#1 and inverter#2 are chosen as CCIs, and inverte#3 and inverter#4 are VCIs. The equivalent circuit model is shown in Fig. 12.7.

12.3.1 Case 1: Source Power Change Under Unity PF

This case aims to verify the independent power regulation capability of each inverter by considering unbalanced DG-source capacities. Figure 12.8 shows experimental

Table 12.1 Experiment parameters

Item	Symbol	Value
Rated grid voltage	V_g^*	311 V
Rated grid angular frequency	ω^*	100 π rad/s
Grid line impedance	Z_{line}	$0.1 + j1.0\Omega$
Series number	N	3/4
Inverter switching frequency	f_{PWM}	10 kHz
PI gains of amplitude-control	K_{Pvi}/K_{Ivi}	0.2/0.6
PI gains of phase-control	$K_{P\omega i}/K_{I\omega i}$	2/0.2
PI gains of DC-link capacitor	k_P/k_I	10/30
DC-link voltage reference	$V_{dci,ref}$	150 V
PF angle reference	φ^*	$0(case1, 3-8)$ $0.128\pi(case2)$

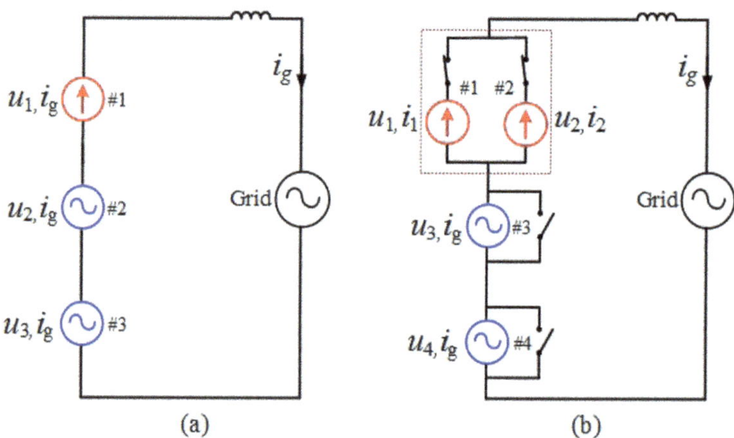

Fig. 12.7 Equivalent circuit of system model. (**a**) Case 1–6 and (**b**) case 7–8

results. The available powers of the three inverters are changed from same values to different values at $t = 0.6$ s.

Before $t = 0.6$ s, output voltages u_1, u_2, u_3 of the three inverters are identical in Fig. 12.8b, and the output real powers P_1, P_2, P_3 are equal to 1.5 kW in Fig. 12.8c. After $t = 0.6$ s, P_2 changes from 1.5 to 1.3 kW, P_3 changes to 1.1 kW, while P_1 is unchanged. As the unity PF is predesigned, the output reactive powers are always zero in Fig. 12.8d. From the steady-state voltage in Fig. 12.8a, output voltages u_1, u_2, u_3 have the same phase than the grid current i_g, which reveals that a predesigned unity PF is realized. Meanwhile, the voltage amplitude V_i is proportional to output real power P_i^* in steady state, which verifies the feasibility of the "varied-amplitude & fixed-phase" voltage control in of Section 1. Moreover, DC-link capacitor voltages of three modules are presented in Fig. 12.8e. The DC-link capacitor voltage of each module is balanced at the desired value 150 V and

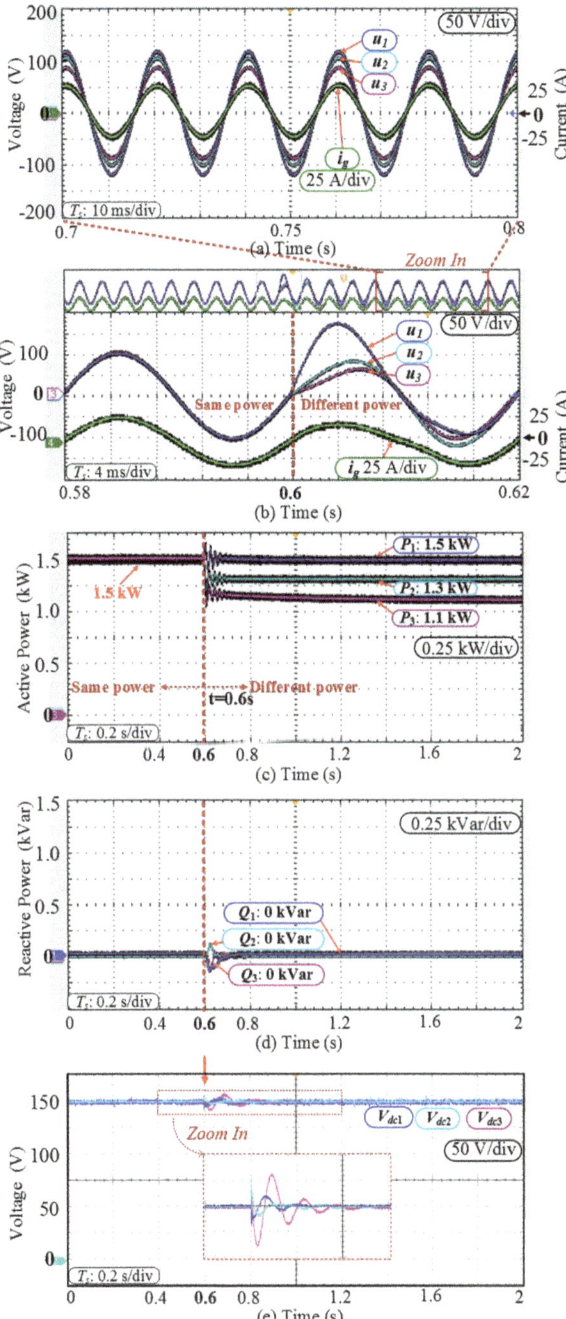

Fig. 12.8 Experiment results of Case 1. (**a**) Steady-state voltage/current, (**b**) three output voltages u_1 u_3, grid current i_g, (**c**) three output real powers P_1 P_3, (**d**) reactive powers Q_1 Q_3, and (**e**) DC capacitor voltage V_{dc1} V_{dc3}

has a satisfactory dynamic response. From Case 1, the method can work in the unbalanced source power conditions and achieve unity PF.

12.3.2 Case 2: Grid Voltage Sag Under No-Unity PF

This case study aims to test the performances of the control under no-unity PF and grid voltage sag. The experimental results of case 2 are shown in Fig. 12.9.

Different from case 1, this case adopts a no-unity PF ($\cos\varphi^* = 0.92$) to show the PF controllability of the method. Under this condition, each inverter could provide reactive power compensation to the utility grid. From the steady-state real/reactive power values in Fig. 12.9b and c, the same power factor angle ($\tan\varphi_i = Q_i/P_i = 0.426$) is ensured for the three inverters. That is, a same no-unity PF $= 0.92$ is realized to ensure power factor angle consistency.

Moreover, to demonstrate the effectiveness of the strategy under grid contingencies, a 15% grid voltage sag is imposed. Before $t = 1$ s, output voltages u_1, u_2, u_3 of the three inverters have the same phase, and the voltage amplitude is proportional to the real power outputs. At $t = 1$ s, a grid voltage sag occurs. To compensate for the grid voltage sag, output voltages u_1, u_2, u_3 react immediately in Fig. 12.9a. After about two cycles, u_1, u_2, u_3 reach the new steady states, and the grid current amplitude I_g increases to guarantee the unchanged real power output. As shown in Fig. 12.9b and c, the real/reactive powers have satisfactory dynamic responses.

In summary, the results of this case indicate that the method achieves a flexible PF regulation and has a disturbance rejection to grid voltage fluctuation.

12.3.3 Case 3: Large Source Power Gap Among Some Inverters

Compared with case 1, a large source power gap is set between VCI#2 and VCI#3 as shown in Fig. 12.10. Before $t = 0.8$ s, the output real powers P_1, P_2, P_3 of the three inverters are equal to 1.2 kW in Fig. 12.10a, and output voltages u_1, u_2, u_3 are identical in Fig. 12.10b. After $t = 0.8$ s, P_2 increases from 1.2 to 1.4 kW, and P_3 rapidly decreases to 0.6 kW, while P_1 is unchanged.

From Fig. 12.10, the active power and reactive power can be quickly adjusted and three output voltages have a fast response. As a result, independent power control for three inverters are achieved under a large, unbalanced source power gap.

12.3.4 Case 4: Grid Frequency Deviation

In this case, a grid frequency deviation from 50 to 49.8 Hz is considered to verify the effectiveness of the hybrid voltage/current control scheme as shown in

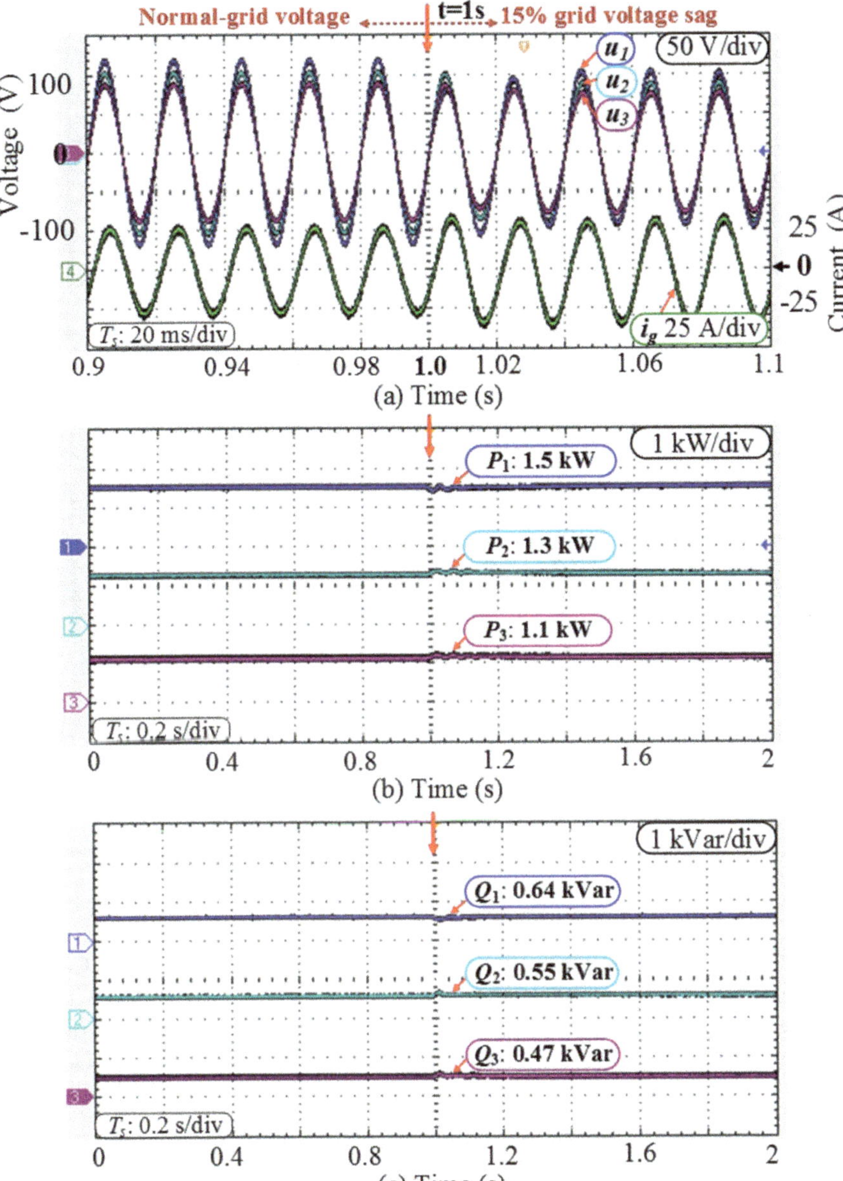

Fig. 12.9 Experimental results of Case 2. (**a**) Steady-state voltage/current, (**b**) three output voltages u_1 u_3, grid current i_g, (**c**) three output real powers P_1 P_3, (**d**) reactive powers Q_1 Q_3, and (**e**) DC capacitor voltage V_{dc1} V_{dc3}

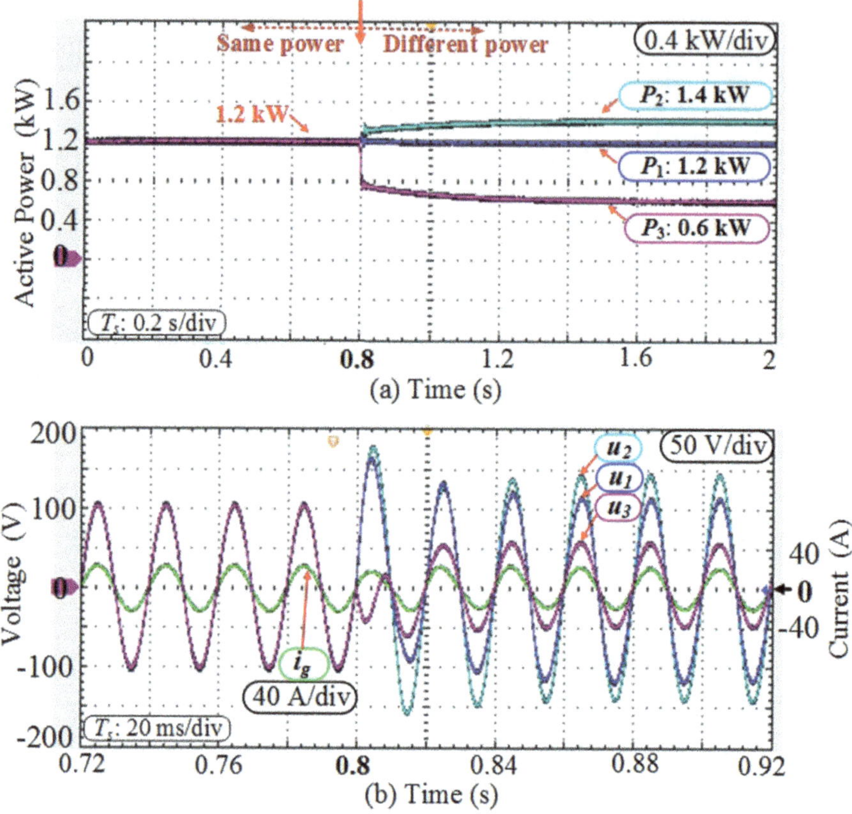

Fig. 12.10 Experimental results of Case 3. (**a**) Three output real powers P_1 P_3, (**b**) three output voltages u_1 u_3, and grid current i_g

Fig. 12.11a. For the dynamic response, only the output voltage u1 of CCI#1 is slightly affected due to the change of grid frequency in Fig. 12.11b. Three inverters can always maintain the synchronization operation with the grid under the frequency disturbance. The output real powers P_1 P_3 are presented in Fig. 12.11c. As seen, the hybrid voltage/current control method is robust against the grid frequency deviation.

12.3.5 *Case 5: Grid Harmonics Condition*

In this case, a distorted grid voltage condition is conducted. Figure 12.12 shows the experimental results. A grid harmonic (THD $= 6.02\%$) is imposed at $t = 1$ s in Fig. 12.12a. As seen from Fig. 12.12a–b, the grid current is unaffected because the CCI#1 just acquires the gird fundamental voltage synchronization signal to regulate the grid current. VCI#2 and VCI#3 achieve voltage synchronization via the sharing

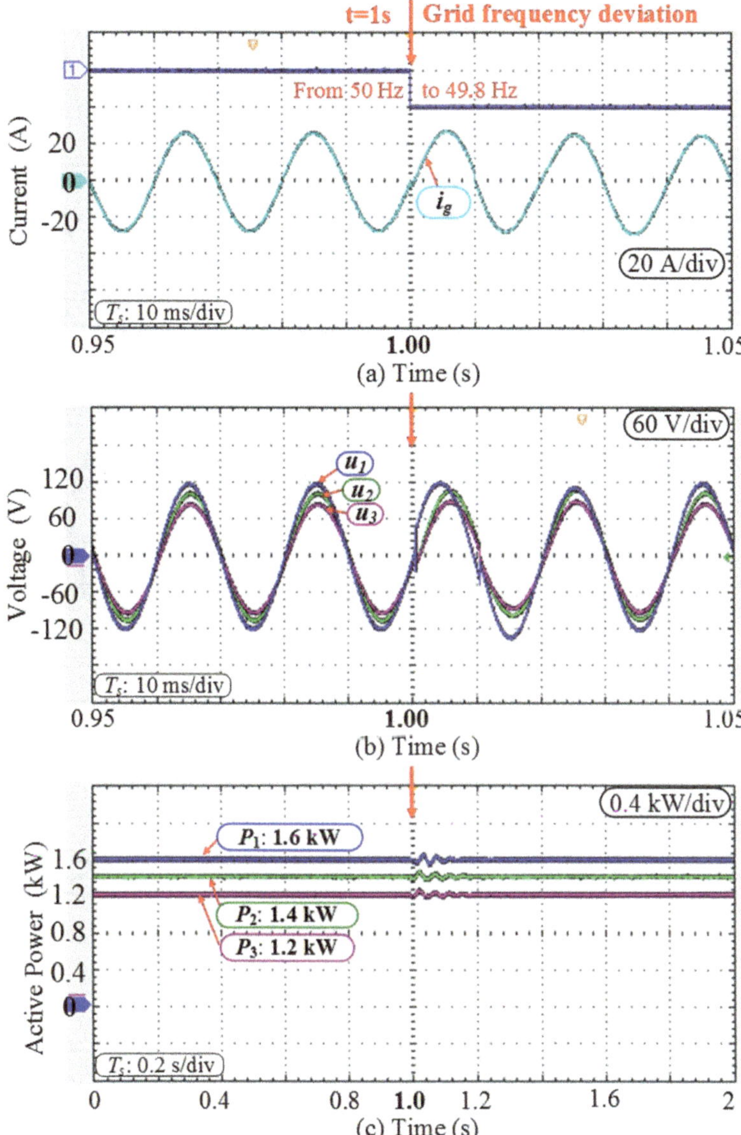

Fig. 12.11 Experimental results of Case 4. (**a**) Grid frequency and grid current i_g, (**b**) three output voltages u_1 u_3, and (**c**) output real powers P_1 P_3

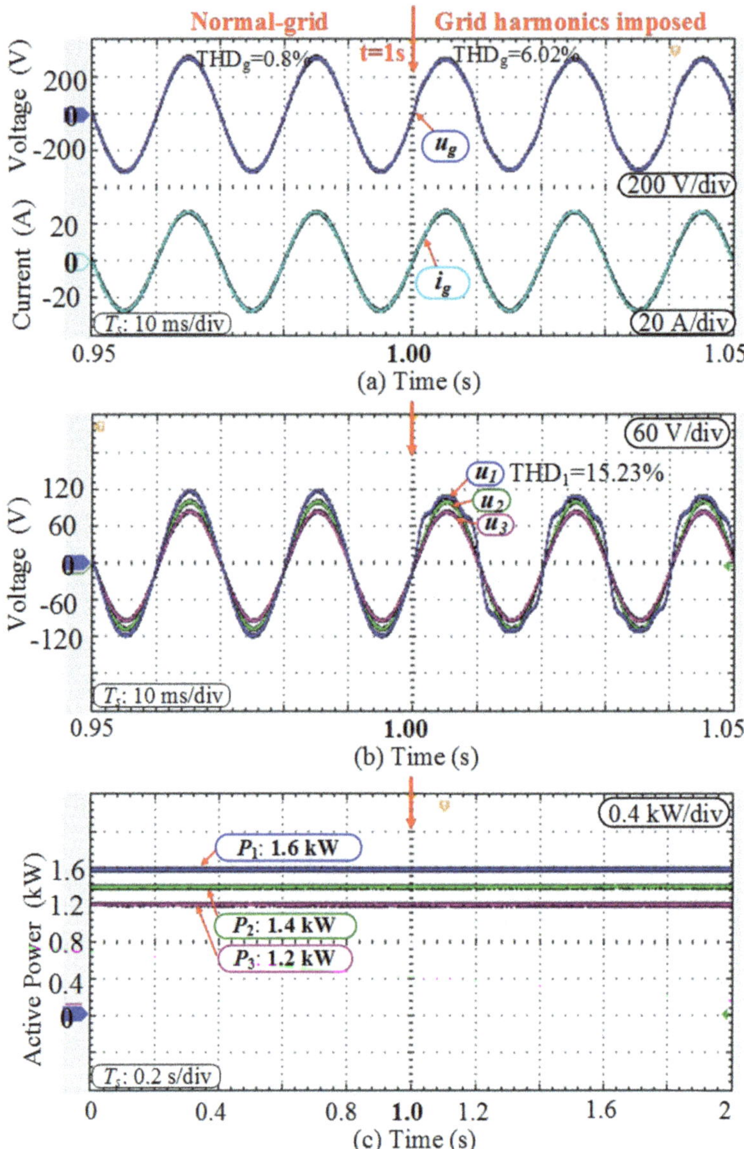

Fig. 12.12 Experiment results of Case 5. (**a**) Grid frequency and grid current i_g, (**b**) three output voltages u_1 u_3, and (**c**) output real powers P_1 P_3

current-phase signal, and then the output voltages of two VCIs are also unaffected. Meanwhile, the CCI#1 suffers the grid voltage harmonics in Fig. 12.12b. The output real powers P_1 P_3 are shown in Fig. 12.12c. As seen, the hybrid voltage/current method is suitable for the grid distortion.

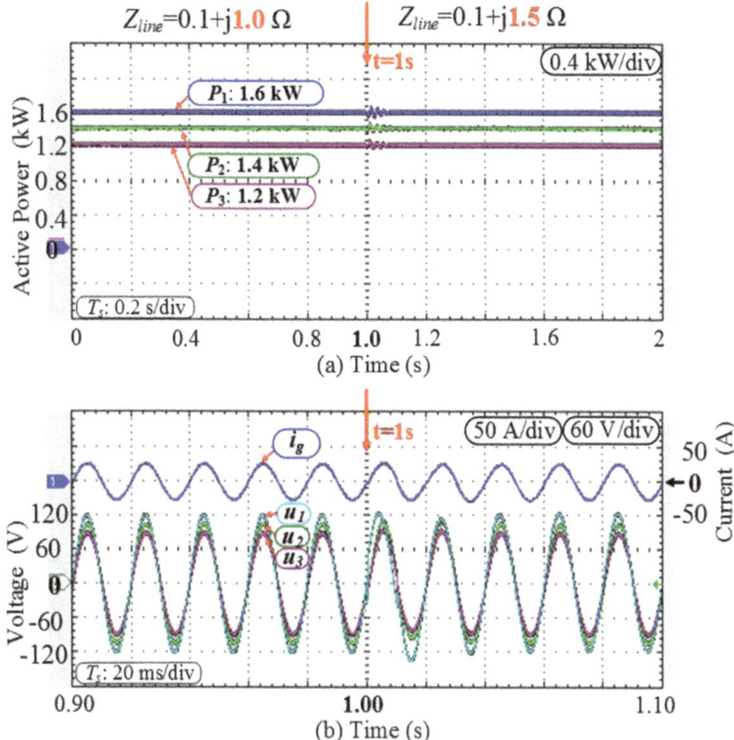

Fig. 12.13 Experimental results of Case 6. (**a**) Output real powers P_1 P_3 and (**b**) three output voltages $u1$ $u3$ and grid current i_g

12.3.6 Case 6: Grid Impedance Variation

To test the effect of grid line impedance variation, an experimental case 6 is performed as shown in Fig. 12.13. The line impedance Z_{line} changes from $0.1 + j1.0\Omega$ to $0.1 + j1.5\Omega$ at $t = 1$ s. From the steady-state results, the active powers of three modules are almost unaffected, and the dynamic response is satisfactory under a line impedance increase. Clearly, the control scheme is feasible for the grid impedance variation.

12.3.7 Case 7: One CCI Unit Fault Redundancy

To further verify the single-point-of-failure redundancy, a system with four inverters is carried out in case 7–8. As shown in Fig. 12.14, inverter#1 and inverter#2 are chosen as CCIs, and inverter#3 and inverter#4 are VCIs. According to the steady-

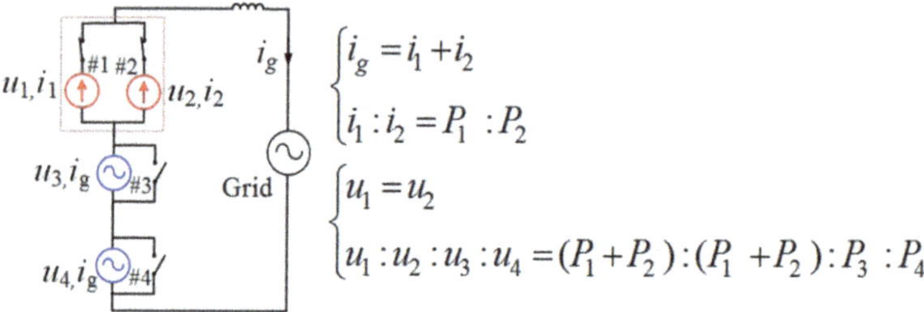

Fig. 12.14 Equivalent circuit of system model and derived equations in case 7–8

state analysis in (12.8)–(12.9), two CCIs can be treated as one equivalent large-capacity CCI unit, and some equations are derived in Fig. 12.14.

Figure 12.15 shows experimental results under CCI#1 failure at $t = 0.4$ s. From Fig. 12.15a, the output real power of CCI#1 changes from 0.6 to 0 kW at $t = 0.4$ s. The clear steady-state results before/after 0.4 s are presented in (12.10), which verifies the above steady-state analysis in (12.8)–(12.9).

Specially, the system can still work well after disconnecting the CCI#1 at $t = 0.4$ s. And the CCI#2 solely regulates the grid current in Fig. 12.15c. Meanwhile, u_2, u_3, u_4 reach the new steady states after about five cycles ($u_1 = 0$) in Fig. 12.15b. As seen, the results indicate that the method achieves a one CCI unit fault redundancy.

$$\begin{cases} t \in [0,\ 0.4]s \begin{cases} P_1 : P_2 : P_3 : P_4 = 0.6 : 1.0 : 1.2 : 1.4 \\ u_1 : u_2 : u_3 : u_4 = 1.6 : 1.6 : 1.2 : 1.4 \\ i_1 : i_2 : i_g == 0.6 : 1.0 : 1.6 \end{cases} \\ t \in [0.4,\ 2]s \begin{cases} P_1 : P_2 : P_3 : P_4 = 0 : 1.0 : 1.2 : 1.4 \\ u_1 : u_2 : u_3 : u_4 = 0 : 1.0 : 1.2 : 1.4 \\ i_1 : i_2 : i_g = 0 : 1.0 : 1.0 \end{cases} \end{cases} \tag{12.10}$$

12.3.8 Case 8: One VCI Unit Fault Redundancy

Figure 12.16 shows experimental results under VCI#3 failure at $t = 0.6$ s. From Fig. 12.16a, the output real power of VCI#3 changes from 1.2 to 0 kW at $t = 0.6$ s. The clear steady-state results before/after 0.6 s are presented in (12.11), which verifies the above steady-state analysis in (12.8)–(12.9). Figure 12.16 reveals that the system can still work well after bypassing the VCI#3 at $t = 0.6$ s. In Fig. 12.16a, each inverter can still achieve independent power regulation. Moreover, as the VCI#3 back out the operation, the rest inverters automatically increase their voltage

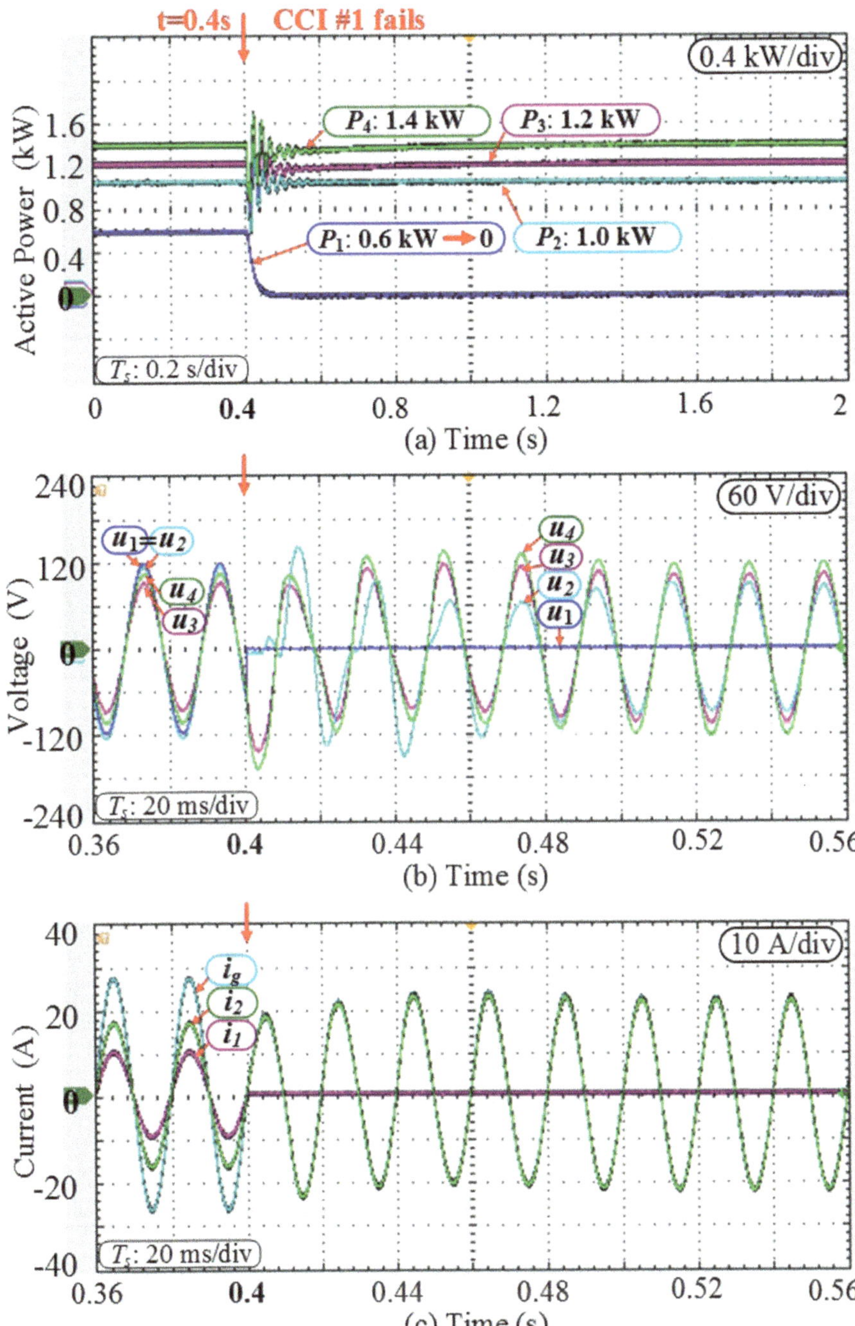

Fig. 12.15 Experimental results of case 7. (**a**) Four output real powers P_1 P_4, (**b**) four output voltages u_1 u_4, and (**c**) two output current i_1 i_2 and grid current i_g

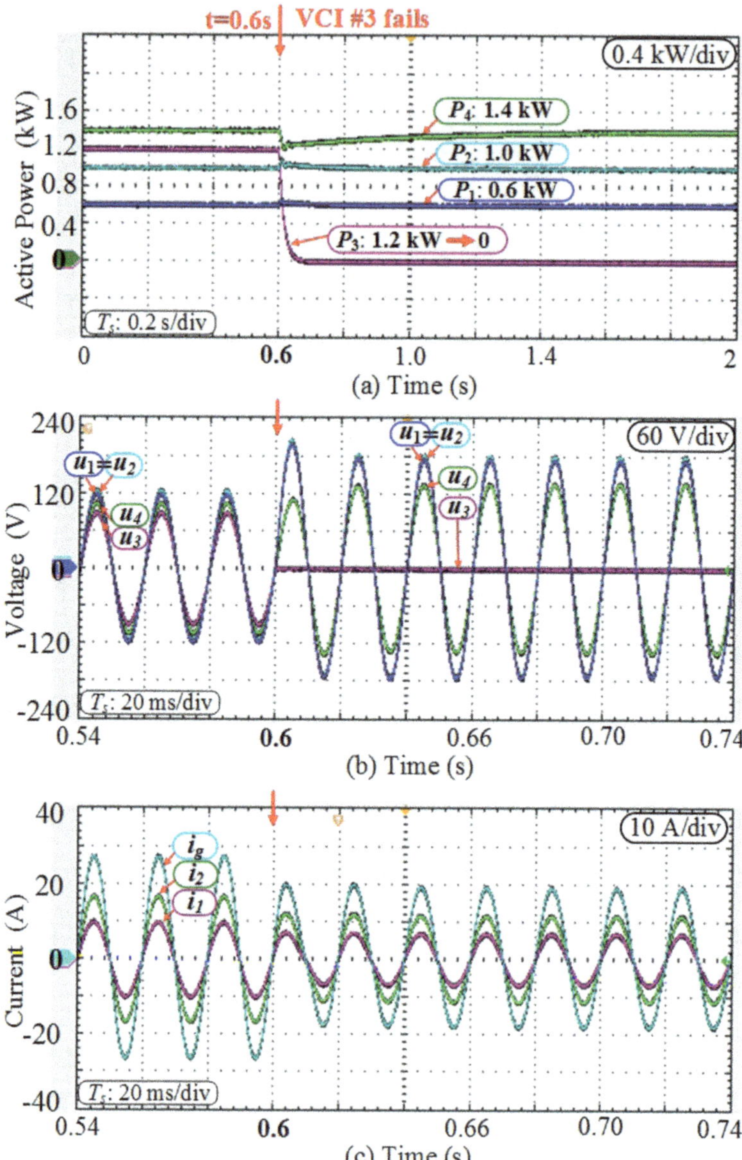

Fig. 12.16 Experimental results of case 8. (**a**) Four output real powers P_1 P_4, (**b**) four output voltages u_1 u_4, and (**c**) two output current i_1 i_2 and grid current i_g

amplitudes to meet the grid voltage requirement in Fig. 12.16b. Meanwhile, due to the decreased grid-injected-power, the grid current is decreased proportionally in Fig. 12.16c. For the dynamic response, u_1, u_2, u_4 reach the new steady states after about four cycles ($u_3 = 0$) in Fig. 12.16b. As seen, this case shows that the method achieves a one VCI unit fault redundancy.

$$\begin{cases} t \in [0,\ 0.6]s \begin{cases} P_1 : P_2 : P_3 : P_4 = 0.6 : 1.0 : 1.2 : 1.4 \\ u_1 : u_2 : u_3 : u_4 = 1.6 : 1.6 : 1.2 : 1.4 \\ i_1 : i_2 : i_g == 0.6 : 1.0 : 1.6 \end{cases} \\ t \in [0.6,\ 2]s \begin{cases} P_1 : P_2 : P_3 : P_4 = 0.6 : 1.0 : 0 : 1.4 \\ u_1 : u_2 : u_3 : u_4 = 1.6 : 1.6 : 0 : 1.4 \\ i_1 : i_2 : i_g = 0.6 : 1.0 : 1.6 \end{cases} \end{cases} \tag{12.11}$$

12.4 Conclusion

To reduce the global communication dependency, this chapter introduces a novel hybrid voltage/current control scheme for series inverters in a decentralized manner. Several CCIs are introduced to jointly adapt the grid variations and take charge of the grid current regulation. For rest VCIs, the voltage regulation adopts a "varied-amplitude & fixed-phase" voltage control to achieve voltage phase self-synchronization and independent power control. The "varied-amplitude" implies that the voltage amplitude is varied according to the available source power of each inverter. The "fixed-phase" refers to a voltage phase consistency of all VCIs by enabling same power factor angle and same grid-current-phase.

Compared with the existing methods, only a few CCI units near the grid acquire the grid phase-angle signal and the most rest inverter units are local self-synchronization based on natural common grid-current-phase signal rather than relying on global real-time grid-phase signal. Thus, the control can be realized with very low communication burden, which is more cost-effective and higher reliability in terms of communication faults. In future work, the method will be expanded to the series PV-storage hybrid systems, in which storages are chosen as CCIs to participate the grid ancillary services and PVs are designed as VCIs to achieve the maximum harvesting of renewable energy.

References

1. J. He, Y. Li, B. Liang, C. Wang, Inverse power factor droop control for decentralized power sharing in series-connected-microconverters-based Islanding microgrids. IEEE Trans. Ind. Electron. **64**, 7444–7454 (2017)
2. L. Zhang, K. Sun, Y.W. Li, X. Lu, J. Zhao, A distributed power control of series-connected module-integrated inverters for PV grid-tied applications. IEEE Trans. Power Electron. **33**, 7698–7707 (2018)
3. S. Kouro, M. Malinowski, K. Gopakumar, J. Pou, L.G. Franquelo, W. Bin, et al., Recent advances and industrial applications of multilevel converters. IEEE Trans. Ind. Electron. **57**, 2553–2580 (2010)
4. Y. Yu, G. Konstantinou, B. Hredzak, V.G. Agelidis, Power balance optimization of cascaded H-bridge multilevel converters for large-scale photovoltaic integration. IEEE Trans. Power Electron. **31**, 1108–1120 (2016)
5. Q. Zhang, K. Sun, A flexible power control for PV-battery hybrid system using cascaded H-bridge converters. IEEE J. Emerg. Select. Topics Power Electron. **7**, 2184–2195 (2019)
6. C. Wang, L. Zhang, L. Xiao, F. Wu, X. Zheng, H. Jiang, Analysis and suppression of the frequency-decrease effect in the capacitor voltage related to the low modulation frequency ratio in an MMC system. IEEE Trans. Power Electron. **35**, 9119–9132 (2020)
7. IEEE standard for interconnection and interoperability of distributed energy resources with associated electric power systems interfaces. IEEE Standard 1547–2018 (2018)
8. J. He, X. Liu, C. Mu, C. Wang, Hierarchical control of series-connected string converter-based Islanded electrical power system. IEEE Trans. Power Electron. **35**, 359–372 (2020)
9. L. Liu, H. Li, Y. Xue, W. Liu, Reactive power compensation and optimization strategy for grid-interactive cascaded photovoltaic systems. IEEE Trans. Power Electron. **30**(1), 188–202 (2015)
10. H. Jafarian, R. Cox, J.H. Enslin, S. Bhowmik, B. Parkhideh, Decentralized active and reactive power control for an AC-stacked PV inverter with singlemember phase compensation. IEEE Trans. Ind. Appl. **54**(1), 345–355 (2018)
11. H. Liu, P.C. Loh, X. Wang, Y. Yang, W. Wang, D. Xu, Droop control with improved disturbance adaption for a PV system with two power conversion stages. IEEE Trans. Ind. Electron. **63**(10), 6073–6085 (2016)
12. X. Hou, Y. Sun, X. Zhang, G. Zhang, J. Lu, F. Blaabjerg, A self-synchronized decentralized control for series-connected H-bridge rectifiers. IEEE Trans. Power Electron. **34**, 7136–7142 (2019)
13. P. Wu, Y. Su, J. Shie, P. Cheng, A distributed control technique for the multilevel cascaded converter. IEEE Trans. Ind. Appl. **55**(2), 1649–1657 (2019)
14. Y. Sun, G.Z. Shi, X. Li, W.B. Yuan, M. Su, H. Han, et al., An f-P/Q droop control in cascaded-type microgrid. IEEE Trans. Power Syst. **33**, 1136–1138 (2018)
15. J. Sun, Impedance-based stability criterion for grid-connected inverters. IEEE Trans. Power Electron. **26**(11), 3075–3078 (2011)
16. X. Hou, K. Sun, X.G. Zhang, Y. Sun, J. Lu, A hybrid voltage/current control scheme with low-communication burden for grid-connected series-type inverters in decentralised manner. IEEE Trans. Power Electron. **37**, 920–931 (2021). https://doi.org/10.1109/TPEL.2021.3093080

Chapter 13
Unified Grid-Connected and Islanded Operation

13.1 Equivalent Models of CMG and Its Control

13.1.1 Equivalent Models of CMG

The structure of the CMG consisting of DG units is shown in Fig. 13.1 [1, 2]. The CMG can operate in grid- connected or islanded mode by switching the static transfer switch (STS). The equivalent circuit of the series DGs is shown in Fig. 13.2. $|Z_{line}| \angle\theta_{line}$ is the generalized transmission impedance, $|Z_{line}| \angle\theta_{line} = \sum z_i$. In both medium and high voltage systems, the line impedance is mainly inductive, i.e., $\theta_{line} \approx \pi/2$.

In the islanded mode, the output active power P_i and reactive power Q_i of the i-th DG are derived as follows

$$P_i + jQ_i = V_i e^{j\delta_i}\left(\sum_{i=1}^{n} V_i e^{j\delta_i} \Big/ |Z'_{load}|\, e^{j\theta'_{load}}\right)^* \tag{13.1}$$

where V_i and δ_i represent the voltage and phase angle of the i-th unit. Z'_{load} and θ'_{load} are the impedance and impedance angle of the generalized load, which includes the generalized transmission line and load, respectively. Then, the power transmission characteristic in the islanded mode is given by,

$$P_i = \frac{V_i}{|Z'_{load}|}\sum_{j=1}^{n} V_j \cos\left(\delta_i - \delta_j + \theta'_{load}\right) \tag{13.2}$$

$$Q_i = \frac{V_i}{|Z'_{load}|}\sum_{j=1}^{n} V_j \sin\left(\delta_i - \delta_j + \theta'_{load}\right) \tag{13.3}$$

Y. Sun et al., *Series-Parallel Converter-Based Microgrids*, Power Systems,
https://doi.org/10.1007/978-3-030-91511-7_13

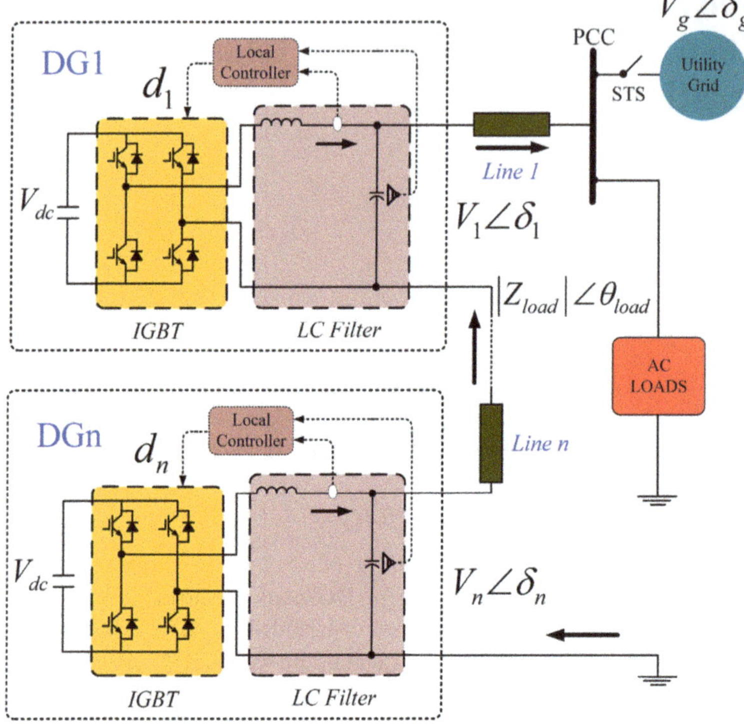

Fig. 13.1 Structure of the CMG

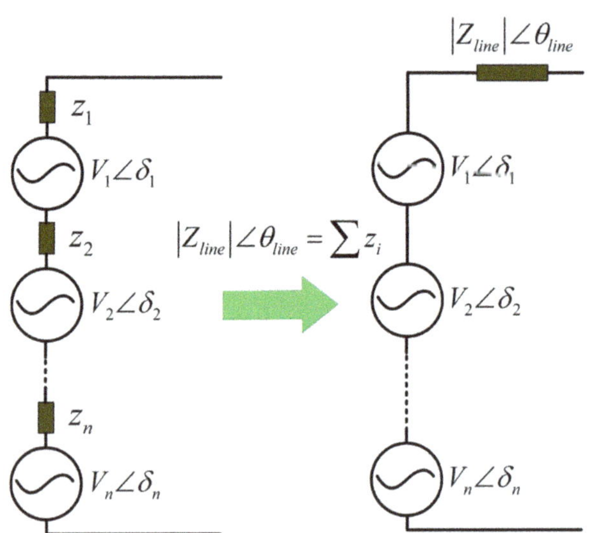

Fig. 13.2 Equivalent circuit of the series DGs

When the output voltages V_i and V_j are set to constant, the active power and reactive power regulation are mainly related to θ'_{load} in the islanded mode.

In the grid-connected mode, P_i and Q_i are expressed as,

$$P_i + jQ_i = V_i e^{j\delta_i} \left(\left(\sum_{i=1}^{n} V_i e^{j\delta_i} - V_g e^{j\delta_g} \right) \Big/ |Z_{line}| \, e^{j\theta_{line}} \right)^* \tag{13.4}$$

where V_g and δ_g represent the voltage amplitude and phase angle of the grid, respectively. The power transmission characteristic in the grid-connected mode is,

$$P_i = \frac{V_i}{|Z_{line}|} \left(V_g \sin\left(\delta_i - \delta_g\right) - \sum_{j=1}^{n} V_j \sin\left(\delta_i - \delta_j\right) \right) \tag{13.5}$$

$$Q_i = \frac{V_i}{|Z_{line}|} \left(\sum_{j=1}^{n} V_j \cos\left(\delta_i - \delta_j\right) - V_g \cos\left(\delta_i - \delta_g\right) \right) \tag{13.6}$$

As seen, the active power and reactive power could be regulated by the output voltage and the phase angle differences in the grid-connected mode.

13.1.2 Review of the Typical Control Schemes

In the islanded mode, the decentralized control scheme is introduced in [3]:

$$\omega_i = \omega^* + m \operatorname{sgn}\left(Q_i\right) P_i \tag{13.7}$$

$$\mathbf{v}_i = \begin{cases} V^* \sin\left(\int \omega_i dt \right), \; P_i \geqslant 0 \\ V^* \sin\left(\int \omega_i dt + \pi \right), \; P_i < 0 \end{cases} \tag{13.8}$$

where ω_i, ω^*, V^* are the angular frequency, the nominal angular frequency, and voltage amplitude, respectively. $\mathbf{v}_i$ is the voltage vector. m is a positive coefficient. sgn() is a signum function:

$$\operatorname{sgn}(x) = \begin{cases} 1, x > 0 \\ 0, x = 0 \\ -1, x < 0 \end{cases} \tag{13.9}$$

In (13.7), the adverse droop control is applied to feeding the RL load ($Q_i > 0$), while the droop control is for the RC load ($Q_i < 0$). Equation (13.8) is to achieve the unique equilibrium point.

In the grid-connected mode, the decentralized control scheme in is expressed as:

$$\begin{cases} \omega_i = \omega^* - mP_i \\ \mathbf{v}_i = V^* \sin\left(\int \omega_i dt\right) \end{cases} \tag{13.10}$$

The droop control is performed to realize the frequency synchronization of the system.

As seen, the adverse droop and droop control are applied in the islanded mode, and the droop control is used in the grid-connected mode. However, the methods in [3] and [4] cannot be served as a unified control scheme for the two modes under the RL and RC loads.

13.2 Unified Decentralized Control

The unified decentralized control strategy for grid-connected and islanded operation of the CMG is expressed as,

$$\omega_i = \omega^* + m\,\mathrm{sgn}\,(Q_i)\,\mathrm{sgn}\,(P_i)\,P_i \tag{13.11}$$

$$\mathbf{v}_i = \begin{cases} V^* \sin\left(\int \omega_i dt\right), \; P_i \geqslant 0 \\ V^* \sin\left(\int \omega_i dt - \left(\pi + 2\arctan\left(Q_i/P_i\right)\right)\right), \; P_i < 0 \end{cases} \tag{13.12}$$

Equation (13.12) is applied to guarantee the unique equilibrium point of system. The output voltage of the ith DG is regulated to track through the inner dual loops controller [5–7], and its control diagram is depicted in Fig. 13.3.

The control parameters in (13.11)–(13.12) are given as (13.13), which will be illustrated in Sect. 13.4 detailedly.

$$\begin{cases} 0 < m < (\omega_{\max} - \omega_{\min})/2P_{Rated} \\ \omega_g < \omega^* < \omega_{\max} \\ \varepsilon V_g/n \leqslant V^* < \cos\bar{\delta} V_g/n \end{cases} \tag{13.13}$$

where P_{Rated} is the rated active power, $\omega_{\max}$ and $\omega_{\min}$ are the allowable maximum and minimum angular frequency of CMG. ε is designed to ensure the load voltage quality (higher than its allowable minimum values in the islanded mode), which is

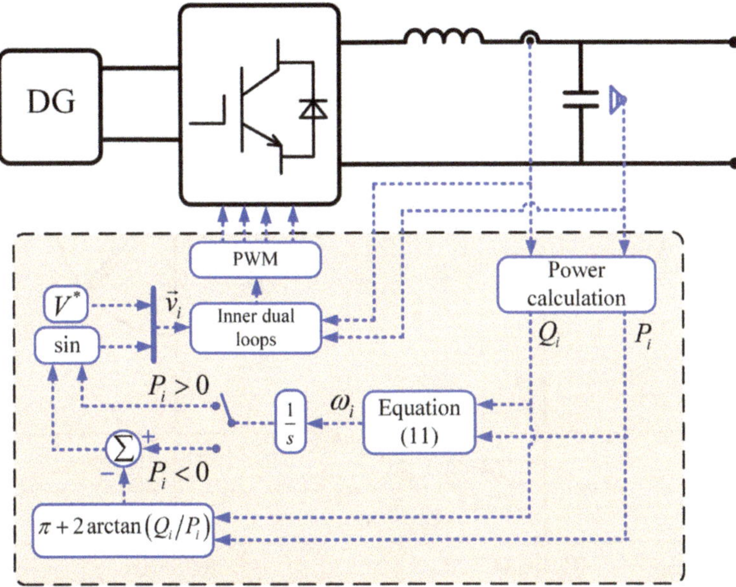

Fig. 13.3 Control diagram of the i-th DG

set at 0.95 in this chapter. The selecting of V^* is the tradeoff between the system stability and load voltage quality.

Compared to the method in [3], the terms $\operatorname{sgn}(P_i)$ and $\omega^* > \omega_g$ are performed to make the system work in the "droop." Then, it is applied to the grid-connected mode. Thus, the unified control for the two modes are realized. Clearly, the scheme (13.11)–(13.12) can be implemented with only local information, thus it is a decentralized control [8–10].

13.3 Stability Analysis

In this section, the small-signal stability near the steady-state operating point of the CMG will be investigated.

13.3.1 Steady-State Analysis

When the system is in the steady state, according to (13.11), we have

$$\operatorname{sgn}(Q_i)\,|P_i| = \operatorname{sgn}\left(Q_j\right)\left|P_j\right|,\ i \neq j \tag{13.14}$$

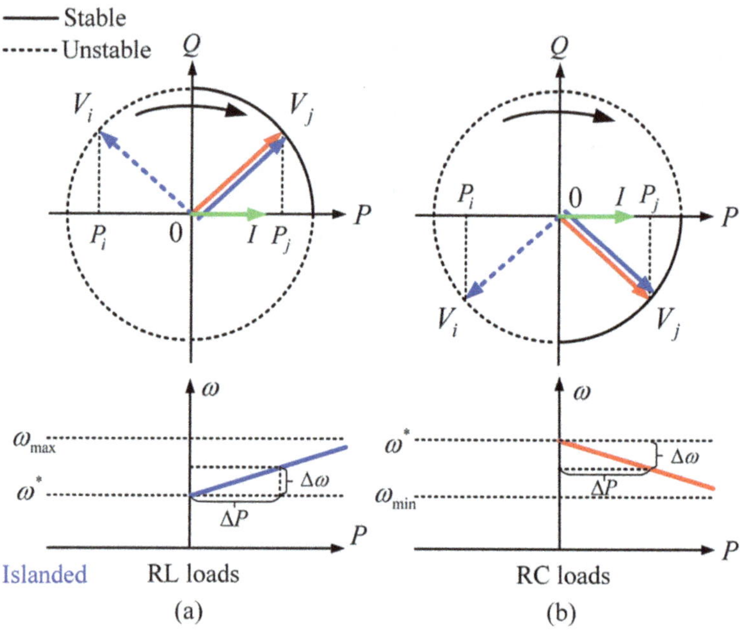

Fig. 13.4 Schematic of the uniqueness of steady-state operation point in the islanded mode (**a**) RL loads, (**b**) RC loads

As the output voltage amplitude and current of all DG units are the same, i.e., $V_i = V_j = V^*$ and $I_i = I_j$, (13.15) is satisfied.

$$|P_i + jQ_i| = |P_j + jQ_j| \tag{13.15}$$

Combine (13.14) and (13.15), if $|P_i| \neq 0$ and $|Q_i| \neq 0$, then

$$\begin{cases} P_i = P_j \ or \ P_i = -P_j \\ Q_i = Q_j \end{cases} \tag{13.16}$$

Thus, the reactive power can be shared accurately in CMG.

In the islanded mode, the undesired steady state $P_i = -P_j$ is removed through (13.12), which has been explained in [3], its detail is omitted here. The system always holds a unique equilibrium point $Q_i = Q_j$ and $P_i = P_j$ when feeding both RL and RC loads, which is depicted in Fig. 13.4a and b, respectively. In other words, $(\delta_i - \delta_j) = 0$ is always achieved in the steady state. From (13.2) and (13.3), we have,

$$P_i = n(V^*)^2 \cos\theta'_{load} \Big/ |Z'_{load}| \tag{13.17}$$

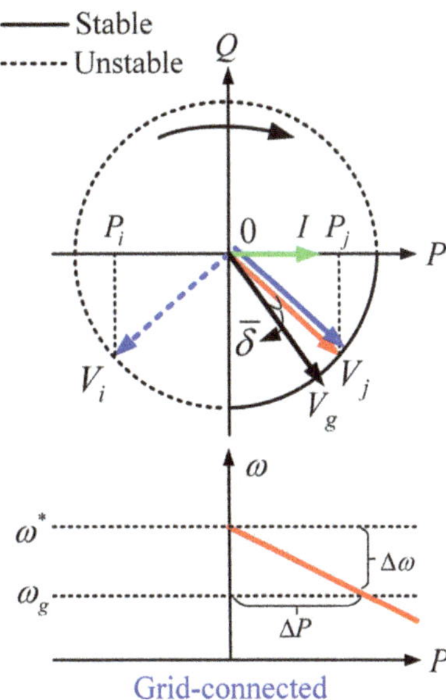

Fig. 13.5 Schematic of the uniqueness of steady-state operation point in the grid-connected mode

$$Q_i = n\left(V^*\right)^2 \sin\theta'_{load} \Big/ \left|Z'_{load}\right| \tag{13.18}$$

Assume that the reactive power losses of transmission line can be neglected. Since the active and reactive power land demands $\sum P_i = P_{load}$ and $\sum Q_i = Q_{load}$, then the unique steady-state operating point is expressed as,

$$\begin{cases} P_i = P_j = P_{load}/n \\ Q_i = Q_j = Q_{load}/n \end{cases} \tag{13.19}$$

In the grid-connected mode, assume that ω_g is the identical grid angular frequency, there is $\omega_i = \omega_g$. Equation (13.11) can be rewritten as,

$$\operatorname{sgn}(P_i)\, P_i = \operatorname{sgn}\left(P_j\right) P_j = \left(\omega_g - \omega^*\right)/m \operatorname{sgn}(Q_i) = \bar{P}^* \tag{13.20}$$

Similarly, the unique equilibrium point $Q_i = Q_j$ and $P_i = P_j = \bar{P}^*$ is obtained, which is depicted in Fig. 13.5.

Assume $\bar{\delta} = \delta_i - \delta_g$ in the steady state, from (13.5)–(13.6), we have,

$$P_i = V^* V_g \sin\bar{\delta} / |Z_{line}| \tag{13.21}$$

$$Q_i = V^* \left(nV^* - V_g \cos\bar{\delta}\right) / |Z_{line}| \tag{13.22}$$

Combine (13.22), then

Table 13.1 Characteristics of the strategy

Operation mode	Load	Characteristic
Islanded mode	RL load	Adverse droop control
Islanded mode	RC load	Droop control
Grid-connected mode	RL and RC load of unit	Droop control

$$nV^* - V_g \cos\bar{\delta} < 0 \tag{13.23}$$

Thus, $V^* < \cos\bar{\delta} V_g/n$ holds in (13.13). Then, $\cos\bar{\delta} > 0$, and $\bar{\delta} \in (-\pi/2, \pi/2)$. Combining $P_i = P_j = \bar{P}^*$, the obtained steady-state operating point of the system is $\bar{\delta} \in (0, \pi/2)$, i.e., $\bar{\delta} = \arcsin\left(\bar{P}^* |Z_{line}|/V^* V_g\right)$.

Based on the analysis above, the active power and reactive power can be shared accurately in both islanded and grid-connected mode. In addition, the characteristics of the control scheme (13.11)–(13.12) are summarized as Table 13.1, which is depicted in Figs. 13.4 and 13.5.

13.3.2 Small-Signal Stability Analysis

For the decentralized control, each DG makes decisions only based on their local information. Thus, the self-synchronization of phase angles is the most essential and crucial problem. Suppose that the inner voltage and current control loops can track the given reference voltage without the steady-state error.

In the islanded mode, assume δ_s is the steady-state synchronized phase angle and denote $\tilde{\delta}_i = \delta_i - \delta_s$, $\dot{\tilde{\delta}}_i = \omega_i$. Linearizing (13.2) and (13.11) near the equilibrium point (13.19), we have

$$\Delta P_i = -\frac{(V^*)^2}{|Z'_{load}|} \sin\left(\theta'_{load}\right) \sum_{j=1}^{n} \left(\Delta\tilde{\delta}_i - \Delta\tilde{\delta}_j\right) \tag{13.24}$$

$$\Delta\dot{\tilde{\delta}}_i = m \,\mathrm{sgn}\left(Q_{load}/n\right) \Delta P_i \tag{13.25}$$

Substituting (13.24) into (13.25) yields

$$\Delta\dot{\tilde{\delta}}_i = -m \,\mathrm{sgn}\left(Q_{load}/n\right) \left(V^*\right)^2 \sin\left(\theta'_{load}\right) \sum_{j=1}^{n} \left(\Delta\tilde{\delta}_i - \Delta\tilde{\delta}_j\right) \Big/ |Z'_{load}| \tag{13.26}$$

Rewriting (13.26) in the matrix form,

$$\dot{\mathbf{X}} = \mathbf{A}\mathbf{X} \tag{13.27}$$

where

$$\mathbf{X} = \left[\, \Delta\tilde{\delta}_1 \cdots \Delta\tilde{\delta}_n \,\right]^T \tag{13.28}$$

$$\mathbf{A} = -K_1 \left(n\mathbf{I}_{\mathbf{n}\times\mathbf{n}} - \mathbf{1}_{\mathbf{n}}\mathbf{1}_{\mathbf{n}}^{\mathbf{T}} \right) \tag{13.29}$$

$$K_1 = m \operatorname{sgn}\left(Q_{load}/n\right)\left(V^*\right)^2 \sin\left(\theta'_{load}\right) \Big/ \left|Z'_{load}\right| \tag{13.30}$$

From (13.29), the eigenvalues of **A** are expressed as

$$\lambda_1(\mathbf{A}) = 0, \lambda_2(\mathbf{A}) = \cdots = \lambda_n(\mathbf{A}) = -nK_1 \tag{13.31}$$

There is one zero eigenvalue, which is corresponding to rotational invariance of the dynamics as explained in [7]. As seen, $m > 0$ (see (13.13)), in case of RL or RC loads, $K_1 > 0$, thus, the system is stable. Therefore, the scheme is feasible for the islanded operation under RL and RC loads. If $\theta'_{load} = 0$ ($Q_i = 0$) or under no-load condition ($P_i = 0$ and $Q_i = 0$), from (13.11), $\omega_i = \omega^*$, the system operates in constant frequency, and it is stable.

In the grid-connected mode, similarly, linearization (13.5) and (13.11) near the equilibrium point yields,

$$\Delta P_i = \frac{V^*}{|Z_{line}|}\left(V_g \cos\bar{\delta}\Delta\tilde{\delta}_i - V^* \sum_{j=1}^{n}\left(\Delta\tilde{\delta}_i - \Delta\tilde{\delta}_j\right)\right) \tag{13.32}$$

$$\Delta\dot{\tilde{\delta}}_i = -m\Delta P_i \tag{13.33}$$

Substituting (13.32) into (13.33), we have

$$\Delta\dot{\tilde{\delta}}_i = -mV^*\left(V_g \cos\bar{\delta}\Delta\tilde{\delta}_i - V^* \sum_{j=1}^{n}\left(\Delta\tilde{\delta}_i - \Delta\tilde{\delta}_j\right)\right) \Big/ |Z_{line}| \tag{13.34}$$

Rewrite (13.34) in the matrix form as (13.35).

$$\dot{\mathbf{X}} = \bar{\mathbf{A}}\mathbf{X} \tag{13.35}$$

where

$$\bar{\mathbf{A}} = K_2\left(\left(V_g \cos\bar{\delta} - nV^*\right)\mathbf{I}_{\mathbf{n}\times\mathbf{n}} + V^*\mathbf{1}_{\mathbf{n}}\mathbf{1}_{\mathbf{n}}^{\mathbf{T}} \right) \tag{13.36}$$

$$K_2 = -mV^*/|Z_{line}| < 0 \tag{13.37}$$

The eigenvalues of the system matrix $\bar{\mathbf{A}}$ are given as follows:

$$\lambda_1\left(\bar{\mathbf{A}}\right) = K_2 V_g \cos\bar{\delta}, \lambda_2\left(\bar{\mathbf{A}}\right) = \cdots = \lambda_n\left(\bar{\mathbf{A}}\right) = K_2\left(V_g \cos\bar{\delta} - nV^*\right) \quad (13.38)$$

Since $\left(V_g \cos\bar{\delta} - nV^*\right) > 0$ (see (13.13), $\cos\bar{\delta} > 0$ and $|Z_{line}| > 0$, the system is stable according to (13.38). As $V_g \geqslant V_g \cos\bar{\delta}$, then $V_g > nV^*$, that is to say, $Q_i < 0$ is needed for the requirement of stable operation. Thus, the system is also feasible for the grid-connected operation.

From the above theoretical analysis, the unified decentralized control is stable under both the grid-connected and islanded mode while the control parameters (13.13) and line impedance $|Z_{line}| > 0$ satisfied. Thus, the seamless transition from grid-connected mode to islanded mode of CMG is obtained via the decentralized manner.

Furthermore, as the expansion of main idea in this chapter, for the active synchronization from the islanded mode to grid-connected mode, we can modify (13.11) as:

$$\omega_i = \omega^* + m\,\text{sgn}\left(Q_i\right)\text{sgn}\left(P_i\right)P_i + \Delta\omega_i \quad (13.39)$$

$$\Delta\omega_i = k_w \int \left(\omega_g - \omega_i\right)dt \quad (13.40)$$

where k_w is a positive constant. $\Delta\omega_i$ is the compensation signal, it is enabled when the CMG needs to switch from the islanded mode to grid-connected mode.

13.4 Simulation Results

The unified decentralized control strategy is verified on MATLAB/Simulink platform. This simulation model of CMG comprises four DGs, which is shown in Fig. 13.6. The associated parameters are listed in Table 13.2.

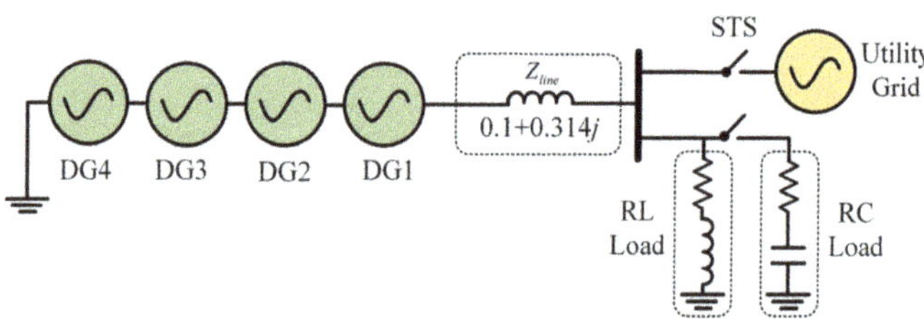

Fig. 13.6 Simulation model of a CMG

Table 13.2 Parameters for simulations

Symbol	Item	Value
L_{line}	Line inductance	1e–3H
R_{line}	Line resistance	0.1Ω
L_f	Filter inductance	1.5e–3H
r_f	Resistance in filter inductance	0.1Ω
C_f	Filter capacitor	20 μF
r_d	Resistance in filter capacitor	3.3Ω
V^*	Voltage reference	77.13 V
f_g	Nominal grid frequency	50 Hz
V_{pcc}	Allowable load voltage ranges	[295.45, 326.55] V
f^*	Frequency reference	50.2 Hz
f	Allowable frequency ranges	[49,51] Hz
m	Sharing coefficients	1e–4rad/(W·s)

13.4.1 Case 1: Islanded Mode Under RL and RC Loads

This case is performed to test the effectiveness of the scheme in the islanded mode. The load demands are shown in Fig. 13.7a, which are fed by RL loads in the interval [0 s, 2 s] and RC loads in the interval [2 s, 4 s], respectively. The corresponding load power factors are 0.89 and 0.95. The frequencies of all DGs are depicted in Fig. 13.7b. As seen, all DGs can realize the self-synchronization and maintain the system stable operation under both the RL loads and RC loads. From Fig. 13.7c, the active power allocations among DGs are always accurate. The reactive power sharing is shown in Fig. 13.7d, in which it is also accurate under the two types of loads. The load voltage and current waveforms are shown in Fig.13.8a and b, in which there are some phase angle differences. From this test, the scheme can realize the accurate power sharing and maintain the system stable in the islanded mode.

13.4.2 Case 2: Grid-Connected Mode Under RL and RC Loads

In this case, the performance of the scheme is verified in the grid-connected mode. The load demands are set as the same as Case 1, which is shown in Fig. 13.7a. The waveforms of grid current and line current are shown in Fig. 13.9a and b, respectively. The injected grid current is decreased as the active power load demands increasing, in which the power balance of system is ensured. The frequencies are shown in Fig. 13.10a, in which it converges to 50 Hz, and the scheme can synchronize with the grid. The active power sharing results are depicted in Fig. 13.10b, in which the DGs can output the desired active power in the grid-connected mode. The accurate reactive power sharing is shown in Fig. 13.10c.

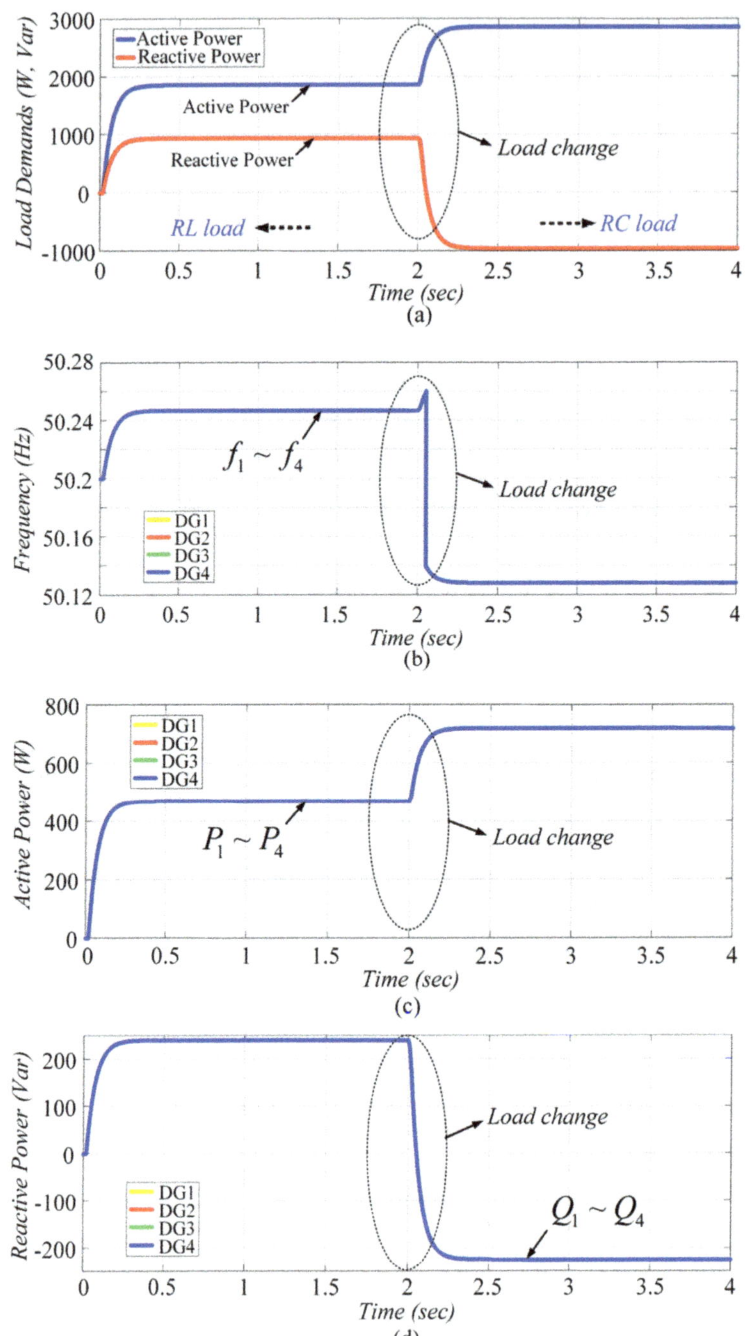

Fig. 13.7 Simulation results of case 1 (**a**) load demands, (**b**) Frequency, (**c**) active power, (**d**) reactive power

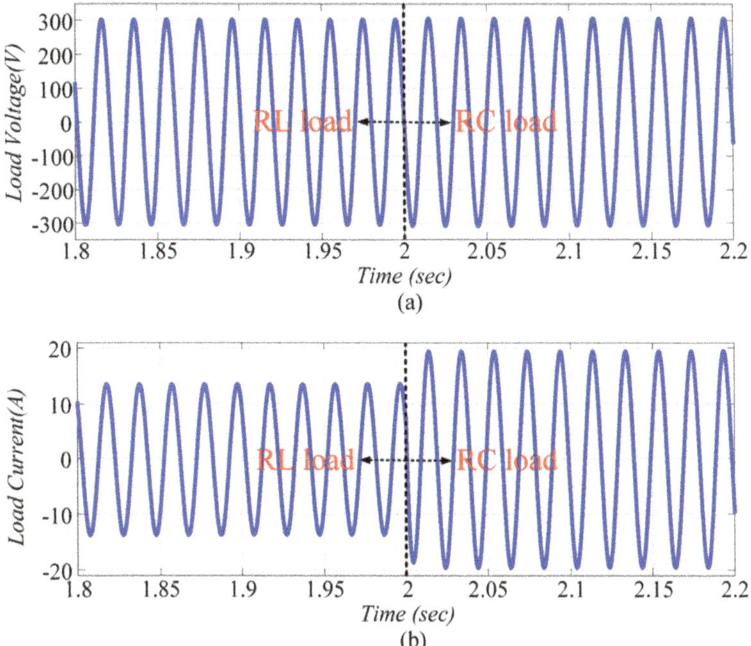

Fig. 13.8 Simulation waveforms of case 1 (**a**) load voltage, (**b**) load current

Therefore, the scheme can output the desired active power and maintain the self-synchronization in the grid-connected mode.

13.4.3 Case 3: Switch from Grid-Connected to Islanded Mode Under RL Loads

This simulation is implemented during switching from the grid-connected to islanded mode under the RL loads. The load power factor is 0.89. The load voltages in the interval [1.8 s, 2.2 s] are shown in Fig. 13.11a, in which the sustainable power supply for the loads is obtained. The grid current and line current are shown in Fig. 13.11b and c, in which the waveforms are smooth while switching from the grid-connected mode to islanded mode. The frequencies of DGs are shown in Fig. 13.12a, in which the system is always stable. The active and reactive power allocations among DGs are accurate, which are shown in Fig. 13.12b and c. Accordingly, it is concluded that the scheme can realize the smooth operation when the system switches from the grid-connected mode to the islanded mode under the RL loads.

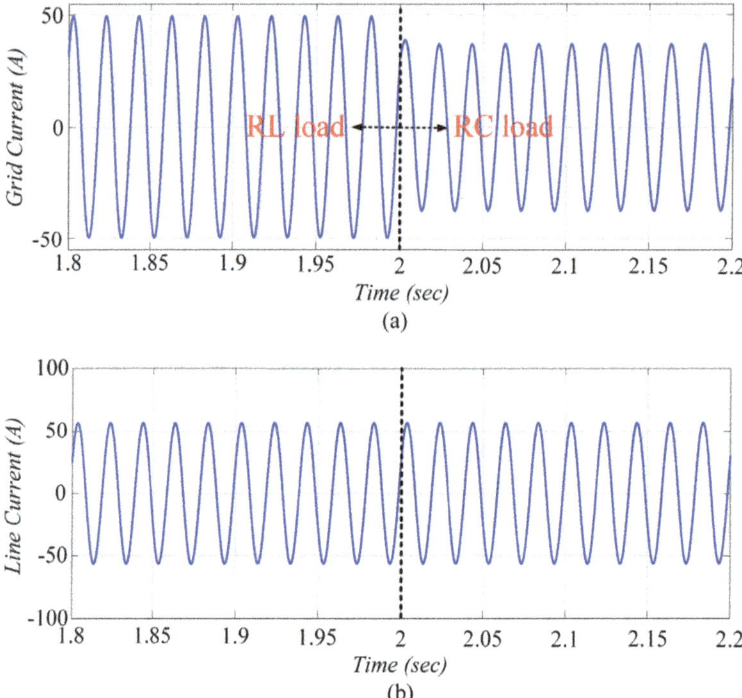

Fig. 13.9 Simulation waveforms of case 2 (**a**) grid current, (**b**) line current

13.4.4 Case 4: Switch from Grid-Connected to Islanded Mode Under RC Loads

This case is carried out during switching from the grid-connected to islanded mode under the RC loads. The load power factor is 0.95. The waveforms of grid current and line current are shown in Fig. 13.13a and b, in which the system can operate smoothly while mode switching under the RC loads. The frequencies are shown in Fig. 13.14a, in which the system is always stable. The active and reactive power allocations among DGs are depicted in Fig. 13.14b and c, respectively. As seen, the scheme is feasible to the CMG during the mode switching from grid-connected mode to islanded mode fed by RC loads.

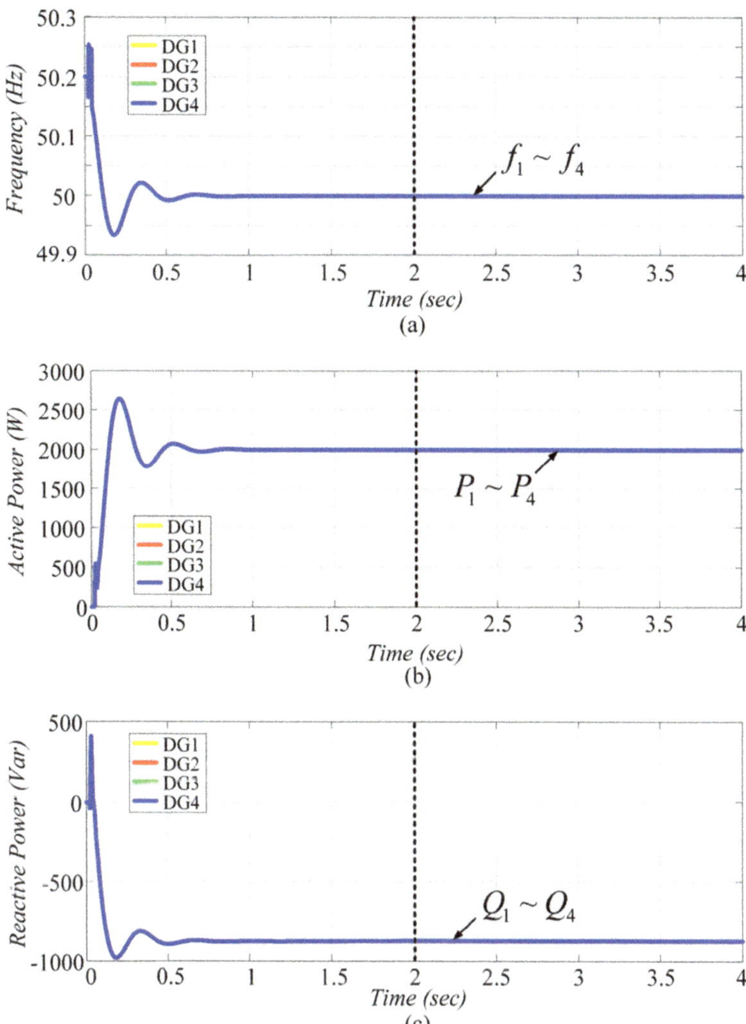

Fig. 13.10 Simulation results of case 2 (**a**) frequency, (**b**) active power, (**c**) reactive power

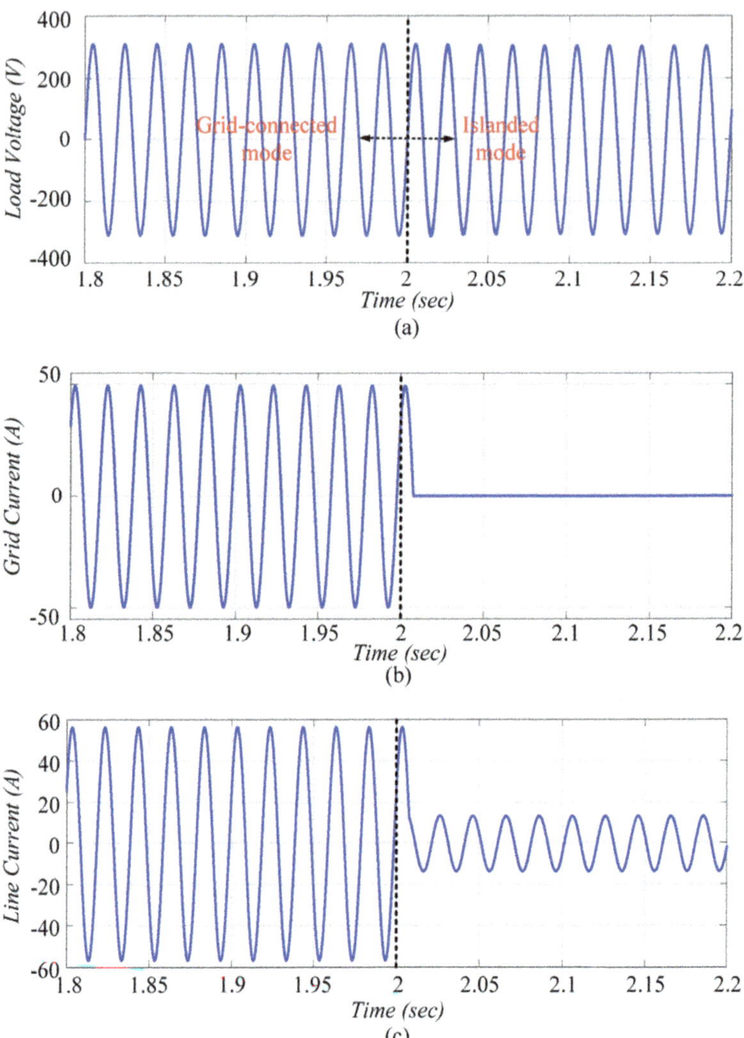

Fig. 13.11 Simulation waveforms of case 3 (**a**) load voltage, (**b**) grid current, (**c**) line current

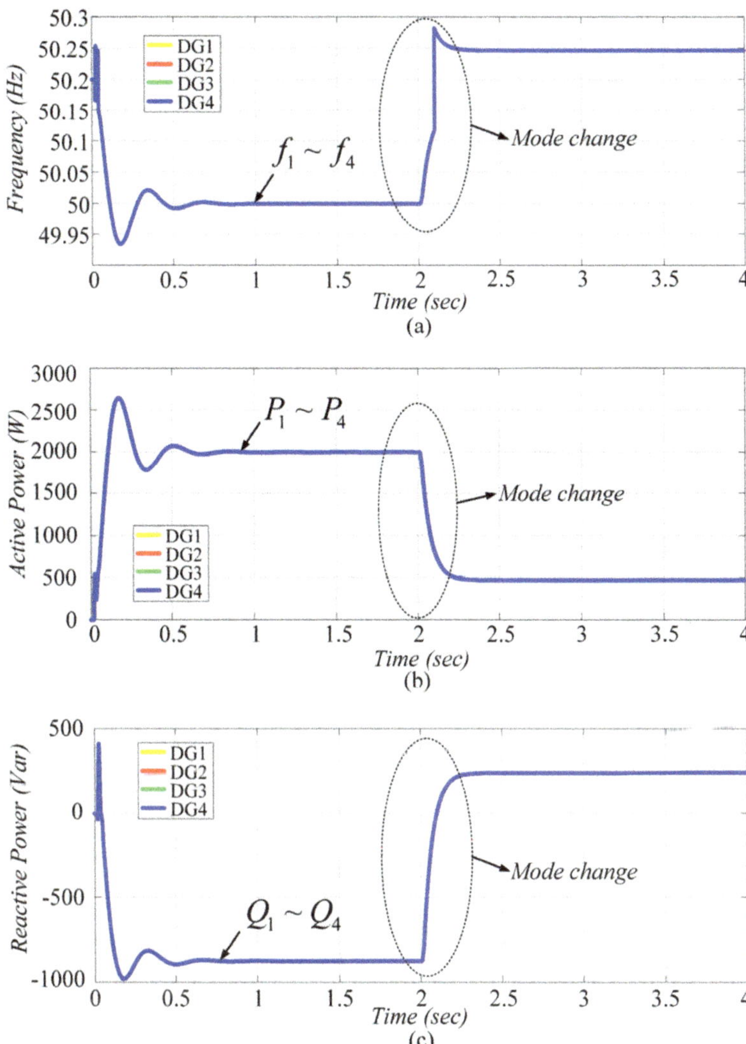

Fig. 13.12 Simulation results of case 3 (**a**) frequency, (**b**) active power, (**c**) reactive power

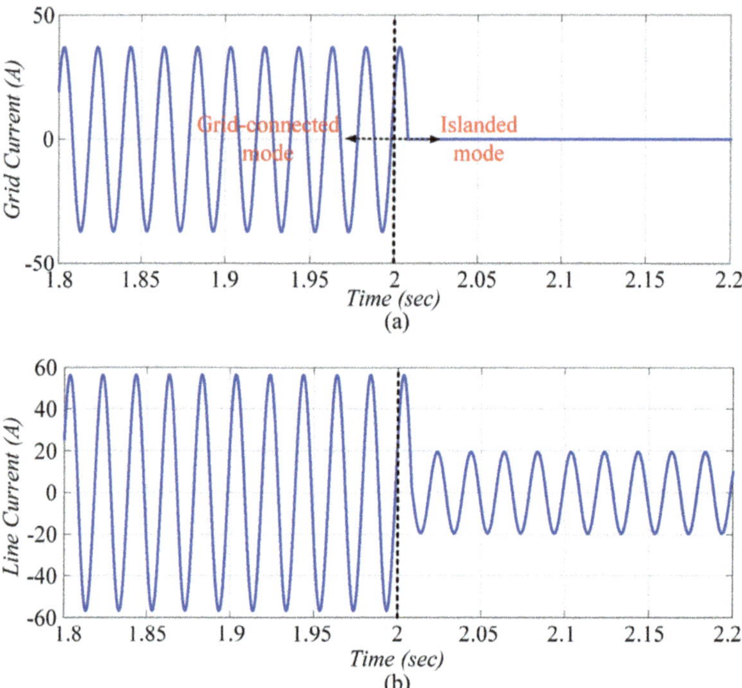

Fig. 13.13 Simulation results of case 4 (**a**) grid current, (**b**) line current

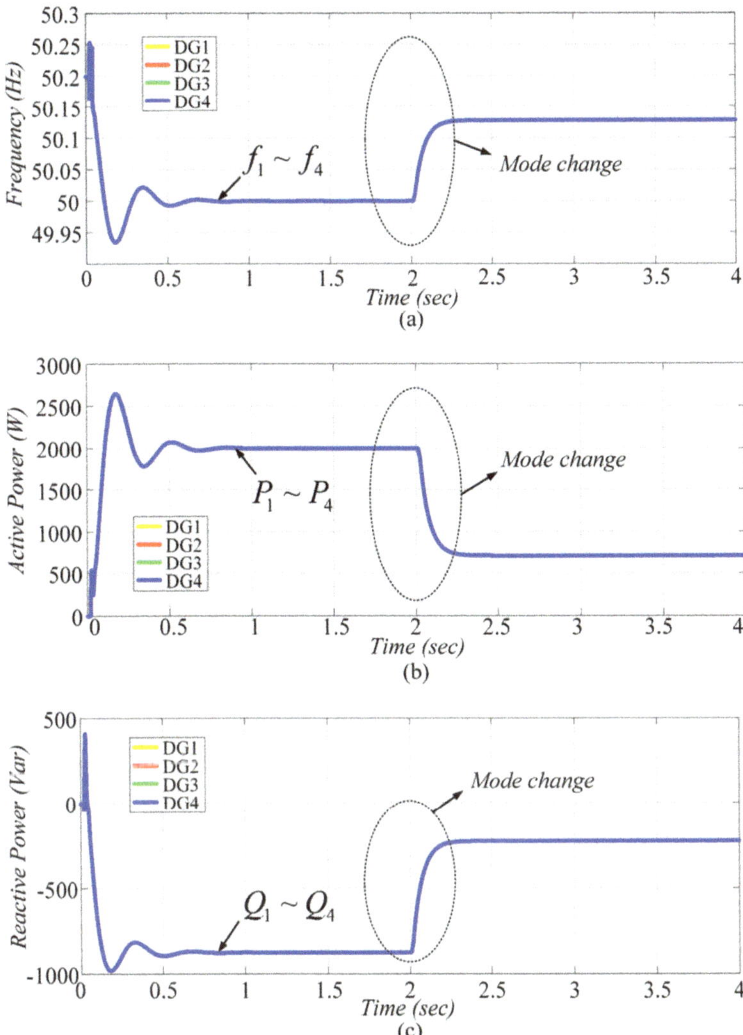

Fig. 13.14 Simulation results of case 4 (**a**) frequency, (**b**) active power, (**c**) reactive power

13.4.5 Case 5: Performance Towards Removal of DG4

This test is carried out with DG4 lost at $t = 2$ s. When DG4 suddenly suffers a fault, the bypass method [8] is applied. The waveforms of active and reactive power in the islanded mode are shown in Fig. 13.15a and b. In the grid-connected mode, the simulation results towards removal of DG4 are depicted in Fig. 13.16a and b. As seen, the CMG can realize the stable operation in both islanded and grid-connected

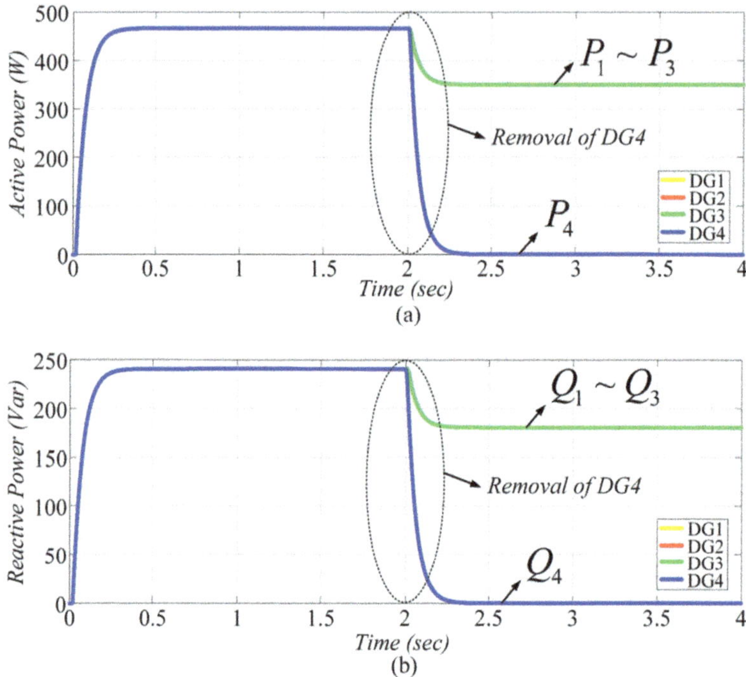

Fig. 13.15 Removal of DG4 in islanded mode (**a**) active power, (**b**) reactive power

mode. Therefore, removal of DG4 will not affect the stability of system for the scheme.

13.4.6 Case 6: Performance Under Single-Phase Asynchronous Machine

This simulation is implemented to verify the performance of the unified decentralized control scheme under single-phase asynchronous machine. The setting mechanical torque is shown in Fig. 13.17a, which is scheduled as 0, 10, 0, 10 in the interval [0 s, 1 s], [1 s, 2 s], [2 s, 3 s], [3 s, 4 s], respectively. The CMG is switched from the grid-connected mode to islanded mode at $t = 2$ s. The speed of single-phase asynchronous machine is shown in Fig. 13.17b. The active power and reactive power allocations are shown in Fig. 13.18a and b. The frequencies of DGs are presented in Fig. 13.18c. As seen, the scheme is feasible for single-phase asynchronous machine load and maintain the stable operation of CMG.

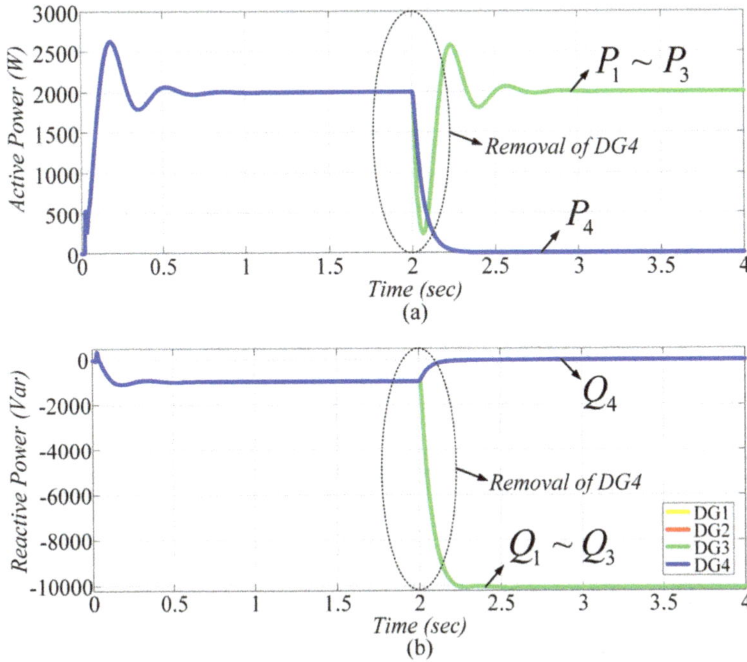

Fig. 13.16 Removal of DG4 in grid-connected mode (**a**) active power, (**b**) reactive power

13.4.7 Case 7: Impact of Line Impedance

This case is carried out to investigate the impact of different line impedance in the islanded mode. The line impedance is scheduled as 1 mH (inductive) in the interval [0 s, 2 s], and 0.3 Ω (resistive) in the interval [2 s, 4 s], respectively. The simulation results are shown in Fig. 13.19a and b. Usually, the line impedance is much less than the load impedance, the impact of inductive or resistive interconnecting line can be almost negligible.

13.4.8 Case 8: Under the Only Resistive Load or No Load Condition

In this case, the generalized load is set as the only resistive load in the interval [0 s, 2 s] and no load condition in the interval [2 s, 4 s], respectively. The active power and reactive power sharing are shown in Fig. 13.20a. The frequency of all DGs is shown in Fig. 13.20b, which is a constant. Therefore, under the only resistive load or no load condition, the system is stable and operates in constant frequency.

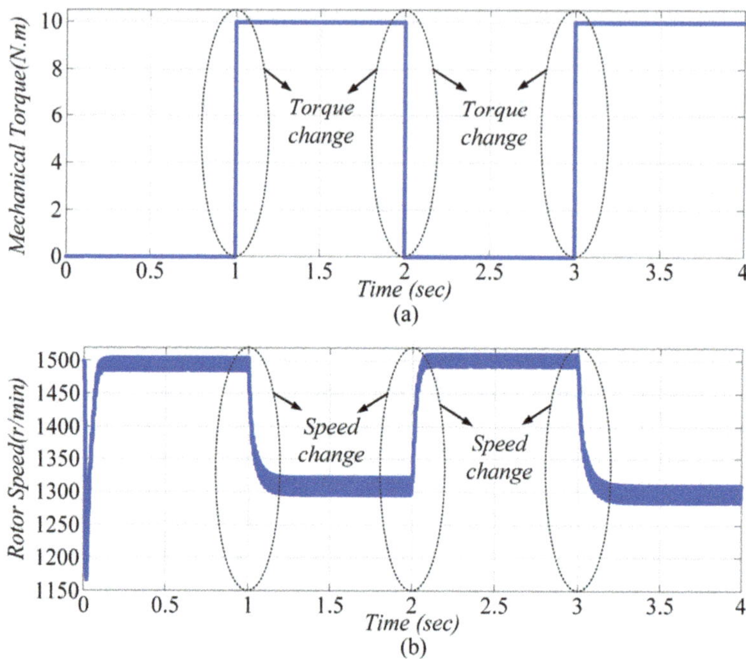

Fig. 13.17 Simulation waveforms of case 6 (**a**) mechanical torque, (**b**) rotor speed

subsectionCase 9: Active Synchronization from The Islanded Mode to Grid-Connected Mode

This case is carried out to verify the performance of active synchronization. At $t = 1$ s, the active synchronization control (13.39) is enabled. The STS is closed at $t = 2$ s. The waveforms of grid current are shown in Fig. 13.21a, which is smooth during mode switching. The frequencies are shown in Fig. 13.21b. Accordingly, the modified active synchronization control (13.39) can realize seamless transition from the islanded mode to grid-connected mode.

13.5 OPAL-RT Based Real-Time Simulation Results

The OPAL-RT based real-time simulation tests are carried out in this section to verify the performance and effectiveness of the scheme. The real-time simulation platform is based on the real-time simulator OPAL-RT 5600, whose time-step is 20 μs. This considered test model is consisting of four DGs, and the associated parameters are listed in Table 13.3.

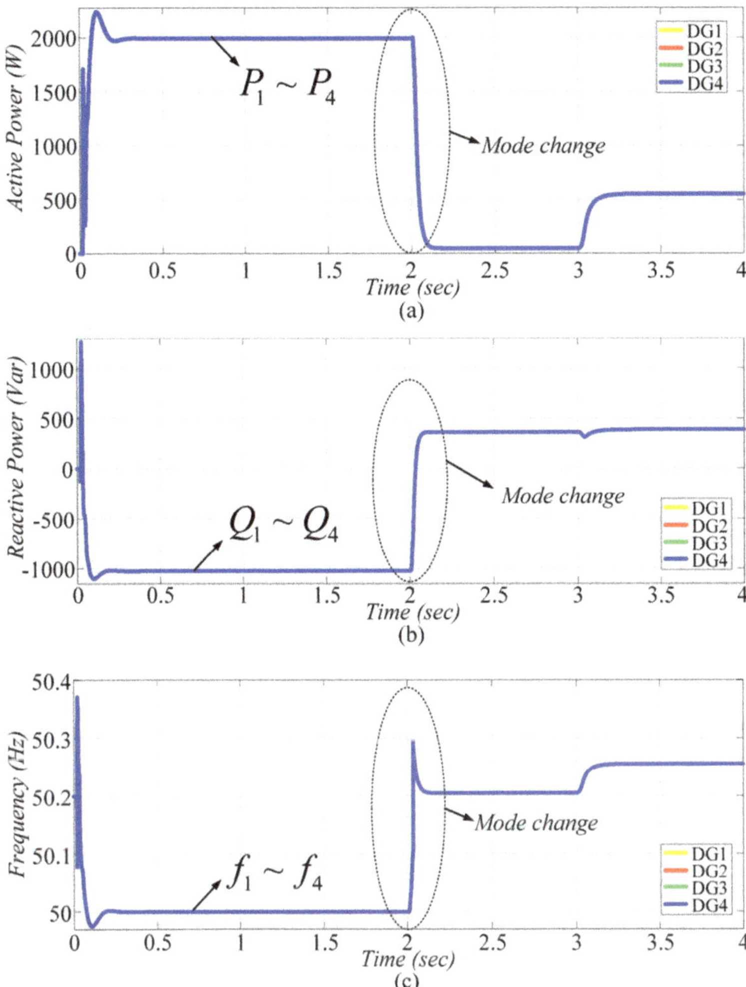

Fig. 13.18 Simulation results of case 6 (**a**) active power, (**b**) reactive power, (**c**) frequency

13.5.1 Case 1: Switch from Grid-Connected to Islanded Mode Under RL Loads

This test is performed when the CMG operates from the grid-connected mode to the islanded mode under the RL loads. The load power factor is 0.89. The waveforms of load voltage, line current, and grid current are shown in Fig. 13.22a, in which the load voltage is smooth regardless of the mode change. After the mode changing, the line current reduces due to only supplying the local RL loads, and the grid current reduces to zero in the islanded mode. The active power allocations among the four

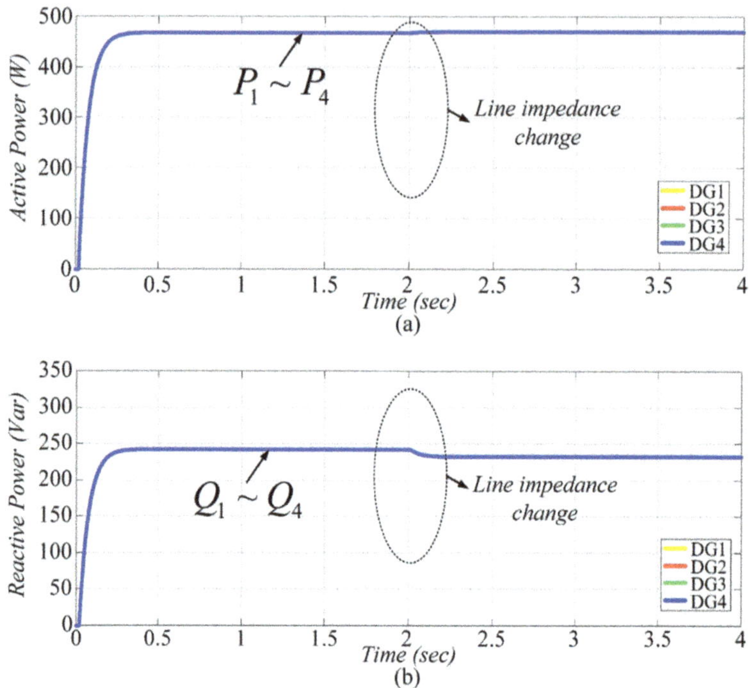

Fig. 13.19 Impact of line impedance in islanded mode (**a**) active power, (**b**) reactive power

DGs are shown in Fig. 13.22b. These DGs can inject the given active powers into the grid, and continuously supply electricity to the local loads. The reactive power sharing is depicted in Fig. 13.22c. The frequencies of DGs are shown in Fig. 13.22d, in which the system can always maintain the stable operation in the grid-connected and islanded mode.

Therefore, it is concluded that the scheme can realize the power sharing and maintain the system table under the RL loads.

13.5.2 Case 2: Switch from Grid-Connected to Islanded Mode Under RC Loads

In this test, the scheme is implemented under RC loads when the CMG operates from the grid-connected mode to islanded mode. The load power factor is set as 0.95. The waveforms of load voltage, line current, and grid current are shown in Fig. 13.23a. The active power and reactive power allocations among DGs are depicted in Fig. 13.23b and c, respectively. As seen, the scheme can inject the desired powers into the grid in grid-connected mode and maintain the stable

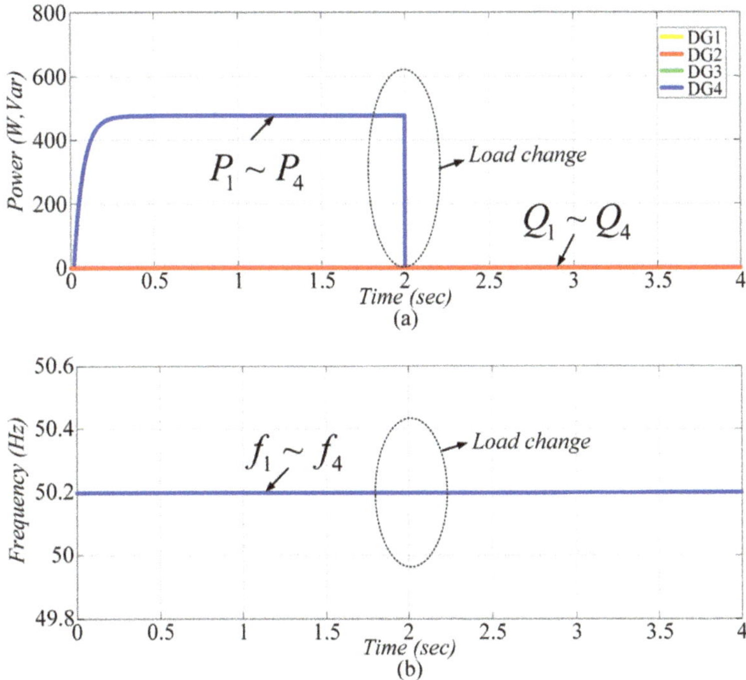

Fig. 13.20 Simulation results of case 8 (**a**) power, (**b**) frequency

operation in islanded mode. The waveform of frequency is depicted in Fig. 13.23d, in which the system switches smoothly from the grid-connected mode to islanded mode.

From the OPAL-RT based real-time simulation test results, the unified decentralized control scheme is feasible for the grid-connected and islanded operation of CMG where the RC loads are fed.

13.6 Conclusion

This chapter introduced a unified decentralized control strategy of CMG for both the grid-connected and islanded operation. The main results of this chapter are summarized as follows: (1) Seamless transition from the grid-connected mode to islanded mode is realized under both RL and RC loads; (2) The CMG always holds a unique steady-state operation point; (3) The scheme is a fully decentralized manner, as it is implemented only with the local information. In addition, the decentralized control scheme of multi-series networked topology will be studied in the future.

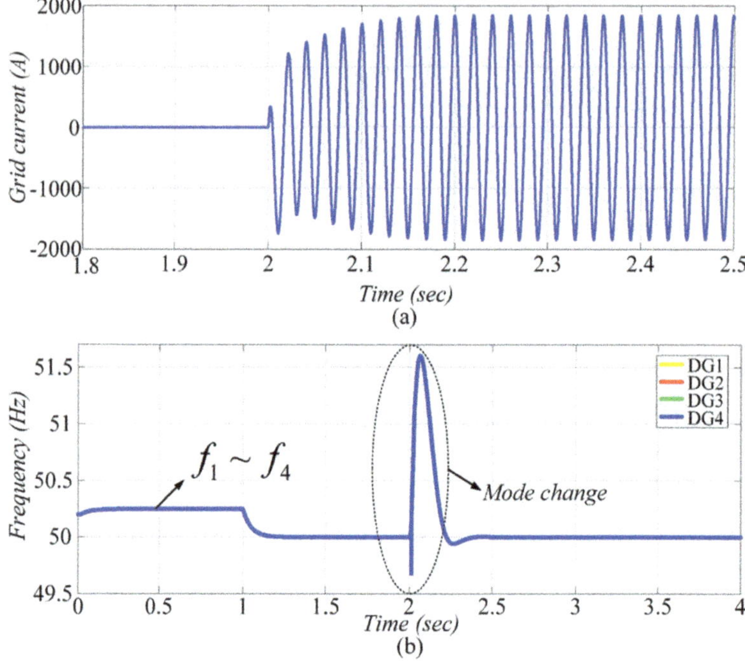

Fig. 13.21 Simulation results of case 9 (**a**) grid current, (**b**) frequency

Table 13.3 Parameters for OPAL-RT based real-time simulation tests

Symbol	Item	Value
L_{line}	Line inductance	1.5e-3H
R_{line}	Line resistance	0.1Ω
L_f	Filter inductance	0.6e-3H
r_f	Resistance in filter inductance	0.1Ω
C_f	Filter capacitor	20μF
r_d	Resistance in filter capacitor	3.3Ω
V^*	Voltage reference	$308.5/4V$
f_g	Nominal grid frequency	50 Hz
f^*	Frequency reference	50.2 Hz
f	Allowable frequency ranges	[49,51] Hz
m	Sharing coefficients	1e-4$rad/(W \cdot s)$

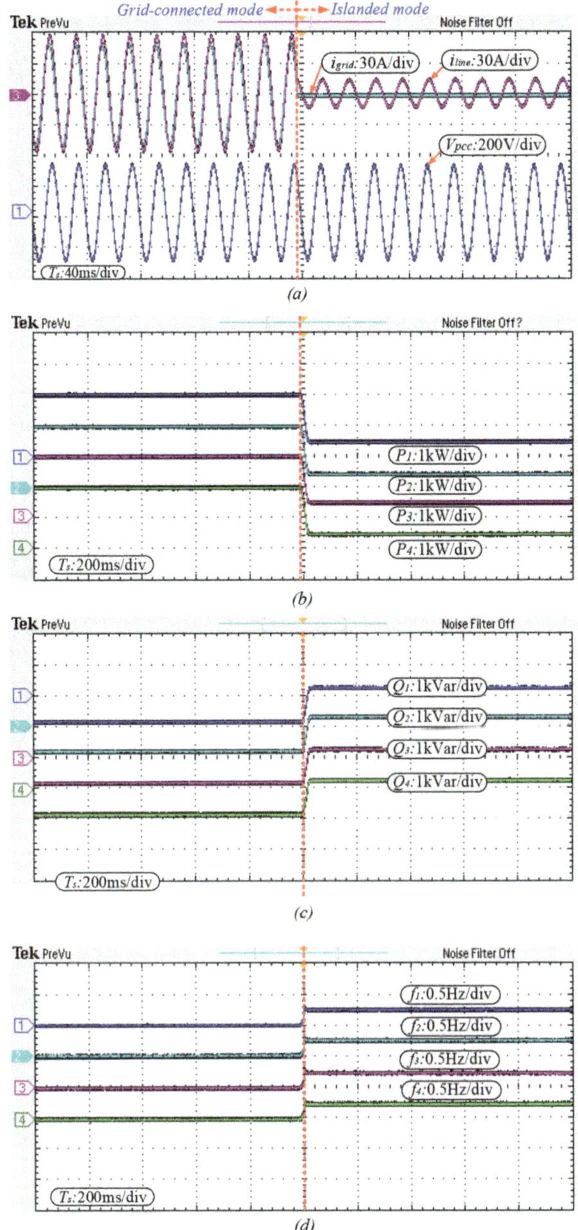

Fig. 13.22 OPAL-RT based real-time simulation tests of case 1 (**a**) Voltage and current waveforms, (**b**) Active power, (**c**) Reactive power, (**d**) Frequency

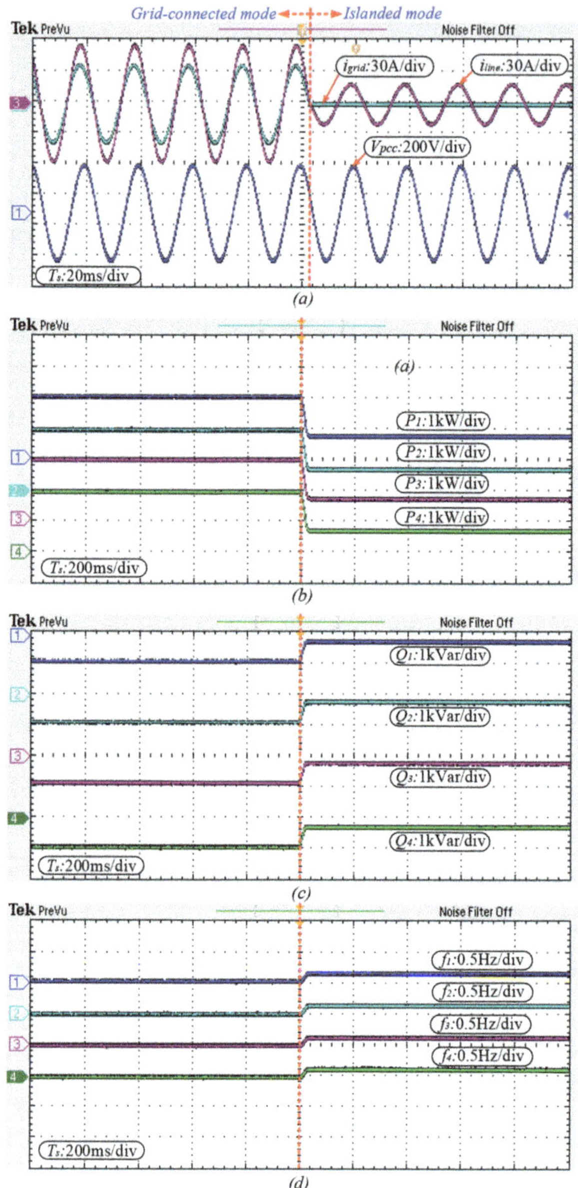

Fig. 13.23 OPAL-RT based real-time simulation tests of case 2 (**a**) Voltage and current waveforms, (**b**) Active power, (**c**) Reactive power, (**d**) Frequency

References

1. J. Lamb, B. Mirafzal, F. Blaabjerg, et al., PWM common mode reference generation for maximizing the linear modulation region of CHB converters in Islanded microgrids. IEEE Trans. Ind. Electron. **65**(7), 5250–5259 (2018)
2. M. Hamzeh, A. Ghazanfari, H. Mokhtari, et al., Integrating hybrid power source into an Islanded MV microgrid using CHB multilevel inverter under unbalanced and nonlinear load conditions. IEEE Trans. Energy Convers. **28**(3), 643–651 (2013)
3. L. Li, Y. Sun, Z. Liu, et al., A decentralized control with unique equilibrium point for cascaded-type microgrid. IEEE Trans. Sustainable Energy **10**(1), 324–326 (2019)
4. X. Hou, Y. Sun, H. Han, et al., A fully decentralized control of grid-connected cascaded inverters. IEEE Trans. Sustainable Energy **10**(1), 315–317 (2019)
5. L. Li, H. Ye, Y. Sun, et al., A communication-free economical-sharing scheme for cascaded-type microgrids. Int. J. Electr. Power Energy Syst. **104**, 1–9 (2019)
6. J.M. Guerrero, M. Chandorkar, T. Lee, et al., Advanced control architectures for intelligent microgrids—Part I: decentralized and hierarchical control. IEEE Trans. Ind. Electron. **60**(4), 1254–1262 (2013)
7. Y. Han, H. Li, P. Shen, et al., Review of active and reactive power sharing strategies in hierarchical controlled microgrids. IEEE Trans. Power Electron. **32**(3), 2427–2451 (2017)
8. J.W. Simpson-Porco, F. Dörfler, F. Bullo, Synchronization and power sharing for droop-controlled inverters in islanded microgrids. Automatica **49**(9), 2603–2611 (2013)
9. L. Maharjan, T. Yamagishi, H. Akagi, J. Asakura, Fault-tolerant operation of a battery-energy-storage system based on a multilevel cascade PWM converter with star configuration. IEEE Trans. Power Electron. **25**(9), 2386–2396 (2010)
10. L. Li, Y. Sun, X. Hou, M. Su, X. Zhang, P. Wang, J.M. Guerrero, Unified decentralized control for both grid-connected and Islanded operation of cascaded-type microgrid. IET Renewable Power Gener. **14**(16), 3138–3148 (2020)

Part III
Hybrid Series–Parallel Microgrid Systems

Chapter 14
Distributed Control Strategy for Hybrid Series–Parallel Microgrid

14.1 Comparative Power Control of Parallel-Series Microgrids

14.1.1 Power Control Characteristic of Parallel-Type Microgrid

Figure 14.1a shows the circuit configuration of parallel-type microgrid. With the assumption of inductive feeder impedance, the output active power P_i and reactive power Q_i of the i-th DG are given as (14.1)–(14.2) [1, 2].

$$P_i = \frac{V_i V_p}{X_i} \sin\left(\delta_{iP}\right) \approx \frac{V_i V_p}{X_i} \delta_{iP} \tag{14.1}$$

$$Q_i = \frac{V_i \left(V_i - V_p\right)}{X_i} \tag{14.2}$$

Where V_i is the output voltage amplitude of i-th converter. V_P is the voltage amplitude of common AC bus. θ_{iP} is the power angle difference between V_i and V_p. X_i is reactance of the feeder impedance between DG#i and AC bus. The relationship between ΔP_i–$\Delta\delta_i$ and ΔQ_i–ΔV_i can be simplified from (14.1)–(14.2) as (14.3)–(14.4).

$$\Delta P_i \propto \Delta\delta_i \tag{14.3}$$

$$\Delta Q_i \propto \Delta V_i \tag{14.4}$$

Where $\propto$ represents "a positive correlation". The mathematical relationships (14.3)–(14.4) indicate that the output active and reactive power can be controlled by

Y. Sun et al., *Series-Parallel Converter-Based Microgrids*, Power Systems,
https://doi.org/10.1007/978-3-030-91511-7_14

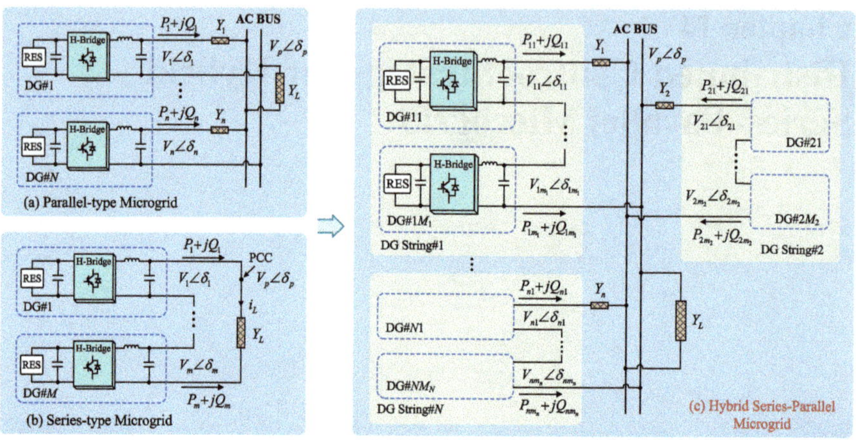

Fig. 14.1 Hybrid series–parallel microgrid

regulating frequency and voltage of the converters. Therefore, the conventional active power-angular frequency (P–ω) and reactive power-voltage (Q–V) droop control can be given as (14.5)–(14.6).

$$\omega_i = \omega^* - m_i P_i \tag{14.5}$$

$$V_i = V^* - n_i Q_i \tag{14.6}$$

where ω_i is the angular frequency, ω^* and V^* represent angular frequency and voltage amplitude without load. m_i and n_i are coefficients of P–ω droop control and Q–V droop control, respectively.

14.1.2 Power Control Characteristic of Series-Type Microgrid

The series-type microgrid is shown in Fig. 14.1b, the output power of the i-th DG is given as (14.7)–(14.8) [3].

$$P_i = V_i\,|Y_L| \sum_{j=1}^{n} V_j \cos\left(\delta_{ij} - \varphi_L\right) \tag{14.7}$$

$$Q_i = V_i\,|Y_L| \sum_{j=1}^{n} V_j \sin\left(\delta_{ij} - \varphi_L\right) \tag{14.8}$$

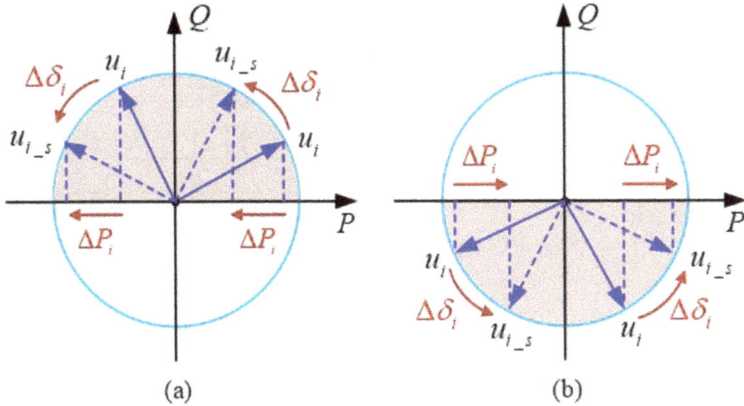

Fig. 14.2 ΔP_i–$\Delta\delta_i$ relationship in series-type microgrid. (**a**) RL load: $\Delta P_i \propto -\Delta\delta_i$. (**b**) RL load: $\Delta P_i \propto \Delta\delta_i$

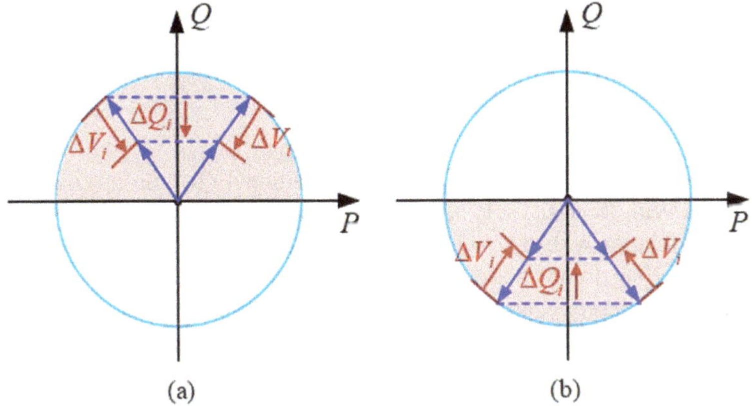

Fig. 14.3 ΔQ_i–ΔV_i relationship in series-type microgrid. (**a**) RL load: $\Delta Q_i \propto \Delta V_i$. (**b**) RL load: $\Delta Q_i \propto -\Delta V_i$

where $|Y_L|$ and φ_L are the amplitude and angle of load admittance. δ_{ij} is the power angle between DG#i and DG#j, which is much smaller than φ_L in steady state. Then, ΔP_i–$\Delta\omega_i$ and ΔQ_i–ΔV_i relationships are given as (14.9) and (14.10)

$$\Delta P_i \propto \sin(\varphi_L)\,\Delta\delta_i \tag{14.9}$$

$$\Delta Q_i \propto -\sin(\varphi_L)\,\Delta V_i \tag{14.10}$$

Figures 14.2 and 14.3 show the ΔP_i–$\Delta\delta_i$ and ΔQ_i–ΔV_i relationships in series-type microgrid. With RL load, there is $\sin(\varphi_L)<0$, which means the ΔP_i has a negative correlation with $\Delta\delta_i$ and ΔQ_i has a positive correlation with ΔV_i. In this

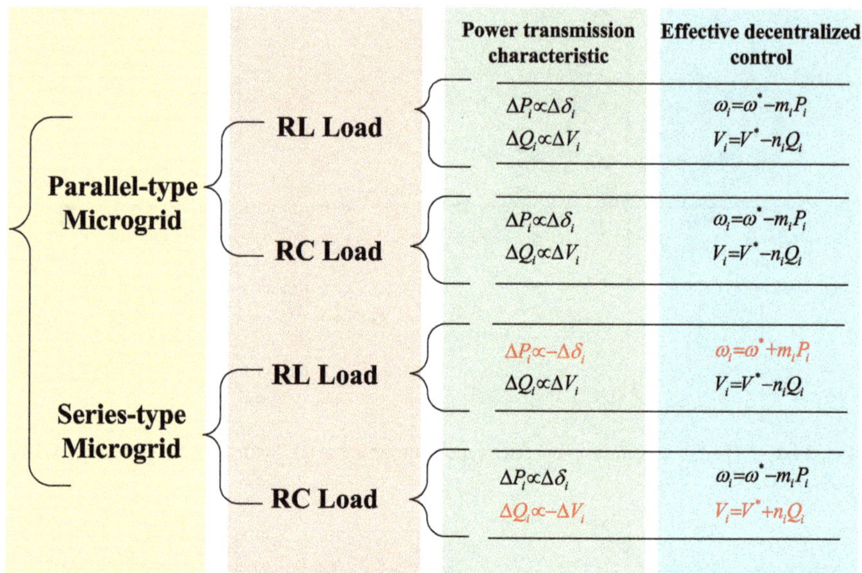

Fig. 14.4 Decentralized power control of microgrid

case, the $P-\omega$ inverse droop control and conventional Q–V droop control should be adopted. With RC load, ΔP_i has a positive correlation with $\Delta\delta_i$ and ΔQ_i has a negative correlation with ΔV_i, which means that the system can be stable under the conventional P–ω droop control and Q–V inverse droop control.

The effective power control strategies of parallel-type and series-type microgrid with RL or RC load are summarized in Fig. 14.4 according to aforementioned analysis. It can be seen that neither droop control nor inverse droop control is compatible among series and parallel microgrid under RL load or RC load, which means these decentralized control methods are not applicable for hybrid series–parallel microgrid. For the decentralized self-synchronous control design of hybrid series–parallel microgrid, there are two unavoidable problems to be overcome:

1. The self-synchronous mechanism of the two subsystems, the series structure and the parallel structure, is difficult to be unified in principle;
2. The control characteristics of the hybrid system are affected by the load properties, and it is difficult to unify the control under resistive load and resistive capacity load.

Obviously, it is difficult to construct the self-synchronization control of the hybrid power electronic system only through the simple combination of the self-synchronization methods of the series-type microgrid and the parallel-type microgrid. However, since the communication links are usually indispensable for fault monitoring and energy management, distributed control using low-bandwidth communication is considered as one of the promising control methods to solve this problem and perform unified power control under RL load and RC load.

14.2 Classification of Distributed Control Methods for Hybrid Series–Parallel Microgrid

Distributed control is widely regarded as an effective way to overcome the defects while retaining the advantages of decentralized control. In the distributed control, each converter adopts a separate control circuit without the need of a central controller and is only equipped with a low-bandwidth communication network that can connect all the sparse micro sources for data exchange. The most significant feature is that each DG is symmetric, and the information needed is only from the adjacent DGs, not the global. Therefore, the distributed control requires far less communication bandwidth than the centralized control while retaining the plug-and-play.

Considering the important application scenarios of microgrid large-scale integration of renewable energy in remote areas, the design of control strategy for island-based hybrid series–parallel microgrid should take communication cost and reliability into full consideration. Therefore, different distributed control structures with low communication dependence should be further discussed in the next chapters to realize coordinated operation of hybrid series–parallel microgrid.

1. **Globally distributed control method**
2. **Locally distributed and globally decentralized control method**
3. **Leader-distributed and follower-decentralized control**

14.3 Globally Distributed Control Strategy of Hybrid Series–Parallel Microgrid

In this section, a distributed control strategy is introduced to perform unified power control for hybrid series–parallel microgrid with RL and RC load, which is able to implement proportional power sharing without frequency deviation [9].

14.3.1 Communication Network Design

Figure 14.1c shows circuit configuration of the hybrid series–parallel microgrid. Several LV DGs are cascaded-connected to provide a medium- or high-voltage. These series DGs are then connected to AC bus in parallel to supply power to the load. Such a physical system can be equipped with a communication network to

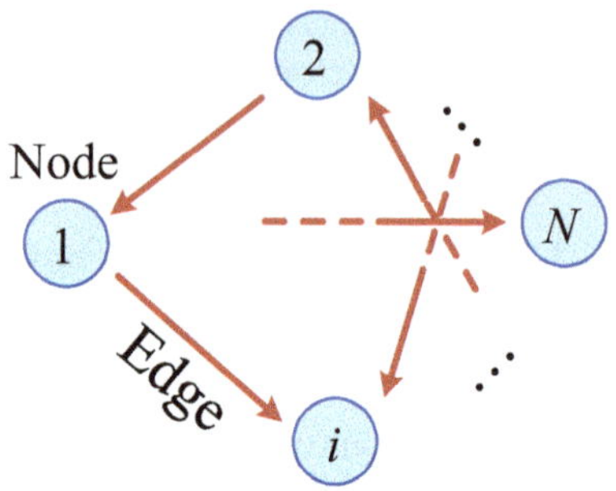

Fig. 14.5 Graphical representation of communication network

facilitate data exchange among DGs for control and monitoring purposes. Figure 14.5 shows the graphical representation of communication network. In the graph, nodes represent DGs and edges connecting notes represent communication links. Each node sends its active power and reactive power information to neighboring nodes. The information transmission can be bidirectional or unidirectional, and the graph is correspondingly called undigraph or digraph. In addition, nodes receive information from neighbors with different gains which are called communication weights. For example, if Node#i receives data x_j from Note#j with weight a_{ij}, it means that the information received by Node#i is $a_{ij} \cdot x_j$. Usually, a_{ij}>0 if Node#i receives information from Node#j and $a_{ij} = 0$, otherwise. In the graph, an adjacency matrix $A = [a_{ij}] \in \mathbb{R}_{N\times N}$ is defined to carry the communication weights. For each node, the weight in degree corresponds to the sum of weights of all ingoing edges into Node#i, where N_i is the set of neighbors of Node#i. The associated in-degree matrix is defined as $D^{in} = diag\left\{d_i^{in}\right\}$. The Laplacian matrix is constructed as $L = D^{in} - A$. In fact, for a system with a large number of DGs, topology of communication network may be various. However, there are some rules for managing a communication network. The mechanism can be explained as follows. Some definitions are first introduced. In the graph, a direct path from Node#i to Node#j is a sequence of edges connecting the two nodes. Root node is defined as a certain node, where there exists at least a direct path between the node and any other nodes. A graph includes a spanning tree if it contains a root node. A communication network is viewed as interconnection if it contains at least a spanning tree. Then, the rules of designing a communication network are given as follows [4, 5]. (1) To ensure connectivity of communication network, the network should contain at least one spanning tree. (2) To ensure system redundancy, the remaining communication network should still be connected in the case of any single communication link failure.

In this section, a bidirectional sparse-network with the minimum redundancy is developed. The graphical diagram of the proposed communication network is shown in Fig. 14.6. It has been proved that the ring structure is one of the most effective structures [5]. Here, the number ij ($i = 1, 2, \ldots$, N; $j = 1, 2, \ldots, M_i$) denotes the j-th DG in the String#i, where N is the number of strings and M_i is the number of DGs in String#i. In the hybrid series–parallel microgrid, information

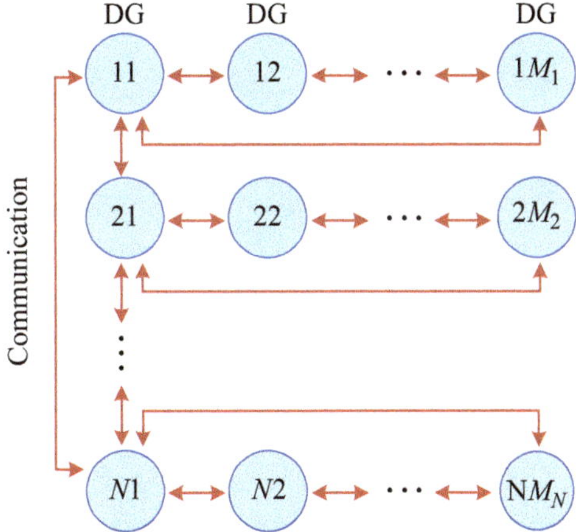

Fig. 14.6 Graphical diagram of the proposed communication network

network includes communication links among series DGs and those among parallel DGs. Considering the sparsity, only one DG in each string is selected as a leader to exchange information with leaders of other strings. In addition, the mechanism for leader selection can be established for different objectives, and here we choose DGs with biggest rated power as leaders for optimum reliability.

14.3.2 Globally Distributed Control Strategy of Hybrid Series–Parallel Microgrid

In autonomous microgrids, the output power can be automatically assigned by controlling frequency and output voltage of DG unit. In hybrid series–parallel microgrid, a unified distributed control strategy is proposed as (14.11) and (14.12).

$$\omega_{ij} = \omega^* - \delta\omega_{ij}^1 + \delta\omega_{ij}^2$$
$$\begin{cases} \delta\omega_{ij}^1 = \text{sgn}\left(Q_{ij}\right) k_{\omega}^{ij} \sum\limits_{h \in N_{ij\_s}} a_{ij\_h}\left(P_h^{pu} - P_{ij}^{pu}\right) \\ \delta\omega_{ij}^2 = k_{\omega\_p}^{ij} \sum\limits_{r \in N_{ij\_p}} a_{ij\_r}\left(P_r^{pu} - P_{ij}^{pu}\right) \end{cases} \tag{14.11}$$

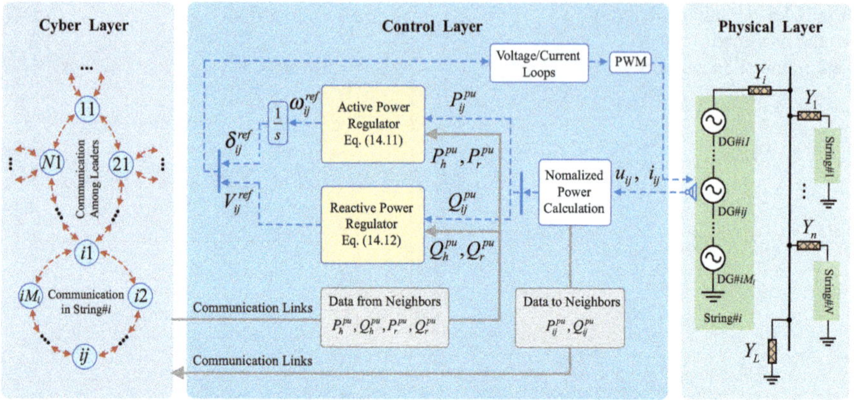

Fig. 14.7 Block diagram of proposed globally distributed control strategy

$$V_{ij} = V_{ij}^{*} + \delta V_{ij}^{1} + \delta V_{ij}^{2}$$

$$\begin{cases} \delta V_{ij}^{1} = \operatorname{sgn}(Q_{ij}) \left(k_{P}^{ij} + \frac{k_{I}^{ij}}{s} \right) \sum\limits_{h \in N_{ij\_s}} a_{ij\_h} \left(Q_{h}^{pu} - Q_{ij}^{pu} \right) \\ \delta V_{ij}^{2} = \left(k_{P\_p}^{ij} + \frac{k_{I\_p}^{ij}}{s} \right) \sum\limits_{r \in N_{ij\_p}} a_{ij\_r} \left(Q_{r}^{pu} - Q_{ij}^{pu} \right) \end{cases} \tag{14.12}$$

where the subscript ij represents the j-th DG of string#i. $h \in N_{ij_s}$is the set of series neighbors of DG#ij.$r \in N_i j_P$ is the set of paralleled neighbors of DG#ij. a_{ij_h}, (a_{ij_r}) is the communication weight between DG#ij and DG#h(DG#r). k_{ω}^{ij} and $k_{\omega\_p}^{ij}$ are the parameters of proportional controllers in the active power regulator (14.11) and are the parameters of proportional-integral (PI) controllers in the reactive power regulator (14.12). k_P^{ij},k_I^{ij},$k_{P\_p}^{ij}$, and $k_{I\_p}^{ij}$ are normalized active and reactive power of DG#ij, which are given as (14.13). The rated voltage amplitude V_{ij}^{*} is given as (14.14).

$$P_{ij}^{pu} = \frac{P_{ij}}{P_{ij}^{rated}}, \ Q_{ij}^{pu} = \frac{Q_{ij}}{Q_{ij}^{rated}} \tag{14.13}$$

$$V_{ij}^{*} = \frac{Q_{ij}^{rated}}{\sum\limits_{l=1}^{m_i} Q_{il}^{rated}} V_{P}^{*} \tag{14.14}$$

where P_{ij}^{rated} and Q_{ij}^{rated} are rated active and reactive power of DG#ij and V_P^{*} is the rated voltage amplitude of AC bus.

Figure 14.7 shows block diagram of the globally distributed control strategy. The distributed controller consists of active power regulator and reactive power regulator. Controller of DG#ij receives information (P_h^{pu}, Q_h^{pu}, P_r^{pu}, Q_r^{pu}) from its neighbors and processes them with local data (P_{ij}^{pu}, Q_{ij}^{pu}) in two regulators. The frequency reference value is calculated by active power regulator according to (14.11). Here, $\delta\omega_{ij}^1$ is added to mitigate the power mismatch among series DGs, so a sign function sgn(Q_{ij}) is introduced to match the different power transmission characteristics under RL load and RC load, ensuring the proposed controller effective under both situations. And $\delta\omega_{ij}^2$ carries the active power mismatch between DG#ij and its paralleled neighbors, which is zero for the non-leaders since the set N_{ij_p} is an empty set. The phase angle of DG#ij is given as (14.15).

$$\delta_{ij} = \int \omega_{ij} dt = \omega^* t + \int \delta\omega_{ij}^1 dt + \int \delta\omega_{ij}^2 dt \tag{14.15}$$

It can be seen from (14.15) that the frequency of DG#ij will synchronize to the rated value in the steady state. And $\delta\omega_{ij}^1$ and $\delta\omega_{ij}^2$ will converge to zero, which means accurate proportional active power sharing is guaranteed. In reactive power regulator, reactive power mismatch is fed to PI controllers, producing correction terms δV_{ij}^1 and δV_{ij}^2 to adjust the voltage amplitude. Similar with the active power regulator, a sign function sgn(Q_{ij}) is introduced in δV_{ij}^1 to match the different reactive power transmission characteristics under RL load and RC load. In the steady state, δV_{ij}^1 and δV_{ij}^2 decay to zero and all normalized reactive power synchronizes, which indicates the proportional reactive power sharing. The proposed distributed control is a globally method for parallel and cascaded converters under RL and RC load. Compared with central control method, DGs only need to exchange data with several neighbors by a low-bandwidth communication network. Therefore, single point of failure can be avoided so that the reliability and flexibility of microgrid can be enhanced.

14.4 Small-Signal Modeling and Stability Analysis

Small-signal stability of microgrids with proposed distributed control strategy is investigated in this section.

14.4.1 Small-Signal Modeling

(1) Normalized power modeling

The power generation of the ij-th DG is presented as (14.16).

$$p_{ij} + jq_{ij} = \frac{V_{ij}e^{j\delta_{ij}}}{2}\left(\left(\sum_{b=1}^{m_i} V_{ib}e^{j\delta_{ib}} - V_P e^{j\delta_P}\right)|Y_i|\, e^{j\varphi_i}\right)^* \tag{14.16}$$

where V_{ij} and δ_{ij} represent the output voltage amplitude and phase angle of ij-th DG. V_P and δ_P are the voltage amplitude and phase angle of AC bus. $|Y_i|$ and φ_i are the amplitude and angle of line admittance of String#i. According to Kirchhoff laws, the voltage of AC bus is obtained as (14.17).

$$V_P e^{j\delta_P} = \sum_{a=1}^{n}\sum_{b=1}^{m_a} \left|Y'_a\right| V_{ab} e^{j\delta_{ab} + \varphi'_a} \tag{14.17}$$

where

$$Y'_a = \frac{Y_a}{\sum\limits_{c=1}^{n} Y_c + Y_L} = \left|Y'_a\right| e^{j\varphi'_a} \tag{14.18}$$

Define $\dot{\tilde{\delta}}_{ij} = \dot{\delta}_{ij} - \dot{\delta}_s = \omega_{ij} - \omega_s$, where ω_s is the frequency in steady state. Then, combining (14.16)–(14.18), the instantaneous power supplied by the ij-th DG is given as (14.19).

$$\begin{cases} p_{ij} = \frac{1}{2}\sum\limits_{a=1}^{n}\sum\limits_{b=1}^{m_a} V_{ij}V_{ab}\left|Y'_a\right||Y_i|\sin\left(\tilde{\delta}_{ij} - \tilde{\delta}_{ab} - \varphi'_a\right) - \frac{1}{2}\sum\limits_{b=1}^{m_i} V_{ij}V_{ib}|Y_i|\sin\left(\tilde{\delta}_{ij} - \tilde{\delta}_{ib}\right) \\ = F\left(\tilde{\delta}_{11},\ \tilde{\delta}_{12},\ \ldots,\ \tilde{\delta}_{nm_n},\ V_{11},\ V_{12},\ \ldots,\ V_{nm_n}\right) \\ q_{ij} = \frac{1}{2}\sum\limits_{b=1}^{m_i} V_{ij}V_{ib}|Y_i|\cos\left(\tilde{\delta}_{ij} - \tilde{\delta}_{ib}\right) - \frac{1}{2}\sum\limits_{a=1}^{n}\sum\limits_{b=1}^{m_a} V_{ij}V_{ab}\left|Y'_a\right||Y_i|\cos\left(\tilde{\delta}_{ij} - \tilde{\delta}_{ab} - \varphi'_a\right) \\ = G\left(\tilde{\delta}_{11},\ \tilde{\delta}_{12},\ \ldots,\ \tilde{\delta}_{nm_n},\ V_{11},\ V_{12},\ \ldots,\ V_{nm_n}\right) \end{cases} \tag{14.19}$$

Instantaneous power is then passed through low-pass filter with the cutoff frequency ω_c. The average active and reactive power are given as (14.20).

$$P_{ij} = \frac{\omega_c}{s + \omega_c} p_{ij},\quad Q_{ij} = \frac{\omega_c}{s + \omega_c} q_{ij} \tag{14.20}$$

Combining (14.13), (14.19), and (14.20), small-signal equations of normalized active and reactive power can be obtained as (14.21).

$$\begin{cases} \Delta\dot{P}^{pu} = diag\{-\omega_c\}\Delta P^{pu} + \omega_c K_{p^{pu}\tilde{\delta}}\Delta\tilde{\delta} + \omega_c K_{p^{pu}V}\Delta V \\ \Delta\dot{Q}^{pu} = diag\{-\omega_c\}\Delta Q^{pu} + \omega_c K_{q^{pu}\tilde{\delta}}\Delta\tilde{\delta} + \omega_c K_{q^{pu}V}\Delta V \end{cases} \tag{14.21}$$

where variable vectors (ΔP^{pu},ΔQ^{pu},$\Delta\tilde{\delta}$ and ΔV) and parameter matrixes ($K_{p^{pu}\tilde{\delta}}$, $K_{p^{pu}\tilde{\delta}}$, $K_{q^{pu}\tilde{\delta}}$ and $K_{q^{pu}V}$) are given in Appendix. (2) Distributed controller modeling

With the proposed distributed control, small-signal models of active and reactive controllers are given as (14.22).

$$\begin{cases} \Delta\dot{\tilde{\delta}} = \left(K_{\omega}diag\left\{\text{sgn}\left(Q_{ij}\right)\right\}L_s + K_{\omega\_p}L_p\right)\Delta P^{pu} = K_1\Delta P^{pu} \\ \Delta\dot{V} = \left(-K_P diag\left\{\text{sgn}\left(Q_{ij}\right)\right\}L_s + K_{P\_p}L_p\right)\Delta\dot{Q}^{pu} \\ \quad + \left(-K_I diag\left\{\text{sgn}\left(Q_{ij}\right)\right\}L_s + K_{I\_p}L_p\right)\Delta Q^{pu} \\ \quad = K_2\Delta\dot{Q}^{pu} + K_3\Delta Q^{pu} \end{cases} \tag{14.22}$$

where K_{ω}, K_{ω_p}, K_P, K_I, K_{P_p}, K_{I_p} are control coefficient matrixes. L_s and L_p are Laplacian matrixes carrying the information of cyber configuration. They are given in Appendix.

(3) Overall small-signal modeling

Small-signal dynamic model of the whole system can be established by combining (14.21)–(14.22) as (14.23).

$$\Delta\dot{x} = A\Delta x \tag{14.23}$$

where

$$\begin{cases} \Delta x = \left[\Delta P^{pu} \ \Delta Q^{pu} \ \Delta\tilde{\delta} \ \Delta V\right]^T \\ A = \begin{bmatrix} diag\{-\omega_c\} & 0_{m_n\times m_n} & \omega_c K_{p^{pu}\tilde{\delta}} & \omega_c K_{p^{pu}V} \\ 0_{m_n\times m_n} & diag\{-\omega_c\} & \omega_c K_{q^{pu}\tilde{\delta}} & \omega_c K_{q^{pu}V} \\ K_1 & 0_{m_n\times m_n} & 0_{m_n\times m_n} & 0_{m_n\times m_n} \\ 0_{m_n\times m_n} & diag\{\omega_c\}K_2 + K_3 & \omega_c K_{q^{pu}\tilde{\delta}}K_2 & \omega_c K_{q^{pu}V}K_2 \end{bmatrix} \end{cases} \tag{14.24}$$

14.4.2 Eigenvalue Analysis

The eigenvalues of matrix A can be calculated to investigate small-signal stability of the system. Eigenvalue trajectory of system (14.23) is analyzed while varying the control parameters, where control parameters are identical for all DG units, defined as k_{ω}, k_{ω_p}, k_P, k_I, k_{P_p}, and k_{I_p}. It is noted that matrix A has one zero eigenvalue which is corresponding to rotational symmetry and only the nonzero eigenvalues are valid for the system dynamic stability [6–8]. Since the system has similar response to the control parameters under RL and RC load, only the dynamic response under

RL load is described, while the stability regions of control parameters under RC load can be found in Table 14.1.

Figures 14.8 and 14.9 show eigenvalue trajectories with k_ω and k_{ω_p} varying. In Fig. 14.7, when k_ω is small, conjugate poles λ_1 and λ_2 are located in right half-plane, which means that system is unstable. The eigenvalue traces will move to left-half plane as increases of k_ω. Meanwhile, λ_3 $\lambda_1 1$ move toward the imaginary axis decreasing the damping ratio of system. In Fig. 14.8, as k_{ω_p} increases, λ_1 and λ_2 move toward unstable region and eventually lie on the right half-plane when $k_{\omega\_p} = 6.4$, making system unstable.

Figures 14.10 and 14.11 show eigenvalue trajectories with k_P and k_I varying. It can be seen from Fig. 14.9 that increasing k_P attracts the conjugate poles λ_1 and λ_2 to the imaginary axis, making the system unstable. It can also be seen from Fig. 14.10 that increasing k_I facilitates λ_1 and λ_2 moving away from the imaginary axis, keeping system stable. Parameters k_{P_p} and k_{I_p} have similar influence with k_P and k_I respectively, as shown in Figs. 14.12 and 14.13.

Figures 14.12 and 14.13 show control parameters k_ω, k_{ω_p}, k_P, k_I, k_{P_p}, and k_{I_p} that have critical influence on system stability and dynamic performance. Taking the overshoot during transients into account, the stability regions of those parameters under RL load and RC load are shown in Table 14.1.

14.5 Simulation Results

To validate effectiveness of the globally distributed control strategy, simulation verification is implemented in a scale-down AC microgrid with 4×4 converters in MATLAB. The circuit configuration of the exemplified microgrid is shown in Fig. 14.14. The ring communication network is shown in Fig. 14.15. The circuit and control parameters applied in simulation verification are given in Table 14.2.

Table 14.1 Stability region of control parameters

RL load		RC load	
Parameter	Stability region	Parameter	Stability region
k_ω	[12.3, 100]	k_ω	[18, 100]
k_{ω_p}	[0, 6.4]	k_{ω_p}	[0, 5.3]
k_P	[0, 82.5]	k_P	[0, 76.2]
k_I	[31, 100]	k_I	[35.8, 100]
k_{P_p}	[0, 44.2]	k_{P_p}	[0, 39.5]
k_{I_p}	[11.5, 50]	k_{I_p}	[13.4, 50]

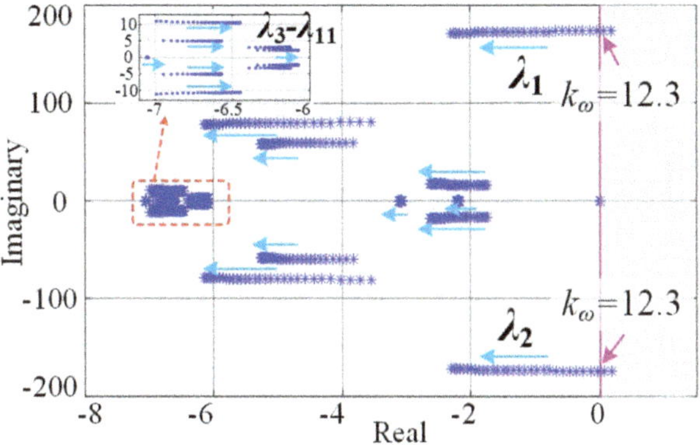

Fig. 14.8 Eigenvalue trajectory as $k_\omega \in [0,\ 100]$

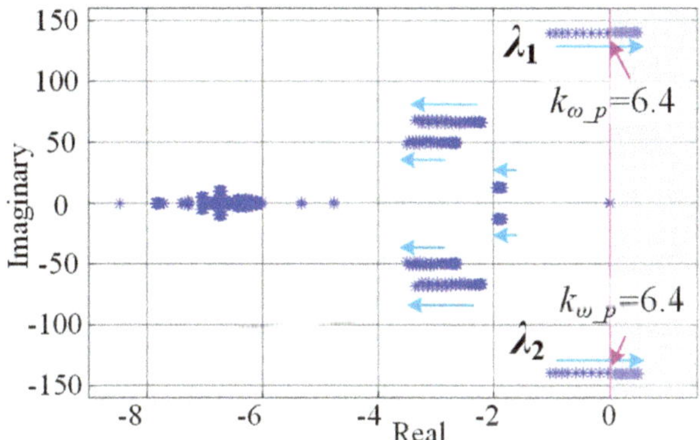

Fig. 14.9 Eigenvalue trajectory as $k_{\omega\_p} \in [0,\ 10]$

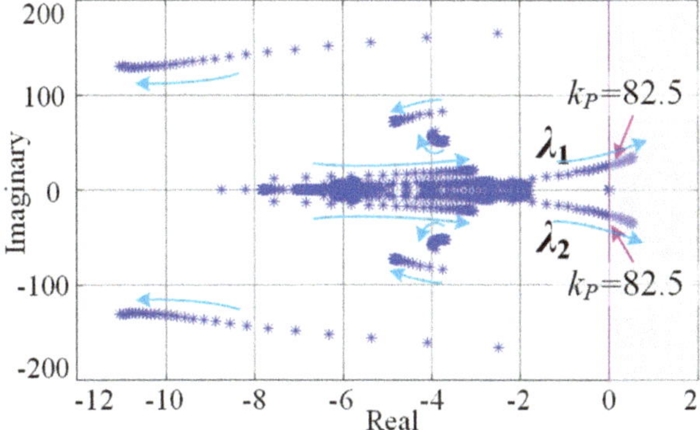

Fig. 14.10 Eigenvalue trajectory as $k_P \in [0,\ 100]$

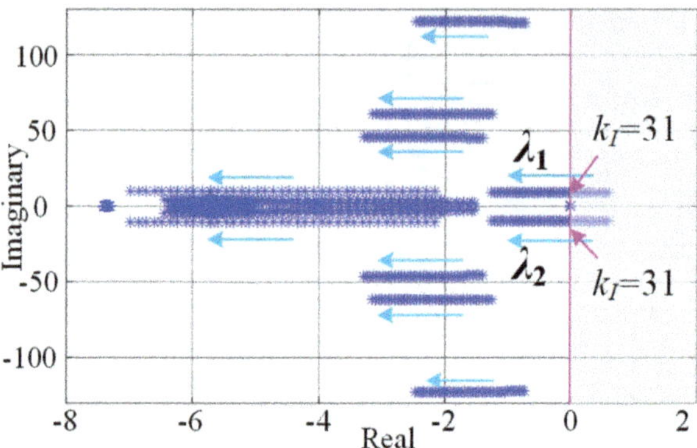

Fig. 14.11 Eigenvalue trajectory as $k_I \in [0,\ 100]$

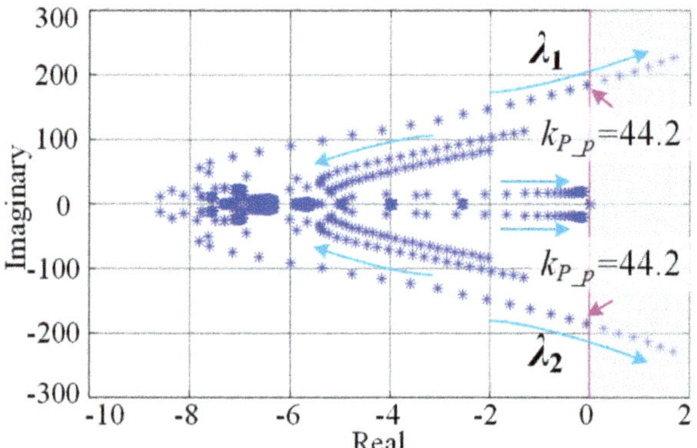

Fig. 14.12 Eigenvalue trajectory as $k_{P\_p} \in [0,\ 50]$

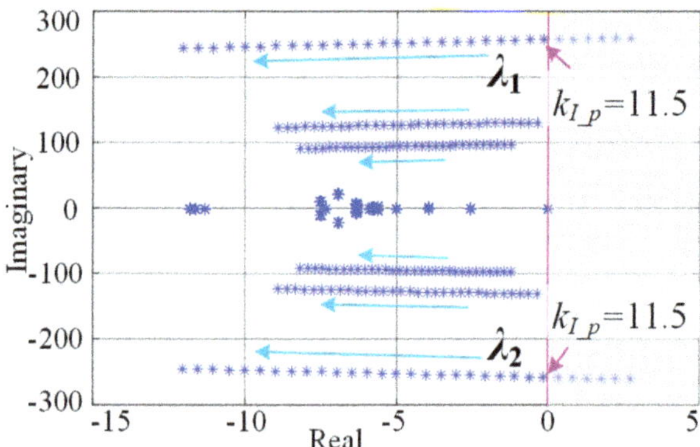

Fig. 14.13 Eigenvalue trajectory as $k_{I\_p} \in [0,\ 50]$

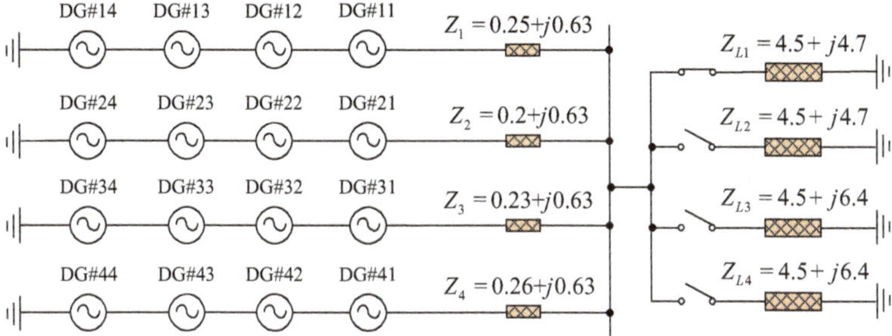

Fig. 14.14 Equivalent circuit of test microgrid

Fig. 14.15 Cyber configuration of test microgrid

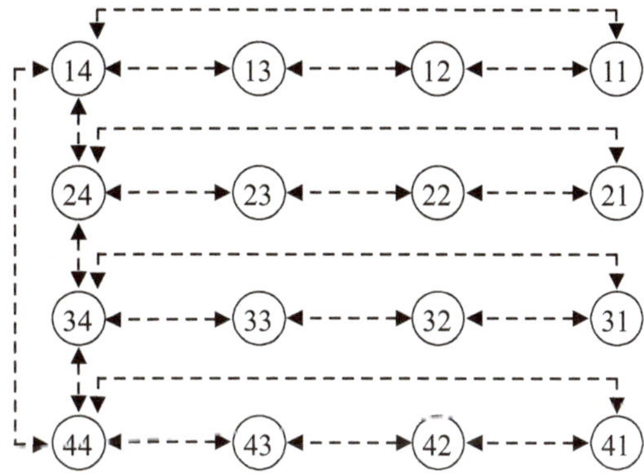

Table 14.2 System parameters

Item	Value
Voltage amplitude reference of AC bus	$V_p^* = 311\,\text{V}$
Rated angular frequency	$\omega^* = 100\,\pi\,\text{rad/s}$
Filter inductance	$L_f = 1.6\,\text{mH}$
Filter capacitance	$C_f = 20\,\mu\text{F}$
Communication weights	$a_{ij\_h,h\in N_{ij\_s}} = 1, a_{ij\_h,h\in N_{ij\_p}} = 1$
Active power control coefficients	$k_\omega = 30, k_{\omega\_p} = 3$
Reactive power control coefficients	$k_P = 35, k_I = 50, k_{P\_p} = 20, k_{I\_p} = 30$

14.5.1 Case 1: Performance of Unified Distributed Controller

In this case, performance of the distributed control method is tested under RL load and RC load. In practical operation, there exist several trivial disturbances such as measurement noise and parameter perturbations [3], etc., which may deteriorate controller performance. Therefore, immunity capability to these disturbances is also tested in this case. Measurement noise is added to voltage/current measurement signals and stochastic disturbance is added to controller parameters. Figure 14.16 shows simulation results under RL load. In order to analyze power sharing performance of proposed controller under different load profiles, Z_{L1} is connected at first and Z_{L2} is plugged at 2 s. It can be seen from Fig. 14.16 that all converters frequencies synchronize to the rated frequency of 50 Hz in the steady state without any deviation. And the system has a fast response with the load change. Figure 14.16b–c show that proportional load sharing is accurately maintained among cascaded and paralleled sources. Figure 14.17 shows simulation results under RC load, where Z_{L3} is connected at first and Z_{L4} is plugged at 2 s. It can be seen that similar performance is achieved under RC load. These simulation results validate the effectiveness of unified power control strategy under RL and RC load. Furthermore, the proposed controller is able to deal with the effect of measurement noise and parameter perturbation as shown in Figs. 14.16 and 14.17.

14.5.2 Case 2: Performance of Droop Controller

Performance of conventional droop controller is tested in this case, where conventional droop controller is enabled during 2–5 s and the distributed controller is activated during 0–2 s and 5–8 s. Figures 14.18 and 14.19 show frequency responses of DGs under RL load and RC load. It can be seen that hybrid series–parallel microgrid with RL and RC load would become unstable once droop controller is activated. The distributed control strategy can ensure the system stability without frequency deviation under both RL and RC load.

14.5.3 Case 3: Communication Delay

To investigate the effect of communication delay on system stability, controller performance with different delay time ($\tau = 50$ ms and $\tau = 100$ ms) is presented in Figs.14.20 and 14.21. Controller performances under RC load are omitted since they are similar with those under RL load. Load change is same as it in case 1. Figure

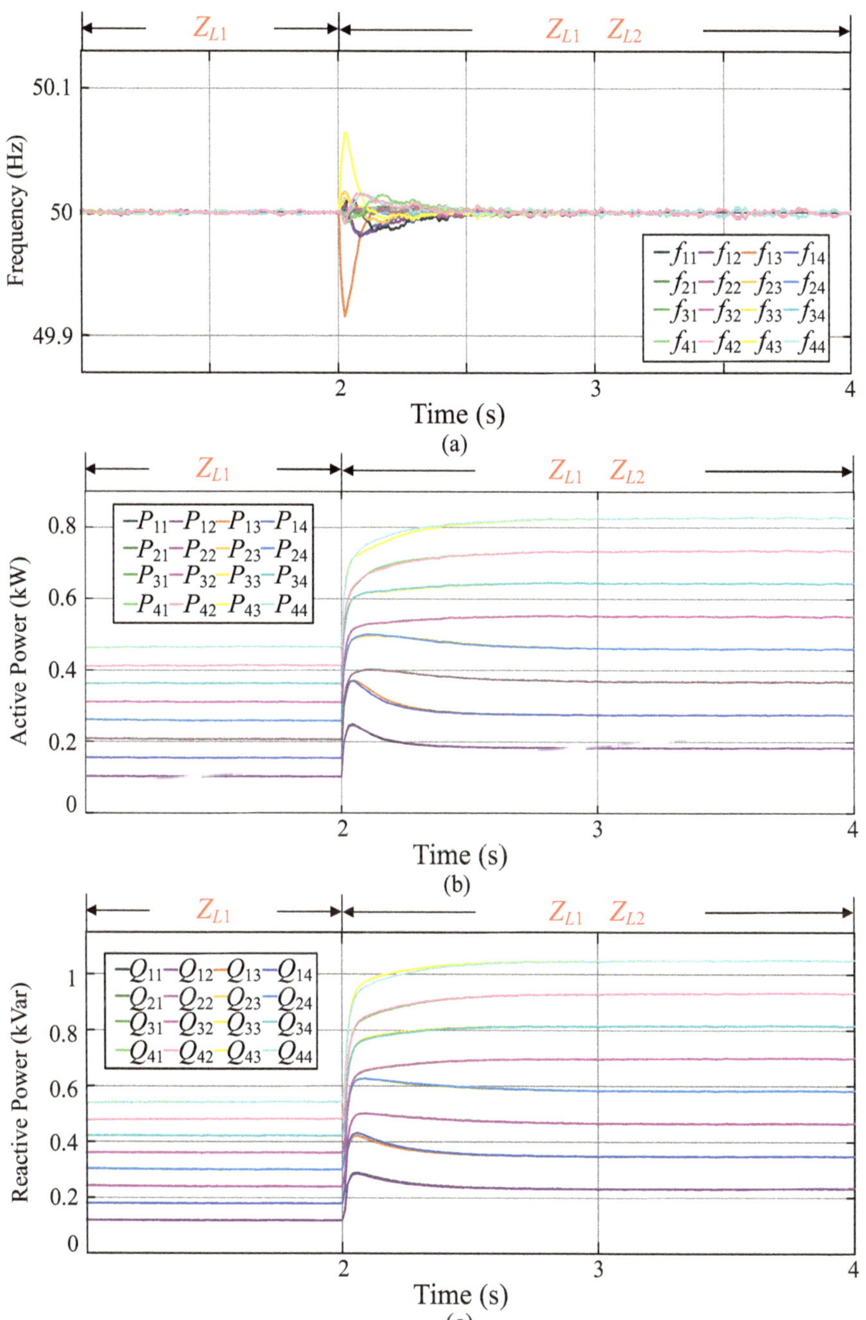

Fig. 14.16 Simulation results of case 1 under RL load. (**a**) Frequency. (**b**) Active power. (**c**) Reactive power

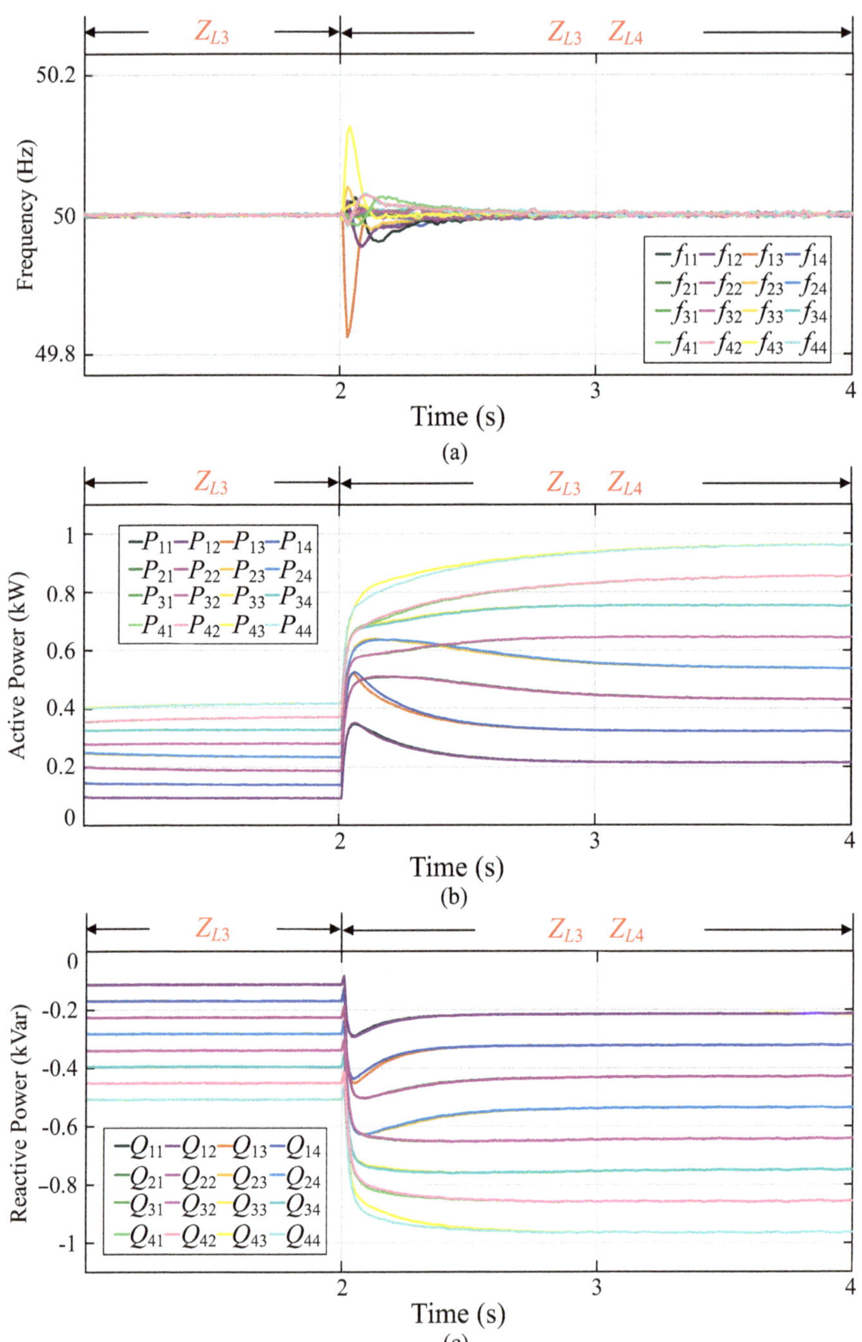

Fig. 14.17 Simulation results of case 1 under RC load. (**a**) Frequency. (**b**) Active power. (**c**) Reactive power

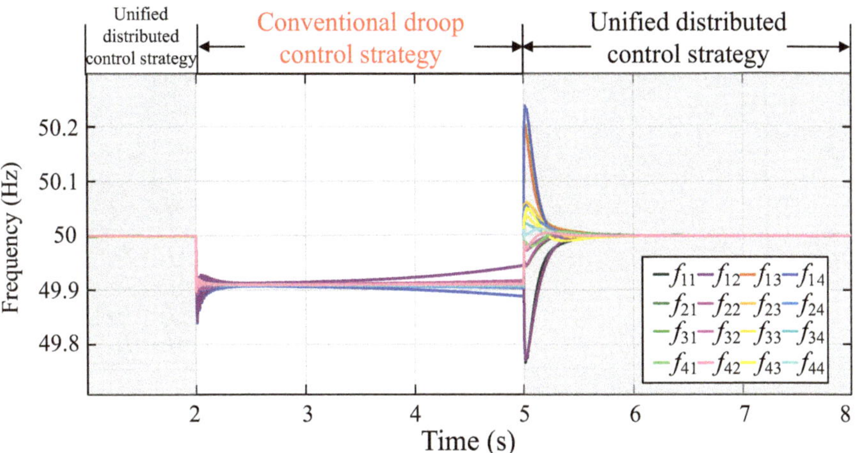

Fig. 14.18 Simulation result of case 2 under RL load

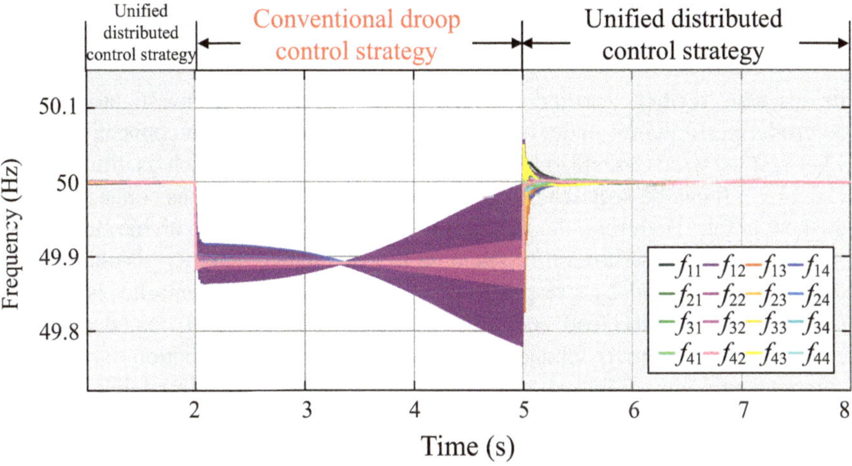

Fig. 14.19 Simulation result of case 2 under RC load

14.20 shows the system frequency with $\tau = 50$ ms. It can be seen that frequency synchronization is achieved at steady state and only little fluctuations are caused during transients. The active and reactive power results are omitted since they are similar with those in case 1. In Fig. 14.21, bigger fluctuations and even the system oscillation during steady state are triggered when $\tau = 100$ ms. From Figs. 14.20 and 14.21, it can be seen that communication delay may mitigate system stability if delay time is higher than critical value of stability region. Here, the controller has a good immunity capability for communication system within delay time 50 ms, which ensures system stability for existing communication protocols including WiFi and ultrawideband (UWB).

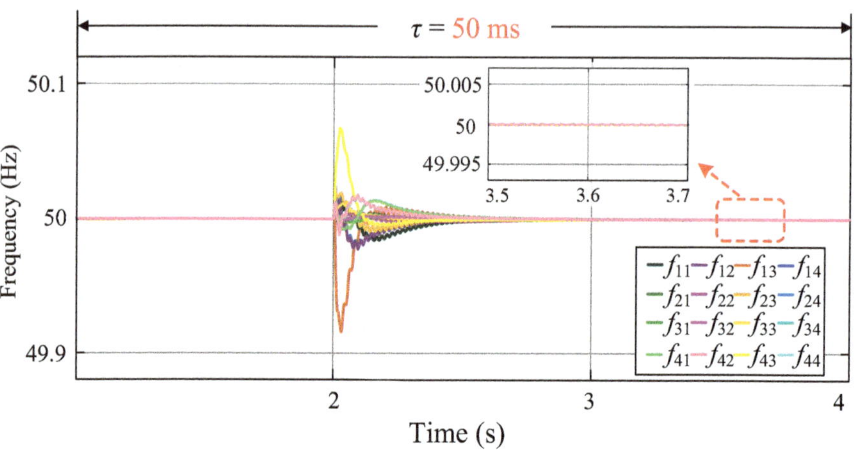

Fig. 14.20 Simulation result as $\tau = 50\,\text{ms}$

14.5.4 Case 4: Communication Link Failure

In this case, resiliency to the communication link failure is investigated. To test controller performance under cyber network with the minimum connectivity, links 13–14, 23–24, 33–34, 43–44, and 14–24 fail at 0.25 s, which is illustrated in Fig. 14.22. It can be seen from Fig. 14.22 that the system remains connected under the new graph. Therefore, the link failure should have no effect on the steady-state performance. In order to test the response to load change, load Z_{L2} is attached and detached at 0.5 s and 2.5 s respectively. The performance of controller is identical under RL load and RC load, and thus we omit the plots under RC load due to space consideration. Figure 14.23 shows frequency and power distribution results under communication link failure. It can be seen that communication link failure does not affect the steady-state performance while it slows down the system dynamics.

14.5.5 Case 5: Plug-and-Play Characteristic

Plug-and-play characteristic of the distributed control method is tested, via unplugging the String#4 at 0.5 s and plugging it back in at 2.5 s. The two physical configurations are illustrated in Fig. 14.24 and simulation results are displayed in Fig. 14.25. It should be noted that when String#4 is unplugged, the communication link 34–44 is disabled at the same time. Nevertheless, the cyber network remains connected and, thus, the steady-state performance is not compromised. From Fig. 14.25, the power supplied by String#4 reduces to zero during 0.5 s–2.5 s and the excess power is proportionately shared among remaining DGs. After String#4 is

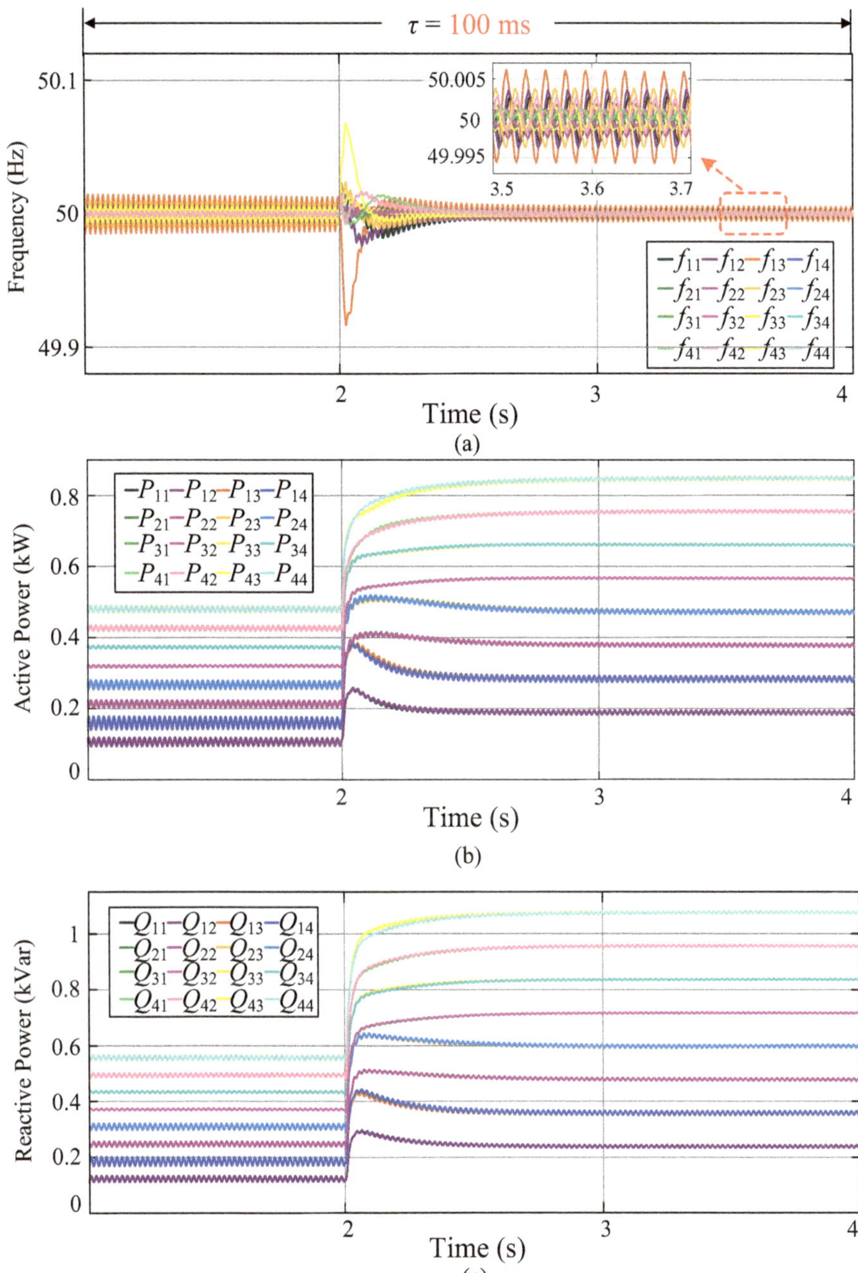

Fig. 14.21 Simulation results as $\tau = 100\,\text{ms}$ (**a**) Frequency. (**b**) Active power. (**c**) Reactive power

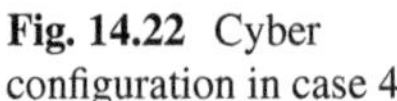

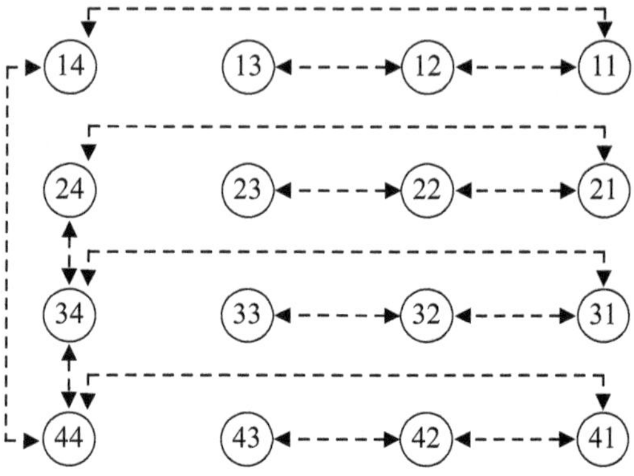

Fig. 14.22 Cyber configuration in case 4

reconnected at 2.5 s, the controllers successfully respond to the converter plugging and share load demand among all DGs.

14.6 Conclusion

This section presents a globally distributed control strategy for hybrid series–parallel microgrid with different load types, where active power and reactive power regulators without frequency drop are developed. A low-bandwidth communication network is designed to support power control and improve system redundancy. Furthermore, small-signal model of hybrid series–parallel microgrid is established. And small-signal stability and dynamic performance is investigated. Eigenvalue analysis shows that parameters of active power regulator and reactive power regulator have critical influences on system stability and dynamic performance. Simulation results show that the globally distributed control strategy is able to implement desirable power sharing under RL load and RC load with superior control performance. Also, the steady-state performance of proposed control strategy is slightly affected in the presence of communication delay and communication link failure. In addition, the distributed control strategy is able to improve system redundancy and support plug-and-play operation of microgrid.

Appendix

The variable vectors (ΔP^{pu}, ΔQ^{pu}, $\Delta\tilde{\delta}$, and ΔV) and parameter matrixes ($K_{p^{pu}\tilde{\delta}}$, $K_{p^{pu}V}$, $K_{q^{pu}\tilde{\delta}}$, and $K_{q^{pu}V}$) in (14.21) are given as (14.25)–(14.26).

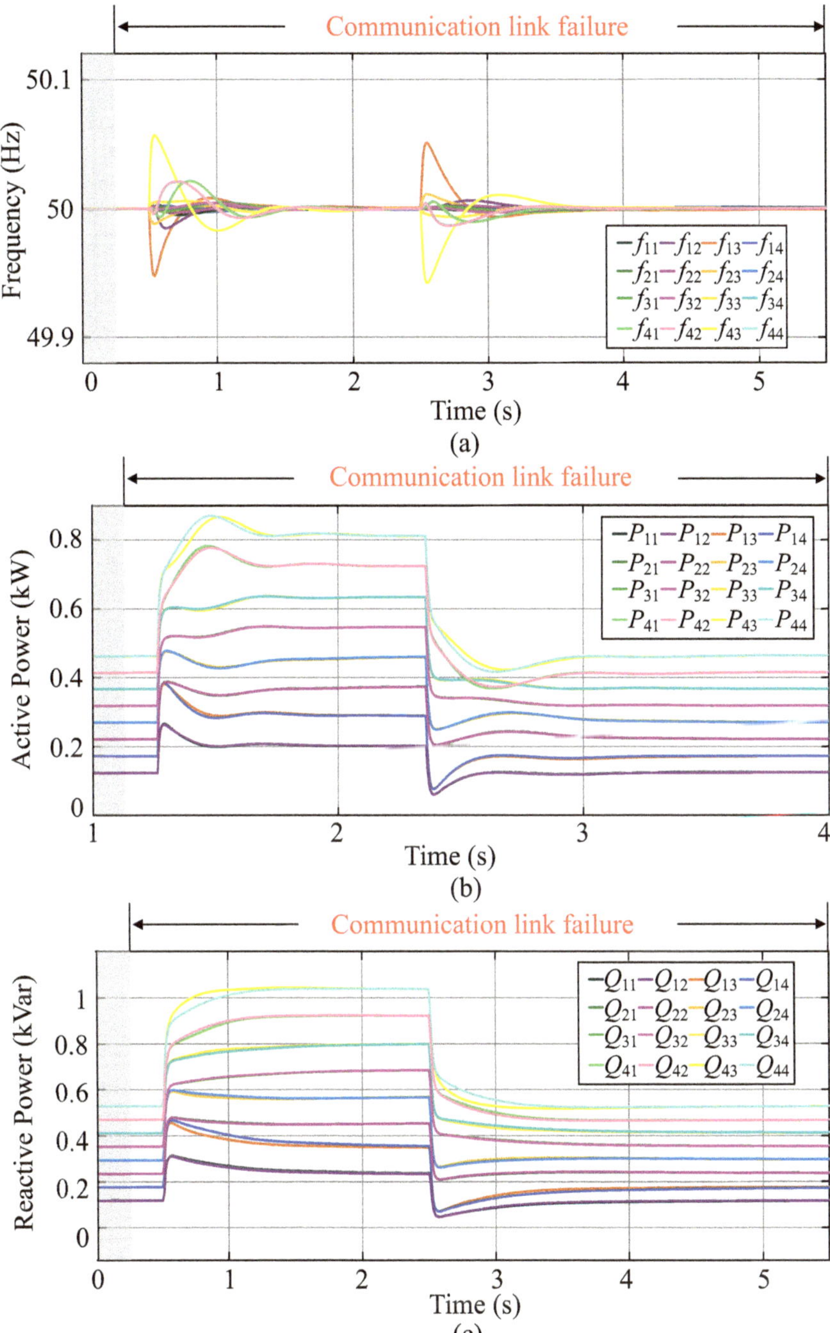

Fig. 14.23 Simulation results of case 4. (**a**) Frequency. (**b**) Active power. (**c**) Reactive power

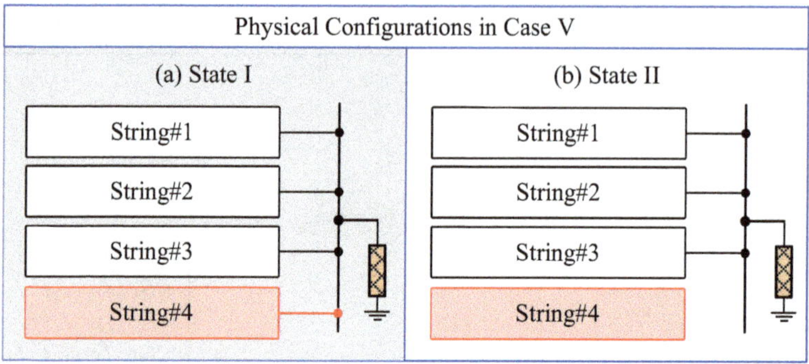

Fig. 14.24 Physical configuration in case 5

$$
\begin{cases}
\Delta P^{pu} = \left[\Delta P^{pu}_{11} \quad \Delta P^{pu}_{12} \quad \cdots \quad \Delta P^{pu}_{nm_n}\right]^T \\
\Delta Q^{pu} = \left[\Delta Q^{pu}_{11} \quad \Delta Q^{pu}_{12} \quad \cdots \quad \Delta Q^{pu}_{nm_n}\right]^T \\
\Delta\tilde{\delta} = \left[\Delta\tilde{\delta}_{11} \quad \Delta\tilde{\delta}_{nm_n} \quad \cdots \quad \Delta\tilde{\delta}_{nm_n}\right]^T \\
\Delta V = \left[\Delta V_{11} \quad \Delta V_{12} \quad \cdots \quad \Delta V_{nm_n}\right]^T \\
K_{p^{pu}\tilde{\delta}} = \begin{bmatrix} k_{p^{pu}_{11}\tilde{\delta}_{11}} & k_{p^{pu}_{11}\tilde{\delta}_{12}} & \cdots & k_{p^{pu}_{11}\tilde{\delta}_{nm_n}} \\ k_{p^{pu}_{12}\tilde{\delta}_{11}} & k_{p^{pu}_{12}\tilde{\delta}_{12}} & \cdots & k_{p^{pu}_{12}\tilde{\delta}_{nm_n}} \\ \vdots & \vdots & \ddots & \vdots \\ k_{p^{pu}_{nm_n}\tilde{\delta}_{11}} & k_{p^{pu}_{nm_n}\tilde{\delta}_{12}} & \cdots & k_{p^{pu}_{nm_n}\tilde{\delta}_{nm_n}} \end{bmatrix} \\
K_{p^{pu}V} = \begin{bmatrix} k_{p^{pu}_{11}V_{11}} & k_{p^{pu}_{11}V_{12}} & \cdots & k_{p^{pu}_{11}V_{nm_n}} \\ k_{p^{pu}_{12}V_{11}} & k_{p^{pu}_{12}V_{12}} & \cdots & k_{p^{pu}_{12}V_{nm_n}} \\ \vdots & \vdots & \ddots & \vdots \\ k_{p^{pu}_{nm_n}V_{11}} & k_{p^{pu}_{nm_n}V_{12}} & \cdots & k_{p^{pu}_{nm_n}V_{nm_n}} \end{bmatrix} \\
K_{q^{pu}\tilde{\delta}} = \begin{bmatrix} k_{q^{pu}_{11}\tilde{\delta}_{11}} & k_{q^{pu}_{11}\tilde{\delta}_{12}} & \cdots & k_{q^{pu}_{11}\tilde{\delta}_{nm_n}} \\ k_{q^{pu}_{12}\tilde{\delta}_{11}} & k_{q^{pu}_{12}\tilde{\delta}_{12}} & \cdots & k_{q^{pu}_{12}\tilde{\delta}_{nm_n}} \\ \vdots & \vdots & \ddots & \vdots \\ k_{q^{pu}_{nm_n}\tilde{\delta}_{11}} & k_{q^{pu}_{nm_n}\tilde{\delta}_{12}} & \cdots & k_{q^{pu}_{nm_n}\tilde{\delta}_{nm_n}} \end{bmatrix} \\
K_{q^{pu}V} = \begin{bmatrix} k_{q^{pu}_{11}V_{11}} & k_{q^{pu}_{11}V_{12}} & \cdots & k_{q^{pu}_{11}V_{nm_n}} \\ k_{q^{pu}_{12}V_{11}} & k_{q^{pu}_{12}V_{12}} & \cdots & k_{q^{pu}_{12}V_{nm_n}} \\ \vdots & \vdots & \ddots & \vdots \\ k_{q^{pu}_{nm_n}V_{11}} & k_{q^{pu}_{nm_n}V_{12}} & \cdots & k_{q^{pu}_{nm_n}V_{nm_n}} \end{bmatrix}
\end{cases} \tag{14.25}
$$

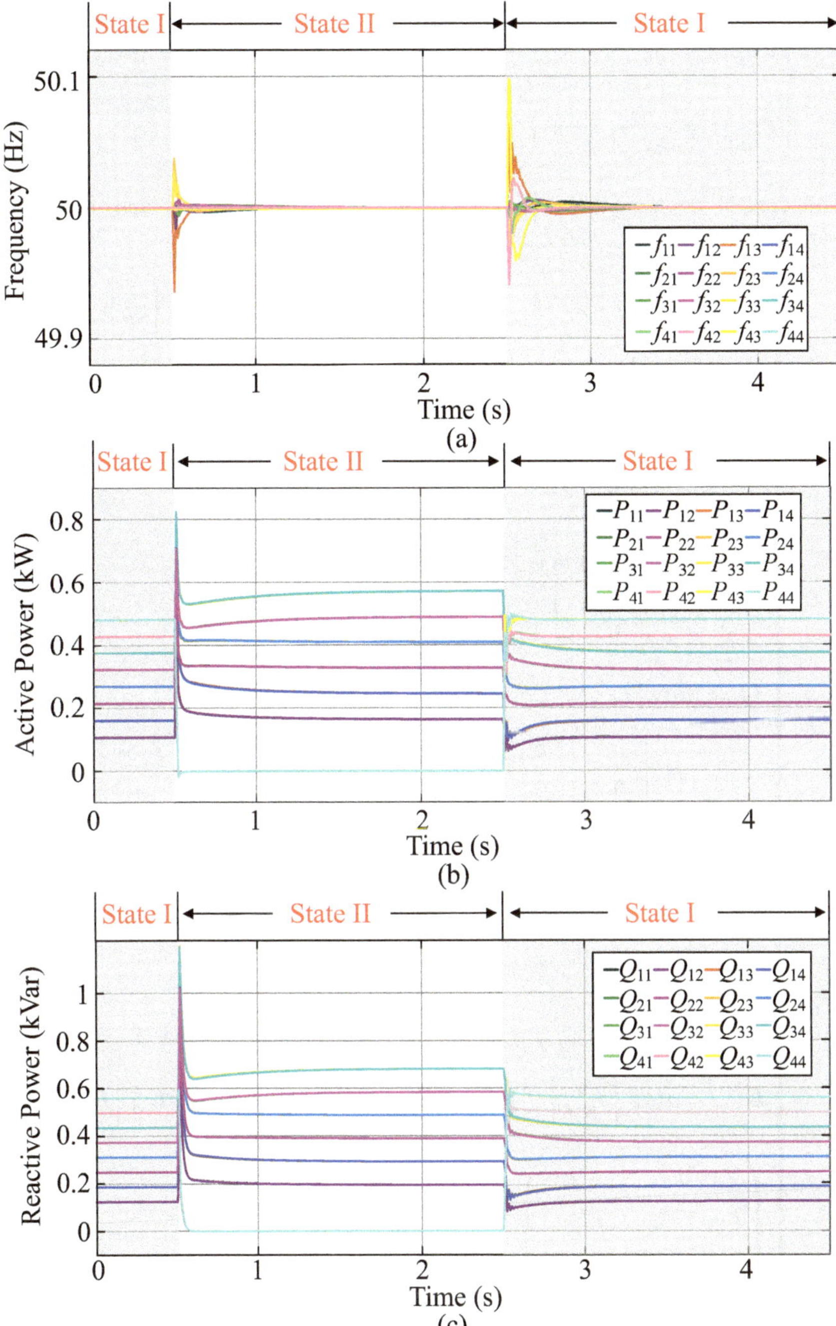

Fig. 14.25 Simulation results of case 5. (**a**) Frequency. (**b**) Active power. (**c**) Reactive power

$$\begin{cases} k_{p_{ij}^{pu}\tilde{\delta}_{ab}} = \dfrac{\partial p_{ij}^{pu}}{\partial \tilde{\delta}_{ab}}, & k_{p_{ij}^{pu}V_{ab}} = \dfrac{\partial p_{ij}^{pu}}{\partial V_{ab}} \\ k_{q_{ij}^{pu}\tilde{\delta}_{ab}} = \dfrac{\partial q_{ij}^{pu}}{\partial \tilde{\delta}_{ab}}, & k_{q_{ij}^{pu}V_{ab}} = \dfrac{\partial q_{ij}^{pu}}{\partial V_{ab}} \end{cases} \begin{pmatrix} i = 1, \cdots, n & j = 1, \cdots, m_i \\ a = 1, \cdots, n & b = 1, \cdots, m_a \end{pmatrix} \tag{14.26}$$

The control coefficient matrixes (K_ω, K_{ω_p}, K_P, K_I, K_{P_p} and K_{I_p}) and Laplacian matrixes (L_s and L_p) in (14.22) are given as (14.27).

$$\begin{cases} K_\omega = diag\left\{k_\omega^{ij}\right\}, \ K_{\omega\_p} = diag\left\{k_{\omega\_p}^{ij}\right\}, \ K_P = diag\left\{k_P^{ij}\right\} \\ K_I = diag\left\{k_I^{ij}\right\}, \ K_{P\_p} = diag\left\{k_{P\_p}^{ij}\right\}, \ K_{I\_p} = diag\left\{k_{I\_p}^{ij}\right\} \\ A_{i\_s} = \begin{bmatrix} 0 & a_{i1\_i2} & \cdots & a_{i1\_im_i} \\ a_{i2\_i1} & 0 & \cdots & a_{i2\_im_i} \\ \vdots & \vdots & \ddots & \vdots \\ a_{im_i\_i1} & a_{im_i\_i2} & \cdots & 0 \end{bmatrix} \\ D_{i\_s}^{in} = diag\left\{d_{ij\_s}^{in}\right\}, \ d_{ij\_s}^{in} = \sum\limits_h a_{ij\_h} \\ L_s = diag\left\{L_{i\_s}\right\}, \ L_{i\_s} = D_{i\_s}^{in} - A_{i\_s} \\ A_{i\_k\_p} = \begin{bmatrix} a_{i1\_l1} & a_{i1\_l2} & \cdots & a_{i1\_lm_l} \\ a_{i2\_l1} & a_{i2\_l2} & \cdots & a_{i2\_lm_l} \\ \vdots & \vdots & \ddots & \vdots \\ a_{im_i\_l1} & a_{im_i\_l2} & \cdots & a_{im_i\_lm_l} \end{bmatrix} \\ A_p = \begin{bmatrix} 0_{m_1\times m_1} & A_{1\_2\_p} & \cdots & A_{1\_n\_p} \\ A_{2\_1\_p} & 0_{m_2\times m_2} & \cdots & A_{2\_n\_p} \\ \vdots & \vdots & \ddots & \vdots \\ A_{n\_1\_p} & A_{n\_2\_p} & \cdots & 0_{m_n\times m_n} \end{bmatrix} \\ D_{i\_p}^{in} = diag\left\{d_{ij\_r}^{in}\right\}, \ d_{ij\_p}^{in} = \sum\limits_r a_{ij\_r}, \ L_p = A_p - diag\left\{D_{i\_p}^{in}\right\} \\ \begin{pmatrix} i = 1, 2, & \ldots, n \\ l = 1, 2, & \ldots, n \\ j = 1, 2, & \ldots, m_n \\ h \in N_{ij\_s} & \\ r \in N_{ij\_p} & \end{pmatrix} \end{cases} \tag{14.27}$$

References

1. Y. Wang, Z. Chen, X. Wang, Y. Tian, Y. Tan, C. Yang, An estimator-based distributed voltage predictive control strategy for AC Islanded microgrids. IEEE Trans. Power Electron. **30**(7), 3934–3951 (2015)
2. W. Yuan, J. Yang, G. Shi, Y. Sun, H. Han, X. Hou, Control design and stability analysis for the cascaded-type AC microgrid, in *IECON 2017 - 43rd Annual Conference of the IEEE Industrial Electronics Society* (2017), pp. 2407–2412
3. Y. Sun, G. Shi, X. Li, W. Yuan, M. Su, H. Han, X. Hou, An f-P/Q droop control in cascaded-type microgrid. IEEE Trans. Power Syst. **33**(1), 1136–1138 (2018)
4. V. Nasirian, Q. Shafiee, J.M. Guerrero, F.L. Lewis, A. Davoudi, Droop-free distributed control for AC microgrids. IEEE Trans. Power Electron. **31**(2), 1600–1617 (2016)
5. A. Bidram, A. Davoudi, F.L. Lewis, S.S. Ge, Distributed adaptive voltage control of inverter-based microgrids. IEEE Trans. Energy Convers. **29**(4), 862–872 (2014)
6. E.A.A. Coelho, P.C. Cortizo, P.F.D. Garcia, Small-signal stability for parallel-connected inverters in stand-alone AC supply systems. IEEE Trans. Ind. Appl. **38**(2), 533–542 (2002)
7. J.W. Simpson-Porco, F. Dörfler, F. Bullo, Synchronization and power sharing for droop-controlled inverters in Islanded microgrids. Automatica **49**(9), 2603–2611 (2012)
8. Z. Liu, M. Su, Y. Sun, W. Yuan, H. Han, J. Feng, Existence and stability of equilibrium of DC microgrid with constant power loads. IEEE Trans. Power Syst. **33**(6), 6999–7010 (2018)
9. W. Yuan, et al., A unified distributed control strategy for hybrid cascaded-parallel microgrid. IEEE Trans. Energy Convers. **34**(4), 2029–2040 (2019)

Chapter 15
A Local-Distributed and Global-Decentralized Control Scheme

15.1 Control Objectives

In Chap. 14, a globally distributed control strategy is presented. There is a DG in each string selected as a leader to exchange information with leaders of other strings. Though accurate power sharing can be guaranteed, the communication network throughout the whole microgrid limits the integration of remote DG units. Therefore, for better flexibility and less communication costs, a local-distributed power sharing control strategy with only in-string communication is introduced in this chapter, whose salient features are given as follows.

1. Proper power sharing. The active power among all DGs and reactive power among in-string DGs can be shared accurately. Furthermore, the approximate reactive power sharing performance among parallel strings can be improved by further decentralized control design, such as virtual impedance [1].
2. Sparse local LBC network. Compared with existing methods, a sparse local LBC network is adopted without central controllers or leader sources. Each DG only needs to exchange data with its local neighbors, which strengthens the reliability and redundancy. The plug-and-play capability of DG strings is also obtained due to no communication among strings.
3. Unified operation under diverse loads. A sign function is introduced to ensure the active power regulator available under diverse loads. Thus microgrid using the proposed control has a unified operation mode under both RL and RC loads.

Y. Sun et al., *Series-Parallel Converter-Based Microgrids*, Power Systems,
https://doi.org/10.1007/978-3-030-91511-7_15

15.2 Local-Distributed Control for Hybrid Series–Parallel Microgrid

15.2.1 Hybrid Series–Parallel Microgrid

The configuration of hybrid series–parallel microgrid is shown in Fig. 15.1. A few of local DG units are cascaded as a DG string and then supply power to the point of common coupling (PCC) in parallel. N and M_i represent the total number of DG strings and DG units in string#i, respectively. Here the interface converters adopt the structure of conventional H-bridge with an output LC filter for easier independent control. If the control method is designed properly, these converters work together to obtain a suitable string output voltage which is equal to the vector sum of all in-string converters' output voltage. Therefore, with this topology, multiple LV sources can be integrated into a higher voltage network without back stage converters, unlike in parallel microgrid where two stage DC/AC power conversion is commonly needed.

The PCC of microgrid is connected to the main grid with a static transfer switch (STS). In the grid-connected operation mode, the PCC voltage is determined by the grid, hence each string inverters in hybrid series–parallel microgrid can be controlled independently as a separate series system. Decentralized power control

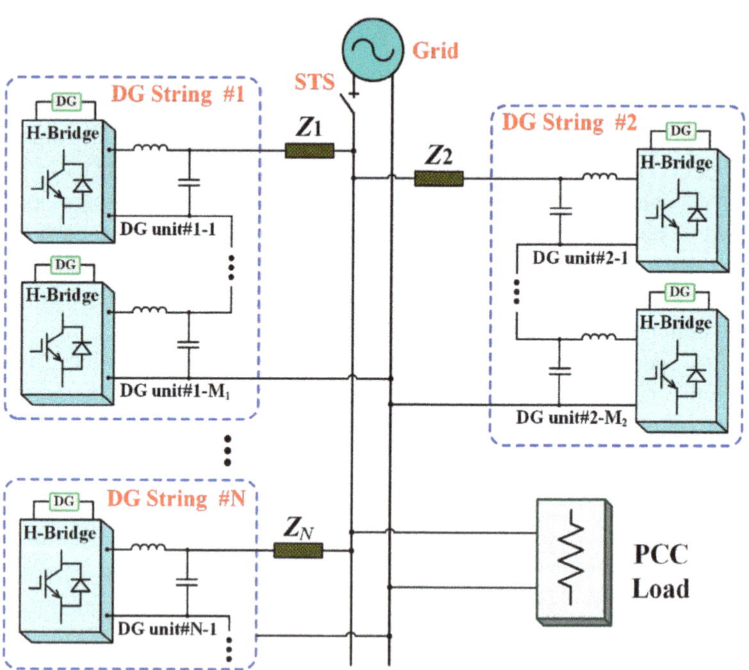

Fig. 15.1 Configuration of hybrid series–parallel microgrid

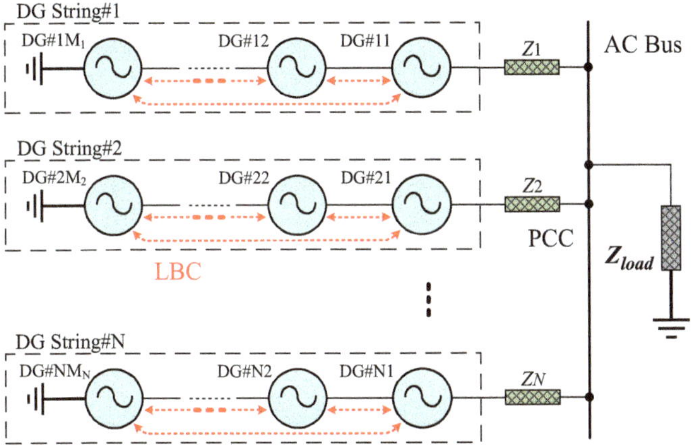

Fig. 15.2 Equivalent circuit of hybrid series–parallel microgrid in island mode

methods proposed for grid-connected series microgrid in [2] and [3] are valid in this hybrid system as well. When the grid is not available and the STS turns off, namely the microgrid runs in islanded mode, PCC load is powered by parallel DG strings. Actually, the hybrid microgrid can be regarded as a more comprehensive system that includes both series and parallel systems. Accordingly, there is not a unified decentralized method to achieve power sharing because of the incompatibility between inverse droop control (RL loads) in series system and droop control in parallel system which are expressed as (15.1) and (15.2), respectively.

$$\omega_i = \omega^* - m_i P_i \tag{15.1}$$

$$\omega_i = \omega^* + m_i \operatorname{sgn}(Q_i) P_i \tag{15.2}$$

where P_i and Q_i are output active power reactive power; ω_i is the angular frequency; ω^* represents the rated angle frequency; m_i is a coefficient of $P - \omega$; $sgn()$ is a signum function. For the purpose of high flexibility and less communication costs, a local-distributed control strategy with only in-string communication is designed.

15.2.2 Communication Network Design

The equivalent circuit of the autonomous series–parallel microgrid is shown in Fig. 15.2 where the adopted LBC links are indicated by red lines. From the control perspective, microgrid can be deemed as a cyber-physical system with a communication network to facilitate data exchange among DGs for different targets [4]. The graphical representation of communication network is illustrated in Fig. 15.3.

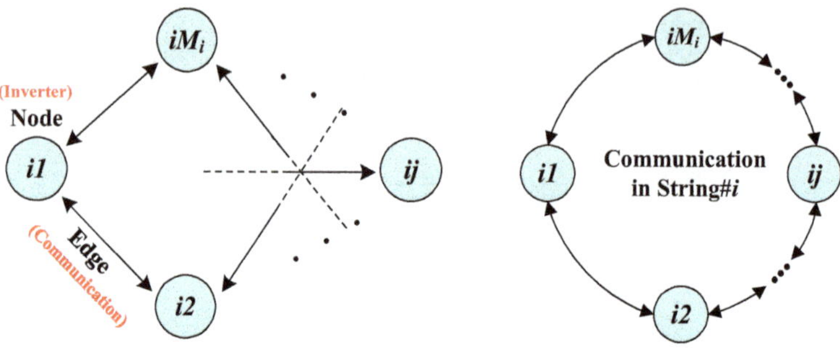

Fig. 15.3 Graphical diagram of the communication network and the designed LBC network in string#i

In the graph, nodes represent DGs and edges represent communication links that can be bidirectional to form a undigraph or not (digraph). Each node broadcasts its information, such as active power measurements and power estimations, to its neighbors with different communication weights. Note ij denotes the j-th DG unit in string#i and $l \in N_{ij}$ is defined as the set of all local in-string neighbors of Node ij, then the communication weight $c_{ij\_l} > 0$ if Node ij receives data from Node l and $c_{ij\_l} = 0$, otherwise. All communication weights in the graph are carried by an associated adjacency matrix $\mathbf{C} = \left[c_{ij\_l}\right] \in \mathbb{R}^{M \times M}$ where M is the number of nodes. The associated in-degree matrix is defined as $\mathbf{D}^{in} = diag\left\{d_{ij}^{in}\right\}$ with $d_{ij}^{in} = \sum_{l \in N_{ij}} c_{ij\_l}$. $\mathbf{L} = \mathbf{D}^{in} - \mathbf{A}$ is constructed as a Laplacian matrix.

The communication network should be designed to improve the reliability and minimize the complexity and costs. In the proposed distributed control strategy, only in-string communication is sufficient for data exchange among local adjacent converters. To simplify the analysis, the local communication network of each string is assumed to be identical. Taking string#i as an example, a bidirectional sparse LBC network with the minimum redundancy is designed as shown in Fig. 15.3. The circular communication graph is turned out to be an effective structure [5] containing enough redundant links to enable system running in the case of any single link failure [6].

15.2.3 Local-Distributed Control

Based on the designed communication network, a local-distributed control strategy is presented. Objectives of the controllers are (1) frequency synchronization, (2) accurate active power sharing, (3) proper reactive power sharing, (4) unified operation in islanded hybrid series–parallel microgrid. Typically, the output active and reactive power can be managed by adjusting frequency and voltage amplitude of each DG, respectively.

$$\omega_{ij} = \omega^* - m_i \bar{P}_{ij} - \mathrm{sgn}(Q_{ij})k_i \sum_{l \in N_{ij}} c_{ij\_l}\left(P_l - P_{ij}\right) \tag{15.3}$$

$$V_{ij} = V_i^* - n\bar{Q}_{ij} \tag{15.4}$$

where ω_{ij} and ω^* are reference angular frequency and nominal angular frequency, respectively. m_i, k_i, and n are control parameters.$\bar{P}_{ij}$ and $\bar{Q}_{ij}$ represent the estimations of local averaged active power and reactive power at Node ij. The rated voltage amplitude V_{ij}, the filtered output active power P_{ij} and reactive power Q_{ij} are expressed as,

$$V_i^* = V_p^*/M_i \tag{15.5}$$

$$\begin{cases} P_{ij} = \dfrac{\omega_c}{s+\omega_c} p_{ij} \\ Q_{ij} = \dfrac{\omega_c}{s+\omega_c} q_{ij} \end{cases} \tag{15.6}$$

where V_p^* is the rated voltage amplitude of PCC, Mi represent the total number of DG units in string# i. p_{ij} and q_{ij} are the measured power, ω_c is the cutoff frequency of the low-pass filter. Eventually, the in-string active power will be shared accurately when all local averaged active power estimations converge to a common value.

Each node has active and reactive power estimators which generate the estimations of averaged power and exchange them with its local neighbors. The average value estimators implement the following dynamics consensus protocol [7].

$$\bar{P}_{ij}(t) = P_{ij}(t) + \int_0^t \sum_{l \in N_{ij}} c_{ij\_l}\left(\bar{P}_l(\tau) - \bar{P}_{ij}(\tau)\right)d\tau \tag{15.7}$$

$$\bar{Q}_{ij}(t) = Q_{ij}(t) + \int_0^t \sum_{l \in N_{ij}} c_{ij\_l}\left(\bar{Q}_l(\tau) - \bar{Q}_{ij}(\tau)\right)d\tau \tag{15.8}$$

The outputs of estimators at each node are updated by processing estimations of adjacent nodes and its output power. Therefore, all other estimations of in-string nodes are affected in case of any power variation. Finally, the local dynamics consensus would be achieved by cooperation among local controllers. By differentiating (15.7)

$$\dot{\bar{P}}_{ij} = \dot{P}_{ij} + \sum_{l \in N_{ij}} c_{ij\_l}\left(\bar{P}_l - \bar{P}_{ij}\right) = \dot{\bar{P}}_{ij} - d_{ij}^{in}\bar{P}_{ij} + \sum_{l \in N_{ij}} c_{ij\_l}\bar{P}_l \tag{15.9}$$

where $d_{ij}^{in} = \sum_{l \in N_{ij}} c_{ij\_l}$ The overall estimator dynamic of string#i can be formulated in matrix form as

$$\dot{\bar{\mathbf{P}}}_i = \dot{\mathbf{P}}_\mathbf{i} - \mathbf{D}_i^{in}\bar{\mathbf{P}}_i + \mathbf{C}_i\bar{\mathbf{P}}_i = \dot{\mathbf{P}}_i - \mathbf{L}_i\bar{\mathbf{P}}_i \ \ (i = 1, \ldots N) \tag{15.10}$$

where $\mathbf{P}_i = [P_{i1}, P_{i2}, \ldots, P_{iM_i}]^T$, $\bar{\mathbf{P}}_i = [\bar{P}_{i1}, \bar{P}_{i2}, \ldots, \bar{P}_{iM_i}]^T$ denote the measured active power vector and the estimated active power vector, respectively. $\mathbf{D}_i^{in} = diag\left\{d_{ij}^{in}\right\}$ is the in-degree matrix of string#i. $\mathbf{C}_i = [c_{ij\_l}] \in \mathbb{R}^{M_i \times M_i}$ is an associated adjacency matrix of string#i. $\mathbf{L}_i = \mathbf{D}_i^{in} - \mathbf{C}_i$ is a Laplacian matrix carrying the information of communication graph in string#i.

With $\bar{P}_{ij}(0) = P_{ij}(0)$ Eqs. (15.7) and (15.11) can be obtained.

$$\bar{\mathbf{P}}_i = s(s\mathbf{I}_i + \mathbf{L}_i)^{-1}\mathbf{P}_i = \mathbf{H}_i\mathbf{P}_i \tag{15.11}$$

where $\mathbf{I}_i \in \mathbb{R}^{m_i \times m_i}$ is the identity matrix, $\mathbf{H}_i$ is the active power estimator transfer function matrix of string#i. It is presented in that all elements of $\bar{\mathbf{P}}_i$ converge to the overall average output active power of string#i in steady state if $\mathbf{L}_i$ is balanced, namely.

$$\bar{P}_{i1}^{ss} = \bar{P}_{i2}^{ss} = \cdots = \bar{P}_{iM_i}^{ss} = \frac{P_{i1}^{ss} + P_{i2}^{ss} + \cdots + P_{iM_i}^{ss}}{M_i} \tag{15.12}$$

where X^{ss} represents the steady-state value of X. Similarly, the averaged reactive power estimations of string#i converge to the true average value in steady state.

$$\bar{Q}_{i1}^{ss} = \bar{Q}_{i2}^{ss} = \cdots = \bar{Q}_{iM_i}^{ss} = \frac{Q_{i1}^{ss} + Q_{i2}^{ss} + \cdots + Q_{iM_i}^{ss}}{M_i} \tag{15.13}$$

15.2.4 Steady-State Analysis

In this chapter, in order to make the analysis procedure more concise, it is assumed that the rated capacities of all sources are equal in steady state. The block diagram of proposed local-distributed control strategy at Node ij is shown in Fig. 15.4. In active power regulator, $\delta\omega_{ij}$ is a correction term carrying the active power mismatch between DG#ij and its neighbors, which is expressed as,

$$\delta\omega_{ij} = \mathrm{sgn}(Q_{ij})k_i \sum_{l \in N_{ij}} c_{ij\_l}\left(P_l - P_{ij}\right) \tag{15.14}$$

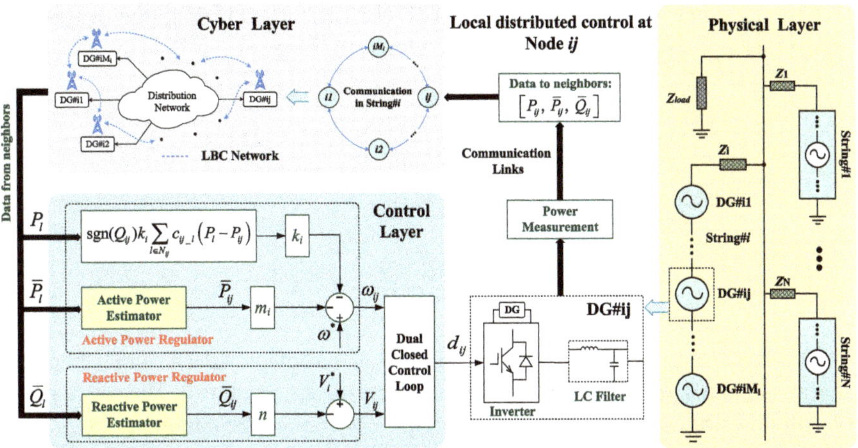

Fig. 15.4 Block diagram of the local-distributed control strategy at Node ij

The sign function $\text{sgn}(Q_{ij})$ is introduced here to ensure that the proposed control strategy is valid under both RL and RC loads. In the steady state, the frequency of DG#ij is derived from (15.3) and (15.14),

$$\omega_{ij} = \omega^* - m_i \bar{P}_{ij}^{ss} - \delta\omega_{ij}^{ss} \tag{15.15}$$

$\bar{P}_{ij}^{ss}$ converges to the true averaged active power of string#i, all frequencies synchronize to the steady-state value, hence

$$\delta\omega_{i1}^{ss} = \delta\omega_{i2}^{ss} = \cdots = \delta\omega_{iM_i}^{ss} \tag{15.16}$$

It can be deduced that $\delta\omega_{ij}^{ss}$ decays to zero, which means in-string accurate active power sharing is guaranteed.

$$P_{i1}^{ss} = P_{i2}^{ss} = \cdots = P_{iM_i}^{ss} \tag{15.17}$$

Combining (15.4), (15.5), and (15.13), the voltage amplitude is same for all DGs in string#i in steady state and they share the same current, so the apparent power of all modules in string#i is equal. Furthermore, (15.18) can be obtained due to the identical active power from (15.17).

$$Q_{i1}^{ss} = Q_{i2}^{ss} = \cdots = Q_{iM_i}^{ss} \tag{15.18}$$

From aforementioned analysis, accurate in-string power sharing is achieved and the steady-state phase angles of all DGs in string#i are equal. So the voltage of PCC can be regulated easily by setting V_P^* in (15.5).

On the other hand, combing (15.12) (15.13), and (15.17), the Eqs. (15.3) and (15.4) in the steady state can be rewritten as follows.

$$\omega_{ij} = \omega^* - \frac{m_i}{M_i} \sum_{j=1}^{M_i} P_{ij}^{ss} \tag{15.19}$$

$$V_{ij} = \frac{1}{M_i} \left(V_p^* - n \sum_{j=1}^{M_i} Q_{ij}^{ss} \right) \tag{15.20}$$

The consistent string could be considered as an integrated DG unit, then load sharing among parallel strings can be ensured by the droop control. The active power of different strings can be shared accurately by designing the parameters as,

$$\frac{m_1}{M_1} = \frac{m_2}{M_2} = \cdots = \frac{m_N}{M_N} \tag{15.21}$$

The reactive power among strings can be approximately shared because of the line impedance mismatch, but the performance can be improved by further decentralized control design, such as virtual impedance method. The proposed distributed control achieves frequency synchronization and load sharing in islanded hybrid series–parallel microgrid with limited local communication. This method is valid under both RL and RC loads. It strengthens the reliability and flexibility of the system and provides a referable application approach to microgrid with complex structure. It is worth noting that proportional load sharing can also be realized if the rated capacities of DGs are unequal. In this case, the output power ought to be normalized and the control parameters should be redesigned as follows.

$$\begin{cases} P_{ij}^{pu} = P_{ij} / P_{ij}^{rated} \\ Q_{ij}^{pu} = Q_{ij} / Q_{ij}^{rated} \end{cases} \tag{15.22}$$

$$V_{ij}^* = \frac{Q_{ij}^{rated}}{\sum_{x=1}^{M_i} Q_{ix}^{rated}} V_P^* \tag{15.23}$$

$$n_{ij} = \frac{Q_{ij}^{rated}}{\sum_{x=1}^{M_i} Q_{ix}^{rated}} n \tag{15.24}$$

Where P_{ij}^{rated} and Q_{ij}^{rated} are the rated active and reactive power of DG#ij.

15.3 Small-Signal Modeling and Stability Analysis

In this part, to study the system stability and dynamic performances with the control strategy, small-signal modeling and eigenvalue analysis are carried out, and reasonable design ranges of parameters are given.

15.3.1 Small-Signal Modeling

(a) Distribution Network Modeling

In the presented hybrid series–parallel microgrid in Fig. 15.1, the output active power p_{ij} and reactive power q_{ij} of DG#ij can be expressed as

$$p_{ij} + jq_{ij} = \frac{V_{ij}e^{j\delta_{ij}}}{2}\left(\left(\sum_{b=1}^{M_i} V_{ib}e^{j\delta_{ib}} - V_P e^{j\delta_P}\right)|Y_i|\, e^{j\varphi_i}\right)^* \tag{15.25}$$

where V_{ij} and δ_{ij} are the output voltage amplitude and phase angle of DG#ij. V_P and δ_P represent the voltage amplitude and phase angle of PCC. $|Y_i|$ and φ_i are amplitude and angle of the equivalent line admittance in string#i. Usually, the line impedance is highly inductive($\varphi_i \approx -\pi/2$) in medium/high voltage power system.

Then $V_P e^{j\delta_P}$ can be derived as follows according to Kirchhoff's laws

$$V_P e^{j\delta_P} = \sum_{a=1}^{N}\sum_{b=1}^{M_a} \frac{Y_a V_{ab} e^{j\delta_{ab}}}{Y_{load} + \sum_{c=1}^{N} Y_c} \tag{15.26}$$

Assume

$$Y_a' = \frac{Y_a}{Y_{\text{load}} + \sum_{c=1}^{N} Y_c} = \left|Y'_a\right| e^{j\varphi'_a} \tag{15.27}$$

Hence the voltage of PCC can be simplified to

$$V_P e^{j\delta_P} = \sum_{a=1}^{N}\sum_{b=1}^{M_a} \left|Y'_a\right| V_{ab} e^{j\left(\delta_{ab} + \varphi'_a\right)} \tag{15.28}$$

Assume that ω_s is the system synchronous angle frequency of steady state. Defining $\dot{\tilde{\delta}}_{ij} = \dot{\delta}_{ij} - \dot{\delta}_s = \omega_{ij} - \omega_s$, then combining (15.25)–(15.28) , the power transmission characteristic is given as,

$$\begin{cases} p_{ij} = \frac{1}{2}|Y_i|\,V_{ij}\left(\sum_{a=1}^{N}\sum_{b=1}^{M_a}|Y'_a|\,V_{ab}\sin\left(\tilde{\delta}_{ij}-\tilde{\delta}_{ab}-\varphi'_a\right)-\sum_{b'=1}^{M_i}V_{ib'}\sin\left(\tilde{\delta}_{ij}-\tilde{\delta}_{ib'}\right)\right) \\ = F_{ij}\left(\tilde{\delta}_{11},\ \tilde{\delta}_{12},\ \ldots,\ \tilde{\delta}_{NM_N},\ V_{11},\ V_{12},\ \ldots,\ V_{NM_N}\right) \\ q_{ij} = \frac{1}{2}|Y_i|\,V_{ij}\left(\sum_{b'=1}^{M_i}V_{ib'}\cos\left(\tilde{\delta}_{ij}-\tilde{\delta}_{ib'}\right)-\sum_{a=1}^{N}\sum_{b=1}^{M_a}|Y'_a|\,V_{ab}\cos\left(\tilde{\delta}_{ij}-\tilde{\delta}_{ab}-\varphi'_a\right)\right) \\ = G_{ij}\left(\tilde{\delta}_{11},\ \tilde{\delta}_{12},\ \ldots,\ \tilde{\delta}_{NM_N},\ V_{11},\ V_{12},\ \ldots,\ V_{NM_N}\right) \end{cases} \tag{15.29}$$

The average active power and reactive power can be obtained by a low-pass filter with the cutoff frequency ω_c which are given as (15.6) . By differentiating (15.6),

$$\begin{cases} \dot{P}_{ij} = \omega_c\left(-P_{ij}+p_{ij}\right) \\ \dot{Q}_{ij} = \omega_c\left(-Q_{ij}+q_{ij}\right) \end{cases} \tag{15.30}$$

Combining (15.29), (15.30) and linearizing them around the steady-state point yield,

$$\begin{cases} \Delta\dot{P}_{ij} = -\omega_c\Delta P_{ij}+\omega_c\sum_{a=1}^{N}\sum_{b=1}^{M_a}k_{p\_\tilde{\delta}_{ab}}^{ij}\Delta\tilde{\delta}_{ab}+\omega_c\sum_{a=1}^{N}\sum_{b=1}^{M_a}k_{p\_V_{ab}}^{ij}\Delta V_{ab} \\ \Delta\dot{Q}_{ij} = -\omega_c\Delta Q_{ij}+\omega_c\sum_{a=1}^{N}\sum_{b=1}^{M_a}k_{q\_\tilde{\delta}_{ab}}^{ij}\Delta\tilde{\delta}_{ab}+\omega_c\sum_{a=1}^{N}\sum_{b=1}^{M_a}k_{q\_V_{ab}}^{ij}\Delta V_{ab} \end{cases} \tag{15.31}$$

where Δ denotes small perturbations around the equilibrium point and.

$$\begin{cases} k_{p\_\tilde{\delta}_{ab}}^{ij} = \frac{\partial p_{ij}}{\partial\tilde{\delta}_{ab}};\ k_{p\_V_{ab}}^{ij} = \frac{\partial p_{ij}}{\partial V_{ab}} \\ k_{q\_\tilde{\delta}_{ab}}^{ij} = \frac{\partial q_{ij}}{\partial\tilde{\delta}_{ab}};\ \ k_{q\_V_{ab}}^{ij} = \frac{\partial q_{ij}}{\partial V_{ab}} \end{cases} \quad \begin{pmatrix} i = 1,\ldots,N;\ j = 1,\ldots,M_i \\ a = 1,\ldots,N;\ b = 1,\ldots,M_a \end{pmatrix} \tag{15.32}$$

Then (15.31) is rewritten in matrix form,

$$\begin{cases} \Delta\dot{\mathbf{P}} = -\omega_c\Delta\mathbf{P}+\omega_c\mathbf{K}_{p\_\tilde{\delta}}\Delta\tilde{\boldsymbol{\delta}}+\omega_c\mathbf{K}_{p\_V}\Delta\mathbf{V} \\ \Delta\dot{\mathbf{Q}} = -\omega_c\Delta\mathbf{Q}+\omega_c\mathbf{K}_{q\_\tilde{\delta}}\Delta\tilde{\boldsymbol{\delta}}+\omega_c\mathbf{K}_{q\_V}\Delta\mathbf{V} \end{cases} \tag{15.33}$$

Where

$$\left\{\begin{array}{l}
\mathbf{\Delta P} = \left[\Delta P_{11}\ \Delta P_{12}\ \cdots\ \ \Delta P_{NM_N}\right]^T;\ \mathbf{\Delta Q} = \left[\Delta Q_{11}\ \Delta Q_{12}\ \ \cdots\ \ \Delta Q_{NM_N}\right]^T \\
\mathbf{\Delta \dot{P}} = \left[\Delta \dot{P}_{11}\ \Delta \dot{P}_{12}\ \cdots\ \ \Delta \dot{P}_{NM_N}\right]^T;\ \mathbf{\Delta \dot{Q}} = \left[\Delta \dot{Q}_{11}\ \Delta \dot{Q}_{12}\ \cdots\ \ \Delta \dot{Q}_{NM_N}\right]^T \\
\mathbf{\Delta \tilde{\delta}} = \left[\Delta \tilde{\delta}_{11}\ \Delta \tilde{\delta}_{12}\ \cdots\ \ \Delta \tilde{\delta}_{NM_N}\right]^T;\quad \mathbf{\Delta V} = \left[\Delta V_{11}\ \ \Delta V_{12}\ \ \cdots\ \ \Delta V_{NM_N}\right]^T \\
\mathbf{K}_{p_\tilde{\delta}} = \begin{bmatrix} k^{11}_{p_\tilde{\delta}_{11}} & k^{11}_{p_\tilde{\delta}_{12}} & \cdots & k^{11}_{p_\tilde{\delta}_{NM_N}} \\ k^{12}_{p_\tilde{\delta}_{11}} & k^{12}_{p_\tilde{\delta}_{12}} & \cdots & k^{12}_{p_\tilde{\delta}_{NM_N}} \\ \vdots & \vdots & \ddots & \vdots \\ k^{NM_N}_{p_\tilde{\delta}_{11}} & k^{NM_N}_{p_\tilde{\delta}_{12}} & \cdots & k^{NM_N}_{p_\tilde{\delta}_{NM_N}} \end{bmatrix} \\
\mathbf{K}_{q_\tilde{\delta}} = \begin{bmatrix} k^{11}_{q_\tilde{\delta}_{11}} & k^{11}_{q_\tilde{\delta}_{12}} & \cdots & k^{11}_{q_\tilde{\delta}_{NM_N}} \\ k^{12}_{q_\tilde{\delta}_{11}} & k^{12}_{q_\tilde{\delta}_{12}} & \cdots & k^{12}_{q_\tilde{\delta}_{NM_N}} \\ \vdots & \vdots & \ddots & \vdots \\ k^{NM_N}_{q_\tilde{\delta}_{11}} & k^{NM_N}_{q_\tilde{\delta}_{12}} & \cdots & k^{NM_N}_{q_\tilde{\delta}_{NM_N}} \end{bmatrix} \\
\mathbf{K}_{q_\tilde{\delta}} = \begin{bmatrix} k^{11}_{q_\tilde{\delta}_{11}} & k^{11}_{q_\tilde{\delta}_{12}} & \cdots & k^{11}_{q_\tilde{\delta}_{NM_N}} \\ k^{12}_{q_\tilde{\delta}_{11}} & k^{12}_{q_\tilde{\delta}_{12}} & \cdots & k^{12}_{q_\tilde{\delta}_{NM_N}} \\ \vdots & \vdots & \ddots & \vdots \\ k^{NM_N}_{q_\tilde{\delta}_{11}} & k^{NM_N}_{q_\tilde{\delta}_{12}} & \cdots & k^{NM_N}_{q_\tilde{\delta}_{NM_N}} \end{bmatrix} \\
\mathbf{K}_{q_V} = \begin{bmatrix} k^{11}_{q_V_{11}} & k^{11}_{q_V_{12}} & \cdots & k^{11}_{q_V_{NM_N}} \\ k^{12}_{q_V_{11}} & k^{12}_{q_V_{12}} & \cdots & k^{12}_{q_V_{NM_N}} \\ \vdots & \vdots & \ddots & \vdots \\ k^{NM_N}_{q_V_{11}} & k^{NM_N}_{q_V_{12}} & \cdots & k^{NM_N}_{q_V_{NM_N}} \end{bmatrix}
\end{array}\right. \tag{15.34}$$

(b) Distributed Controller Modeling

With the proposed distributed control, the controllers of adjacent DGs in one string process and exchange information to update the frequency and voltage set points, which is shown in Fig. 15.4. In the frequency domain,

$$\left\{\begin{array}{l}
\dot{\tilde{\delta}}_{ij} = \omega_{ref} - \omega_s - m_i \bar{P}_{ij} - \mathrm{sgn}(Q_{ij}) k_i \sum\limits_{l \in N_{ij}} c_{ij_l} \left(P_l - P_{ij}\right) \\
\dot{V}_{ij} = -n \dot{\bar{Q}}_{ij}
\end{array}\right. \tag{15.35}$$

Linearizing them around the steady-state point yields,

$$
\begin{cases}
\Delta\dot{\bar{\delta}}_{ij} = -m_i\Delta\bar{P}_{ij} - \text{sgn}(Q_{ij})k_i \sum\limits_{l\in N_{ij}} c_{ij\_l}\left(\Delta P_l - \Delta P_{ij}\right) \\
\dot{V}_{ij} = -n\Delta\dot{\bar{Q}}_{ij}
\end{cases}
\tag{15.36}
$$

Combining (15.11) and (15.36), writing them in matrix form, the small-signal models of controllers are obtained.

$$
\begin{cases}
\mathbf{\Delta}\dot{\bar{\boldsymbol{\delta}}} = \left(diag\left\{\text{sgn}(Q_{ij})\right\}\mathbf{kL} - \mathbf{mH}\right)\mathbf{\Delta P} \\
\mathbf{\Delta\dot{V}} = -n\mathbf{H\Delta\dot{Q}}
\end{cases}
\tag{15.37}
$$

where **L** is a Laplacian matrix containing the information of system communication graph topology. H is the entire active/reactive power estimator transfer function matrix. **m** and **k** are matrixes of controller parameters. They can be expressed as,

$$
\begin{cases}
\mathbf{m} = \begin{bmatrix} diag\{m_1\} & & & \\ & diag\{m_2\} & & \\ & & \ddots & \\ & & & diag\{m_n\} \end{bmatrix}; \\
\mathbf{k} = \begin{bmatrix} diag\{k_1\} & & & \\ & diag\{k_2\} & & \\ & & \ddots & \\ & & & diag\{k_n\} \end{bmatrix}; \\
\mathbf{L} = \begin{bmatrix} \mathbf{L}_1 & & & \\ & \mathbf{L}_2 & & \\ & & \ddots & \\ & & & \mathbf{L}_n \end{bmatrix} \mathbf{L}_i = \mathbf{D}_i^{in} - \mathbf{C}_i; \\
\mathbf{H} = \begin{bmatrix} \mathbf{H}_1 & & & \\ & \mathbf{H}_2 & & \\ & & \ddots & \\ & & & \mathbf{H}_n \end{bmatrix} \mathbf{H}_i = s(s\mathbf{I}_i + \mathbf{L}_i)^{-1}
\end{cases}
\begin{pmatrix} i = 1, \ldots, N; \\ j = 1, \ldots, M_i \end{pmatrix}
\tag{15.38}
$$

(c) Dynamic Model of the Entire Microgrid

Combining the small-signal models of the distribution network (15.33) and distributed controllers (15.37), the small-signal dynamic model of the entire series–parallel microgrid is constructed as follows.

$$
\dot{\mathbf{x}} = Ax \tag{15.39}
$$

Where

$$\begin{cases} \mathbf{x} = \begin{bmatrix} \Delta\mathbf{P} \\ \Delta\mathbf{Q} \\ \Delta\tilde{\boldsymbol{\delta}} \\ \Delta\mathbf{V} \end{bmatrix} \\ \mathbf{A} = \left[\begin{array}{c|c|c|c} -\omega_c\mathbf{I}_M & 0_{M\times M} & \omega_c\mathbf{K}_{p\_\tilde{\delta}} & \omega_c\mathbf{K}_{p\_V} \\ \hline 0_{M\times M} & -\omega_c\mathbf{I}_M & \omega_c\mathbf{K}_{q\_\tilde{\delta}} & \omega_c\mathbf{K}_{q\_V} \\ \hline diag\left\{\mathrm{sgn}(Q_{ij})\right\}\mathbf{kL}-\mathbf{mH} & 0_{M\times M} & 0_{M\times M} & 0_{M\times M} \\ \hline 0_{M\times M} & n\omega_c\mathbf{H} & -n\omega_c\mathbf{HK}_{q\_\tilde{\delta}} & -n\omega_c\mathbf{HK}_{q\_V} \end{array}\right] \end{cases} \tag{15.40}$$

and $M = \sum_{i=1}^{N} M_i$ denotes the total number of DGs in the microgrid.

15.3.2 Eigenvalue Analysis

To evaluate the stability and dynamic performance of the system, the eigenvalue analysis of matrix A in (15.39) is adopted [8]. Based on the simulation model described in the next section which consists of 3×3 DGs with only Z_{l1}, the system eigenvalues were analyzed while varying the control parameters: m_i, k_i, and n. And the corresponding eigenvalue trajectory diagrams are illustrated. To simplify the analysis, it is assumed that the control parameters of all DGs are equal. The dynamic response under RC loads is omitted due to space constraints.

(1) Eigenvalue trajectory with respect to m. Figure 15.5 shows the eigenvalue trajectory diagram as m increases from 2e–5 to 1e–3, with k = 8e–3 and n = 5e–4. It can be seen that the conjugate poles λ_1 and λ_2 lie on the right half-plane when m is small, which means that the system is unstable. λ_1 and λ_2 move to the left half-plane of when m is greater than 0 and gradually move away from the imaginary axis as m increases. Meanwhile, the dominant poles move away from the imaginary axis to increase the damping ratio of system. In other words, the system becomes stable in this case.

(2) Eigenvalue trajectory with respect to k. The active power regulator coefficient k has a significant influence on the system stability. Figure 15.6 shows the eigenvalue trajectory diagram as k changes from 2e–3 to 2 with m = 4e–4 and n = 5e–4. In Fig. 15.6, the conjugate poles λ_1 and λ_2 move from the left half-plane to the right half-plane when k = 0.35, making the system unstable. So a too large k is not conductive to the stability of the system.

(3) Eigenvalue trajectory with respect to n. The droop coefficient n should be designed properly to ensure the system stability. Figure 15.7 shows the eigen-

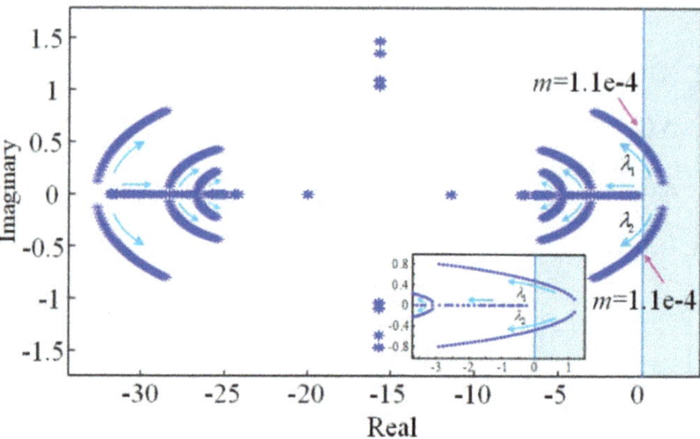

Fig. 15.5 Eigenvalue trajectory as m increases from 2e–5 to 1e–3

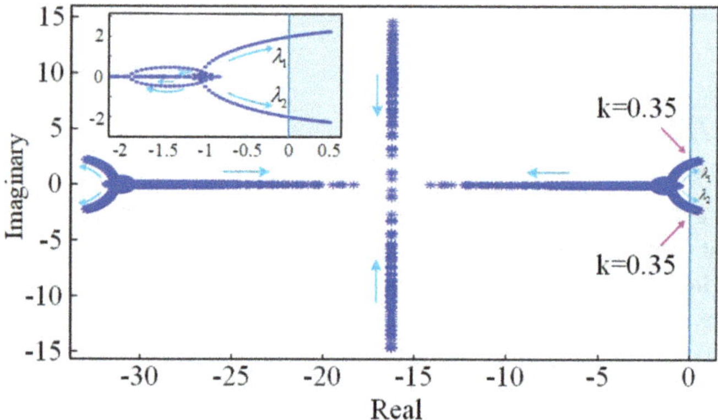

Fig. 15.6 Eigenvalue trajectory as k increases from 2e–3 to 2

value trajectory diagram as n varies from 1e–5 to 1e–3, with $m = 4e{-}4$ and $k = 8e{-}3$. In the beginning, λ_1 lies in the right half-plane, indicating that the system is unstable. When n increases to 6e–5, all eigenvalues stay in the left half-plane so the system becomes stable.

To conclusion, the control parameters m_i, k_i, and n have crucial effect on the system stability and dynamic performance. Considering the voltage and frequency deviations in steady state as well as the overshoot during transients, a reasonable design region of those parameters under RL loads and RC loads is given as shown in Table 15.1.

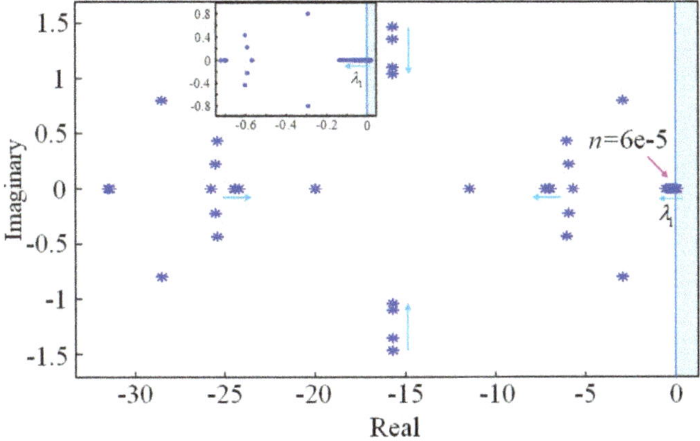

Fig. 15.7 Eigenvalue trajectory as n increases from 1e–5 to 1e–3

Table 15.1 Design regions of control parameters

Parameters	RL loads	RC loads
m	[1.1e–4, 2e–3]	[1.3e–4, 2e–3]
k	[2e–3, 5e–2]	[2e–3, 5e–2]
n	[6e–5, 2e–3]	[9e–5, 2e–3]

15.4 Simulation Results

In this section, to verify the feasibility of the distributed control strategy in this paper, a series–parallel single-phase AC microgrid model consisting of 3×3 DGs is built in MATLAB/Simulink. The equivalent circuit of the simulation model is presented in Fig. 15.8. The physical and control parameters are listed in Table 15.2, and four simulation cases are carried out.

15.4.1 Case 1: Simulation with the Same Rated Capacity

In this case, the rated capacity of each DG is equal and power sharing performance of the proposed local-distributed control method is tested under RL loads and RC loads. Z_{l1} is connected to the PCC at first and Z_{l2} is added at 2 s, then the PCC load changes to Z_{l3} at 4 s and Z_{l4} is plugged at 6 s. The simulation results are illustrated in Fig. 15.9. It can be seen from Fig. 15.9a that all frequencies synchronize quickly with a slight deviation to the rated frequency. Figure 15.9b shows that accurate active power sharing is guaranteed under both RL and RC loads. In Fig. 15.9c, reactive power of DGs in the same string is controlled to be equal in steady state. The reactive power among different strings is shared approximately due to the line impedance mismatch. Besides, when the load suddenly changes, the system has a

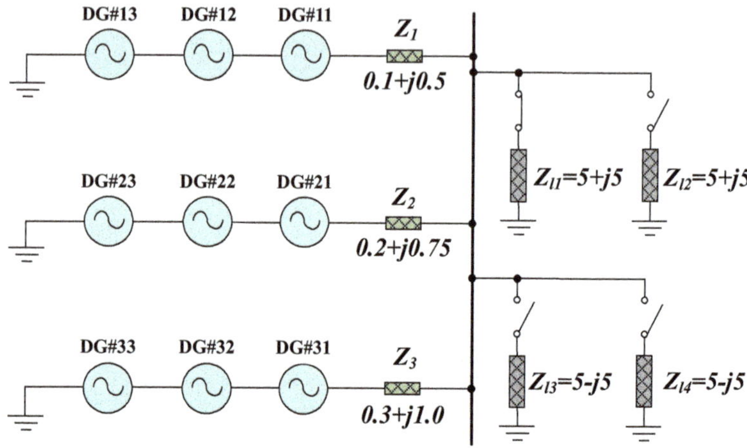

Fig. 15.8 Equivalent circuit of the simulation model

Table 15.2 Simulation parameters

Items	Symbol value
Voltage reference of PCC	$V_P^* = 311\,\text{V}$; $f_P^* = 50\,\text{Hz}$
Rated angle frequency	$\omega^* = 100\pi\,\text{rad/s}$
Filter inductance	$L_f = 1.6\,\text{mH}$
Filter capacitance	$C_f = 20\,\mu\text{F}$
Communication weights	$c_{ij\_l, l\in N_{ij}} = 1$

fast response and the active power and reactive power can be quickly adjusted. The desired control targets are achieved with the proposed method.

15.4.2 Case 2: Simulation with Different Rated Capacities

In this case, the PCC load change is the same as case 1. The corresponding simulation results are illustrated in Fig. 15.10. It shows that the system stabilizes rapidly after running and has a fast response to load change. In the steady state, the active power among all converters is shared exactly in proportional to the rated active power. The output active power is equal for DGs with the same rated value, such as P_{33} and P_{21}. Similarly, the in-string reactive power is shared proportionally and the output reactive power of DGs in different strings with the same rated value has a small difference, such as Q_{33} and Q_{21}. As a result, proportional load sharing can also be realized under both RL and RC loads if the rated capacities of DGs are unequal.

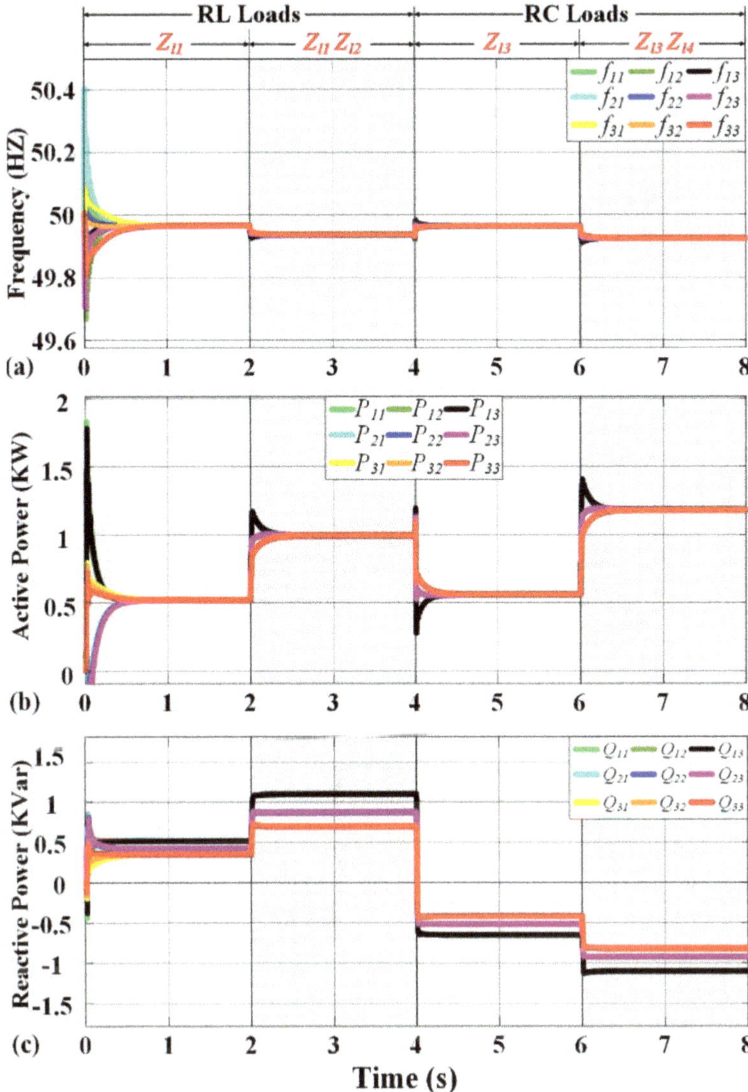

Fig. 15.9 Simulation results of case 1. (**a**) Frequency, (**b**) active power, (**c**) reactive power

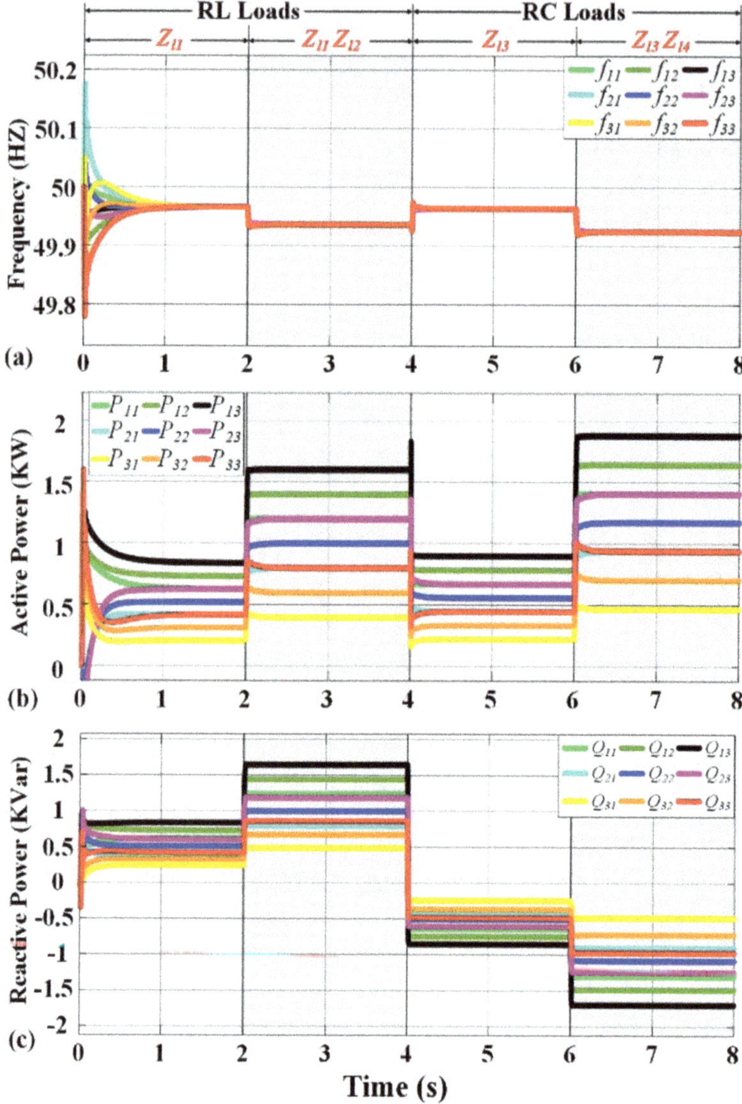

Fig. 15.10 Simulation results of case 2. (**a**) Frequency, (**b**) active power, (**c**) reactive power

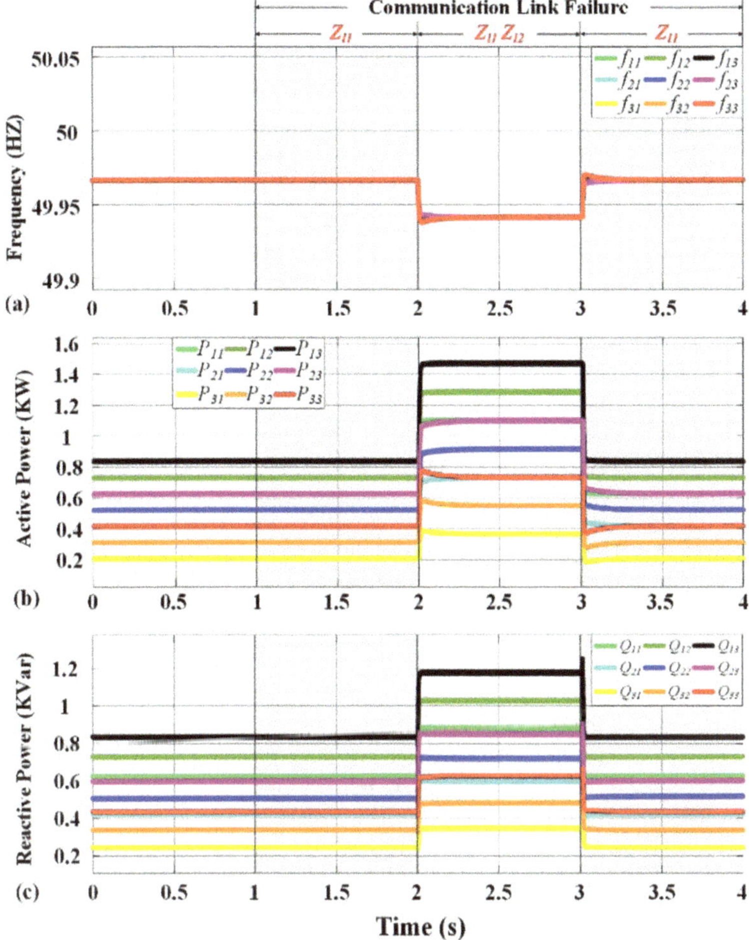

Fig. 15.11 Simulation results of case 3. (**a**) Frequency, (**b**) active power, (**c**) reactive power

15.4.3 Case 3: Simulation with Communication Link Failure

The in-string circular communication network with the minimum redundancy is adopted to deal with any single communication link failure in this paper. In this case, to study the system resiliency to the communication link failure, controller performance with sudden communication links failure is presented in Fig. 15.11 (results under RC loads are omitted). Links 11–12, 22–23, and 33–31 fail at 1 s, 2 s and 3 s respectively, which is illustrated in Fig. 15.12a. Load $Z_{l5} = 10 + j5$ Ω is attached at 2 s and detached at 3 s to test the response to load change. From Fig. 15.10, it can be seen that the steady-state performance is not affected when the link 11–12 is disconnected at 1 s. The steady-state power sharing performance in

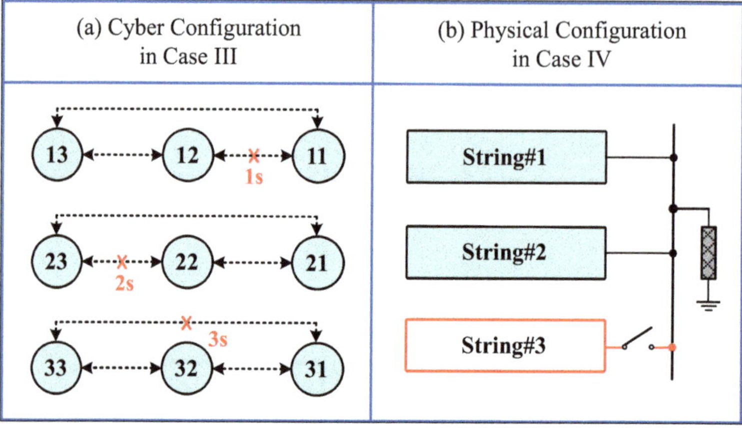

Fig. 15.12 (**a**) Cyber configuration in case 3 (**b**) Physical configuration in case 4

the interval [3 s, 4 s] gradually returns to the inceptive state when Z_{l5} is detached. However, the frequency synchronization and power balance are achieved more slowly than case 2 after load changes. In a word, single link failure does not affect the steady-state performance while it slows down the system dynamics to some extent.

15.4.4 Case 4: Simulation with Plug-and-Play

In this case, the system is tested via unplugging the string#3 at 1 s and plugging it back at 3 s, whose physical configuration is depicted in Fig. 15.12b. The simulation results are shown in Fig. 15.13. When string#3 is unplugged, the system quickly reaches a new steady state. The output power of DGs in string#3 reduces to zero while proper power sharing is still maintained among the remaining DGs. After string#3 is plugged back, the controllers have a fast response and the steady-state performance is consistent with the interval [0 s,1 s]. The load demand is shared among all DGs again. So the plug-and-play capability of DG strings can be obtained with the control method.

15.5 Conclusion

In this chapter, a local-distributed control is presented for the hybrid series–parallel microgrid. Only a local sparse LBC network is used for limited data exchange among in-string converters. The general control scheme is presented in Sect. 15.2. The frequency synchronization and proper power sharing are achieved in islanded

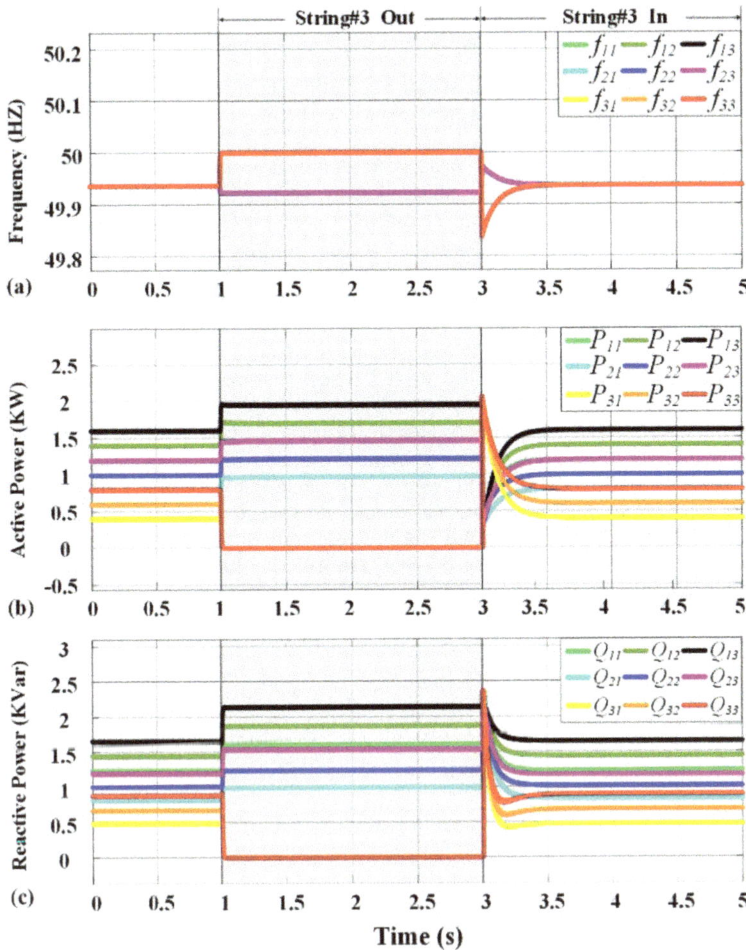

Fig. 15.13 Simulation results of case 4. (**a**) Frequency, (**b**) active power, (**c**) reactive power

mode. Meanwhile, a sign function is introduced to ensure the controller available under both RL loads and RC loads. With the proposed control, the reliability and redundancy of the system are improved and the communication costs are reduced comparing with existing methods. Small-signal modeling and eigenvalue analysis are carried out to study the system stability and dynamic performances, and reasonable design ranges of parameters are given. Finally, the power sharing performance is verified by simulation results with the same and different rated capacities under both RL and RC loads. The resiliency to communication link failure and plug-and-play operation are proved as well.

References

1. J.M. Guerrero, L.G. de Vicuna, J. Matas, M. Castilla, J. Miret, Output impedance design of parallel-connected UPS inverters with wireless load-sharing control. IEEE Trans. Ind. Electron. **52**(4), 1126–1135 (2005)
2. X. Hou, Y. Sun, H. Han, Z. Liu, W. Yuan, M. Su, A fully decentralized control of grid-connected cascaded inverters. IEEE Trans. Sustainable Energy **10**(1), 315–317 (2019)
3. M. Su, C. Luo, X. Hou, W. Yuan, Z. Liu, H. Han, J.M. Guerrero, A communication-free decentralized control for grid-connected cascaded PV inverters. Energies **11**, 1375 (2018)
4. Y. Wang, Y. Xu, Y. Tang, M.H. Syed, E. Guillo-Sansano, G.M. Burt, Decentralised-distributed hybrid voltage regulation of power distribution networks based on power inverters. IET Gener. Transmiss. Distribut. **13**(3), 444–451 (2019)
5. H. Han, H. Wang, Y. Sun, J. Yang, Z. Liu, Distributed control scheme on cost optimisation under communication delays for DC microgrids. IET Gener. Transmiss. Distribut. **11**(17), 4193–4201 (2017)
6. A. Bidram, A. Davoudi, F.L. Lewis, S. Sam Ge, Distributed adaptive voltage control of inverter-based microgrids. IEEE Trans. Energy Convers. **29**(4), 862–872 (2014)
7. D.P. Spanos, R. Olfati-Saber, R.M. Murray, Dynamic consensus for mobile networks, in *Proceedings of the 16th International Federation of Automatic Control* (2005), pp. 1–6
8. X. Ge, et al., Locally-distributed and globally-decentralized control for hybrid series-parallel microgrids. Int. J. Electri. Power Energy Syst. **116**, 105537 (2020)

Chapter 16
SoC Balancing Control Strategy for Hybrid Series–Parallel Storage System

16.1 Control Objectives

A local-distributed and global-decentralized SoC balancing scheme is introduced for the hybrid series–parallel ESS. In a local ESU string, a distributed SoC balancing algorithm based on low-bandwidth communication is designed to balance the SoC of ESUs [1–6]. A modified droop control based on SoC and power estimators is presented for dispersed ESU strings. The SoC of all ESUs in strings with different capacities converge to the global average SoC of ESS under both the charging mode and discharging mode. Compared with SoC balancing method in [7, 8], the proposed method significantly reduces the communication cost and improves the reliability of the system. Besides, the proposed method shows a superiority in resisting communication failure. Without communication links between dispersed ESU strings, the scalability of the ESS is also improved, which can achieve the string plug-and-play. Moreover, the small-signal stability and steady-state performances of the hybrid series–parallel ESS are analyzed in theory. And the simulation results of some cases validate the feasibility of the proposed control. It is of great significance for SoC balance and power coordination of complex distributed energy storage systems.

16.2 SOC Balancing Control Strategy

16.2.1 ESS Structure for Hybrid Series–Parallel Microgrid

The configuration of hybrid series–parallel ESS is shown in Fig. 16.1. It consists of N parallel strings, which is formed by series-connected ESUs, and local loads. The ESU interface converters adopt the conventional H-bridge structure with LC filter for independent control. Low voltage ESUs can be integrated without additional

Y. Sun et al., *Series-Parallel Converter-Based Microgrids*, Power Systems,
https://doi.org/10.1007/978-3-030-91511-7_16

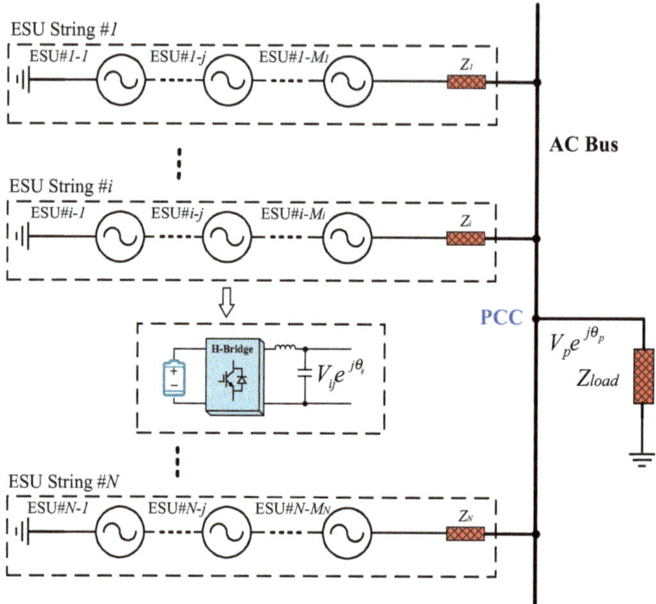

Fig. 16.1 Equivalent model of hybrid series–parallel ESS in islanded mode

converters for a high voltage ESS, promoting ESS's flexibility and efficiency. In this section, ESU strings can supply controllable high-level voltage, which means the ESU serves as a controlled voltage source.

16.2.2 Relationship of SOC and Active Power

There are many techniques to measure or monitor the SOC. Since this section focuses mainly on achieving SOC balance instead of any specific sophisticated SOC estimation methods, the SOC value of ESU #$i - j$ is obtained as follows by its definition:

$$SOC_{ij} = SOC_{ij0} - \frac{1}{C_{ij}} \int i^{out\_i} dt \tag{16.1}$$

where, SOC_{ij0} and C_{ij} are the initial SOC value and capacity of ESU#$i - j$, respectively. i^{out_i} is the output current of ESU string #i. In this work, the power loss in the converter is neglected and the input voltages of the converter are assumed to be the same, the relationship of the SoC and the active power can be obtained

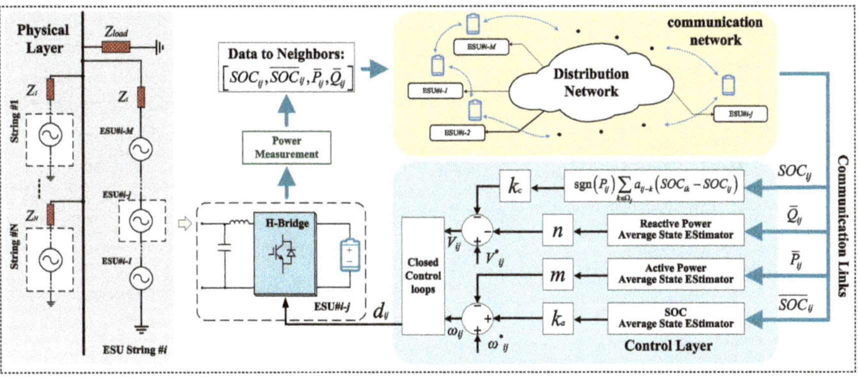

Fig. 16.2 Control diagram of the SOC balancing control method

$$SOC_{ij} = SOC_{ij0} - \frac{\int p_{ij} dt}{V_{in} C_{ij}} \tag{16.2}$$

where V_{in} is the input voltage of the converter.

16.2.3 SOC Balancing Control Method

The proposed SOC balancing control scheme is shown in Fig. 16.2 and it is expressed as (16.3) and (16.4):

$$\omega_{ij} = \omega^* - \frac{m}{C_{ij}} \bar{P}_{ij} + k_a \overline{SOC}_{ij} \tag{16.3}$$

$$V_{ij} = V^* - n\bar{Q}_{ij} - \text{sgn}\left(P_{ij}\right) k_c \sum_{k \in \Omega_j} a_{ij\_k} \left(SOC_{ik} - SOC_{ij}\right) \tag{16.4}$$

where ω^* and V^* are angular frequency reference and voltage amplitude reference, respectively. $V^* = V_P^*/M_i$, V_p^* represents voltage amplitude of PCC at no load. ω_{ij}, V_{ij}, and P_{ij} are the angular frequency, voltage amplitude, active power of ESU $\#i - j$, respectively. Moreover, m, k_a, n, and k_c are modified droop control coefficients, respectively. Sign function in (16.4) is designed to distinguish the charging mode and discharging mode of the ESS. $\bar{P}_{ij}$, $\bar{Q}_{ij}$, and $\overline{SOC}_{ij}$ represent the estimated average value of active power, reactive power, and SOC of ESU $\#i - j$, respectively. Each ESU has the average state estimators of active power, reactive power, and SOC, which adopt local measured value and neighbors' information to update average estimated value. The distributed average consensus protocol is implemented as follows

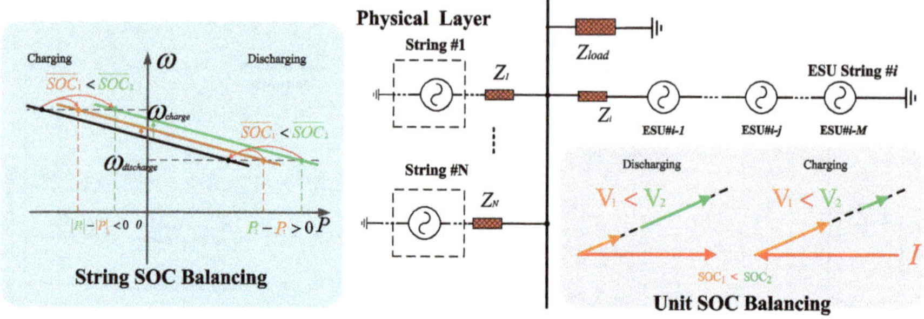

Fig. 16.3 The basic principle of the presented method

$$\bar{P}_{ij}(t) = P_{ij}(t) + \int_0^t \sum_{k\in\Omega_j} a_{ij\_k}(\bar{P}_{ik}(\tau) - \bar{P}_{ij}(\tau))d\tau \tag{16.5}$$

$$\bar{Q}_{ij}(t) = Q_{ij}(t) + \int_0^t \sum_{k\in\Omega_j} a_{ij\_k}(\bar{Q}_{ik}(\tau) - \bar{Q}_{ij}(\tau))d\tau \tag{16.6}$$

$$\overline{SOC}_{ij}(t) = SOC_{ij}(t) + \int_0^t \sum_{k\in\Omega_j} a_{ij\_k}(\overline{SOC}_{ik}(\tau) - \overline{SOC}_{ij}(\tau))d\tau \tag{16.7}$$

where $a_{(ij\_k)}$ is the communication weight. In order to express the proposed scheme clearly, the basic principle of method is shown in Fig. 16.3. According to the proposed control function (16.3) (16.4), the method is divided into two parts: the unit SOC balancing and the string balancing.

Unit SOC Balancing Within one string, every ESU shares the same current. The output power is determined by the voltage. In this paper, the local and neighbor's SOC information are utilized to adjust the voltage amplitude based on distributed consensus algorithm. In the discharging mode, higher voltage amplitude energy storage units output greater power and lower ones output less power. The SOC difference will be decreased and the SOC balancing will be achieved in the string ultimately.

String SOC Balancing When ESUs in one string are coordinated, the whole string can be regarded as a high voltage level ESU with large capacity. The hybrid series–parallel ESS can be simplified as a parallel ESS. Based on the traditional decentralized SOC balancing method [6, 9] and the average state estimators, the drop control bias is modified by the average SOC. In the discharging mode, higher average SOC strings output greater power and lower SOC ones output less power. The global SOC balancing is achieved with string SOC balanced.

The SOC balancing of each ESU in the same string is realized by regulating voltages, which is shown in (16.4). Clearly, it relies on distributed communication in the same string. By summing the voltages of the ESUs in i-th string, we have

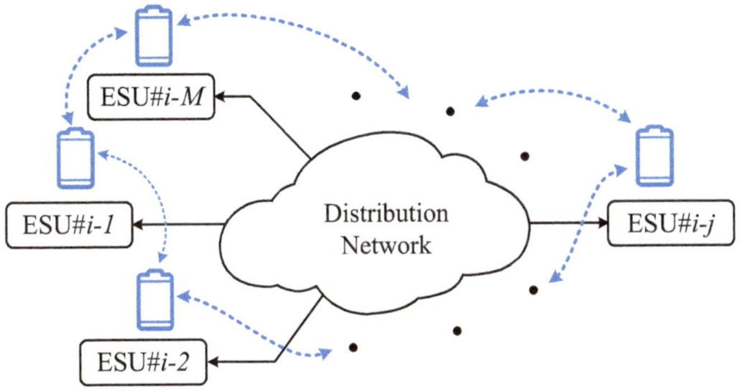

Fig. 16.4 Communication network in ESU string #*i*

$$\sum V_{ij} = M_i V^* - \sum n\bar{Q}_{ij} \tag{16.8}$$

From (16.8), it can be found that there is no total voltage deviation due to the SOC balancing, which is beneficial to load voltage regulation.

16.2.4 Steady-State Analysis

Figure 16.4 illustrates the communication network is in ESU string #i. In Fig. 16.4, nodes represent ESUs, and edges represent communication links between two ESUs. Node $i - j$ represents the ESU #$i - j$, and $k \in \Omega_j$ denotes the set of all in-string neighbors of ESU #$i - j$. If communication weight $a_{ij\_k} = 1$, ESU #$i - j$ receives data from ESU #$i - k$, otherwise $a_{ij\_k} = 0$. In the communication network, each node has in-degree $d_{ij} = \sum_{k\in\Omega_j} a_{ij\_k}$ and out-degree $d_{ij}^0 = \sum_{k\in\Omega_j} a_{ik\_j}$. When $d_{ij} = d_{ij}^0$ for each node, communication could come to a balanced status. For ESU string #i, the SOC adjacency matrix is given by $\mathbf{A}_i = [a_{ij\_k}]$ and in-degree matrix is written as $\mathbf{D}_i = diag[d_{ij}]$. So, the Laplacian matrix can be obtained as $\mathbf{L}_i = \mathbf{D}_i - \mathbf{A}_i$. The dynamics of the distributed average consensus protocol express as follows:

$$\begin{aligned}
&\dot{\bar{X}}_i = \dot{X}_i - D_i\bar{X}_i + A_i\bar{X}_i = \dot{X}_i - L_i\bar{X}_i \\
&(i = 1, 2, \ldots N) \\
&\bar{X}_i = \left[\bar{X}_{i1}, \bar{X}_{i2}, \ldots \bar{X}_{iM_i}\right]^T \\
&X_i = \left[X_{i1}, X_{i2}, \ldots, X_{iM_i}\right]^T
\end{aligned} \tag{16.9}$$

where X can be used to indicate active power P, reactive power Q, and SoC. Applying the Laplace transform yields the following transfer function matrix:

$$H_i = \frac{\bar{X}_i}{X_i} = s(sI_M + L_i)^{-1} \tag{16.10}$$

where $I_M \in \mathbb{R}^{M_i \times M_i}$ is the identity matrix. When communication comes to a balanced status, elements of $\bar{X}_i$ will converge to a value, which is the actual average value of all X. In other words,

$$\lim_{s \to 0} H_i = E_i \tag{16.11}$$

$$\lim_{t \to 0} \bar{X}_i(t) = E_i \times \lim_{s \to 0} sX_i = E_i x_i^{ss} = \langle x_i^{ss} \rangle \underline{1} \tag{16.12}$$

where elements of the averaging matrix E_a equal $1/M_i$. x_i^{ss} is the steady value of X_i. $\langle x_a^{ss} \rangle$ represents the average value of elements of vector X_a. $\underline{1}$ is the vector being all ones. Based on aforementioned analysis, the estimated average value of active power, reactive power, and SOC in steady state can be written as

$$\bar{X}_{i1}^{ss} = \bar{X}_{i2}^{ss} = \cdots = \bar{X}_{iM_i}^{ss} = \frac{X_{i1}{}^{ss} + X_{i2}{}^{ss} + \cdots + X_{iM_i}{}^{ss}}{M_i} = \bar{X}_i^{ss} \tag{16.13}$$

$\bar{Q}_{ij}^{ss}$, $\bar{P}_{ij}^{ss}$, and $\bar{S}OC_{ij}^{ss}$ will converge to the average value $\bar{Q}_i^{ss}$, $\bar{P}_i^{ss}$, and $\bar{S}OC_i^{ss}$ of ESU string #i. Within ESU string #i, the dynamics are given as:

$$\delta V_i = -L_i SOC_i \tag{16.14}$$

where $SOC_i = [SOC_{i1} \cdots SOC_{iM}]^T$. The closed-loop dynamics depends on the graph Laplacian matrix $\mathbf{L}$. $\mathbf{L}$ has the rank of $N - 1$, then $c\underline{1}$ is the only vector in the null-space of $\mathbf{L}$ and one has $\mathbf{SOC}_i^{ss} = c\underline{1}$ for some constant c. At steady state, $-\mathbf{L}_i \mathbf{SOC}_i^{ss} = 0$.

$$SOC_{ij}^{ss} = SOC_{ik}^{ss} \tag{16.15}$$

SOC balancing in one ESU string is proved to be achieved. Equation (16.13) of the ESU in two ESU strings is written as

$$\omega_i - \omega^* = -\frac{m}{C_i} \bar{P}_i + k_a \overline{SOC_i} \tag{16.16}$$

$$\omega_k - \omega^* = -\frac{m}{C_k} \bar{P}_k + k_a \overline{SOC_k} \tag{16.17}$$

All frequencies synchronize to the steady-state value. (16.18) can be given by (16.15), (16.16), and (16.12).

$$\begin{aligned} 0 &= -\frac{m}{C_i}\bar{P}_i + \frac{m}{C_k}\bar{P}_k + k_a\left(\overline{SOC}_i - \overline{SOC}_k\right) \\ &= -\frac{m}{C_i}\bar{P}_i + \frac{m}{C_k}\bar{P}_k + k_a\left(SOC_{i0} - \frac{\bar{P}_i}{sC_i}\right) - k_a\left(SOC_{k0} - \frac{\bar{P}_k}{sC_k}\right) \end{aligned} \tag{16.18}$$

where SOC_{i0} and SOC_{k0} can represent the initial SOC value of any ESUs in ESU string #i and ESU string #k, respectively. Then, multiplying sC_iC_k to each side of Eq. (16.18),

$$\begin{aligned} & sC_iC_kk_a\left(SOC_{k0} - SOC_{i0}\right) \\ & = sm\left(C_i\bar{P}_k^{ss} - C_k\bar{P}_i^{ss}\right) + k_a\left(C_i\bar{P}_k^{ss} - C_k\bar{P}_i^{ss}\right) \end{aligned} \tag{16.19}$$

When $t \to \infty$, $\bar{P}_k^{ss} : \bar{P}_i^{ss} = C_k : C_i$ can be obtained. The active power of different cascaded string shares accurately following the capacity of ESU. Substituting it into Eq. (16.18), $SOC_i^{ss} = SOC_k^{ss}$. In steady state, $SOC_{ij}^{ss} = SOC_i^{ss}$.

$$SOC_{ij}^{ss} = SOC_{kl}^{ss} \tag{16.20}$$

Thus, it is proved that global SoC balancing will be achieved under the proposed method if the system is stable [10].

16.3 Small-Signal Modeling and Stability Analysis

16.3.1 Small-Signal Modeling

To simplify the analysis, it is supposed that each cascaded string consists of M ESUs. The power transmission characteristic is given as

$$p_{ij} = V_{ij}\left|Y_i\right|\left(\sum_{l=1}^{N}\sum_{k=1}^{M}\left|Y_l'\right|V_{lk}\sin\left(\tilde{\theta}_{ij} - \tilde{\theta}_{lk} - \theta_l'\right) - \sum_{k=1}^{M}V_{ik'}\sin\left(\tilde{\theta}_{ij} - \tilde{\theta}_{ik'}\right)\right) \tag{16.21}$$

$$q_{ij} = V_{ij}\left|Y_i\right|\left(-\sum_{l=1}^{N}\sum_{k=1}^{M}\left|Y_l'\right|V_{lk}\cos\left(\tilde{\theta}_{ij} - \tilde{\theta}_{lk} - \theta_l'\right) + \sum_{k=1}^{M}V_{ik'}\cos\left(\tilde{\theta}_{ij} - \tilde{\theta}_{ik'}\right)\right) \tag{16.22}$$

Where, $\left|Y_l'\right| e^{j\theta_l'} = Y_l/(Y_{load} + \sum_{c=1}^{M} Y_c) \cdot Y_{load}$ is the admittance of load. Let $\tilde{\theta} = \theta - \theta_s$, the small-signal variations of active and reactive power are given as follows:

$$\Delta\dot{P}_{ij} = -\omega_c \Delta P_{ij} + \omega_c \sum_{l=1}^{N}\sum_{k=1}^{M} T_{pv-lk}^{ij} \Delta V_{lk} + \omega_c \sum_{l=1}^{N}\sum_{k=1}^{M} T_{p\theta-lk}^{ij} \Delta\tilde{\theta}_{lk} \tag{16.23}$$

$$\Delta\dot{Q}_{ij} = -\omega_c \Delta Q_{ij} + \omega_c \sum_{l=1}^{N}\sum_{k=1}^{M} T_{qv-lk}^{ij} \Delta V_{lk} + \omega_c \sum_{l=1}^{N}\sum_{k=1}^{M} T_{q\theta-lk}^{ij} \Delta\tilde{\theta}_{lk} \tag{16.24}$$

where ω_c is the cutoff frequency of the low-pass filter.

Defining $\dot{\tilde{\theta}}_{ij} = \dot{\theta}_{ij} - \dot{\theta}_s = \omega_{ij} - \omega_s \approx \omega_{ref} - \omega_s$, the control strategy can be written as

$$\dot{\tilde{\theta}}_{ij} = \omega_{ref} - \omega_s - \frac{m}{C_{ij}} \bar{P}_{ij} + k_a \overline{SOC}_{ij} \tag{16.25}$$

$$\dot{V}_{ij} = -n\dot{\bar{Q}}_{ij} - \text{sgn}\left(P_{ij}\right) k_c \sum_{k\in\Omega_j} a_{ij\_k} \left(S\dot{O}C_{ik} - S\dot{O}C_{ij}\right) \tag{16.26}$$

Linearizing (16.25) and (16.26), it yields

$$\dot{\tilde{\theta}}_{ij} = \omega_{ref} - \omega_s - \frac{m}{C_{ij}} \bar{P}_{ij} + k_a \overline{SOC}_{ij} \tag{16.27}$$

$$\Delta\dot{V}_{ij} = -n\Delta\dot{\bar{Q}}_{ij} - \text{sgn}\left(P_{ij}\right) k_c \sum_{k\in\Omega_j} a_{ij\_k} \left(\Delta S\dot{O}C_{ik} - \Delta S\dot{O}C_{ij}\right) \tag{16.28}$$

The disturbances of SOC can be obtained by perturbing (16.2)

$$\Delta SOC_{ij} = -\frac{1}{sV_{in}C_{ij}} \Delta P_{ij} \tag{16.29}$$

Then, matrix of aforementioned state equation is achieved as follows:

$$\begin{cases} \Delta\dot{\mathbf{P}} = -\omega_c \mathbf{\Delta P} + \omega_c \mathbf{T}_{p\theta} \mathbf{\Delta}\tilde{\boldsymbol{\theta}} + \omega_c \mathbf{T}_{pv} \mathbf{\Delta V} \\ \Delta\dot{\mathbf{Q}} = -\omega_c \mathbf{\Delta Q} + \omega_c \mathbf{T}_{q\theta} \mathbf{\Delta}\tilde{\boldsymbol{\theta}} + \omega_c \mathbf{T}_{qv} \mathbf{\Delta V} \\ \mathbf{\Delta}\dot{\tilde{\boldsymbol{\theta}}} = -\mathbf{MH\Delta P} + \mathbf{K_a H \Delta SOC} \\ \Delta\dot{\mathbf{V}} = \mathbf{DK_c LC\Delta P} + \omega_c \mathbf{NH\Delta Q} - \omega_c \mathbf{NHT}_{q\theta} \mathbf{\Delta}\tilde{\boldsymbol{\theta}} - \omega_c \mathbf{NHT_{qv}\Delta V} \\ \mathbf{\Delta S\dot{O}C} = \mathbf{C\Delta P} \end{cases} \tag{16.30}$$

Where $\mathbf{\Delta P}$, $\mathbf{\Delta Q}$, $\mathbf{\Delta V}$, $\mathbf{\Delta \tilde{\theta}}$, and $\mathbf{\Delta SOC}$ are the state variable vectors. $\mathbf{M}$, $\mathbf{N}$, $\mathbf{K_a}$, and $\mathbf{K_c}$ are the control coefficients matrixes. $\mathbf{H}$ is the power transfer function matrix. $\mathbf{D} = diag\{\text{sgn}(P_{ab})$ is the sign matrix of active power. $\mathbf{L}$ is the Laplacian matrix that represents the communication relation of each ESU. $\mathbf{T}_{pv}$, $\mathbf{T}_{p\theta}$, $\mathbf{T}_{qv}$, and $\mathbf{T}_{q\theta}$ are the partial derivative matrixes that indicate the correlation between active/reactive power and voltage magnitude/angle. The state-space form can be obtained as

$$\dot{\mathbf{X}} = \mathbf{A} \cdot \mathbf{X} \tag{16.31}$$

$$\begin{cases} \mathbf{X} = \left[\, \mathbf{\Delta P} \; \mathbf{\Delta Q} \; \mathbf{\Delta \tilde{\theta}} \; \mathbf{\Delta V} \; \mathbf{\Delta SOC} \,\right]^T \\ \mathbf{A} = \left[\begin{array}{c|c|c|c|c} -\omega_c \mathbf{I}_M & \mathbf{0}_{M\times M} & \omega_c \mathbf{T}_{p\theta} & \omega_c \mathbf{T}_{pv} & \mathbf{0}_{M\times M} \\ \mathbf{0}_{M\times M} & -\omega_c \mathbf{I}_M & \omega_c \mathbf{T}_{q\theta} & \omega_c \mathbf{T}_{qv} & \mathbf{0}_{M\times M} \\ -\mathbf{MH} & \mathbf{0}_{M\times M} & \mathbf{0}_{M\times M} & \mathbf{0}_{M\times M} & \mathbf{K}_a \mathbf{H} \\ \mathbf{DK}_c \mathbf{LC} & \omega_c \mathbf{NH} & -\omega_c \mathbf{NHT}_{q\theta} & -\omega_c \mathbf{NHT}_{qv} & \mathbf{0}_{M\times M} \\ \mathbf{C} & \mathbf{0}_{M\times M} & \mathbf{0}_{M\times M} & \mathbf{0}_{M\times M} & \mathbf{0}_{M\times M} \end{array} \right] \end{cases} \tag{16.32}$$

Where

$$\begin{cases} \mathbf{\Delta P} = \left[\, \Delta P_{11} \cdots \Delta P_{1M} \; \Delta P_{21} \cdots \Delta P_{2M} \cdots \Delta P_{N1} \cdots \Delta P_{NM} \,\right]^T; \\ \mathbf{\Delta Q} = \left[\, \Delta Q_{11} \cdots \Delta Q_{1M} \; \Delta Q_{21} \cdots \Delta Q_{2M} \cdots \Delta Q_{N1} \cdots \Delta Q_{NM} \,\right]^T; \\ \mathbf{\Delta \tilde{\theta}} = \left[\, \Delta \theta_{11} \cdots \Delta \theta_{1M} \; \Delta \theta_{21} \cdots \Delta \theta_{2M} \cdots \Delta \theta_{N1} \cdots \Delta \theta_{NM} \,\right]^T; \\ \mathbf{\Delta V} = \left[\, \Delta v_{11} \cdots \Delta v_{1M} \; \Delta v_{21} \cdots \Delta v_{2M} \cdots \Delta v_{N1} \cdots \Delta v_{NM} \,\right]^T; \\ \mathbf{\Delta SOC} = \left[\, \Delta SOC_{11} \cdots \Delta SOC_{21} \cdots \Delta SOC_{N1} \cdots \Delta SOC_{NM} \,\right]^T; \end{cases} \tag{16.33}$$

16.3.2 Eigenvalue Analysis

Supposing control parameters and all modules are the same, and ESS works in discharging mode. The stability and dynamic performance of the hybrid series–parallel ESS are analyzed by varying the control parameters to get the corresponding eigenvalue trajectory diagrams. Figure 16.5a shows the eigenvalue trajectory diagram of m/C increasing from 1e−5 to 2e−4, in which blue part shows $m \in$ [1e−5,1e−4) and red parts show m ∈[1e−4,2e−4]. Other parameters are set: $n = 2.8\text{e} - 4$, $k_a = 1$, and $k_c = 500$. It can be seen that some of the dominant poles move closer to the imaginary with the increasing of m, which means the dynamic character of the system may be harmed when m/C is too large. All eigenvalues have a negative real part, so that the system is stable with the m/C changing. And the eigenvalue trajectory diagram representing the effect of the variation of n is shown

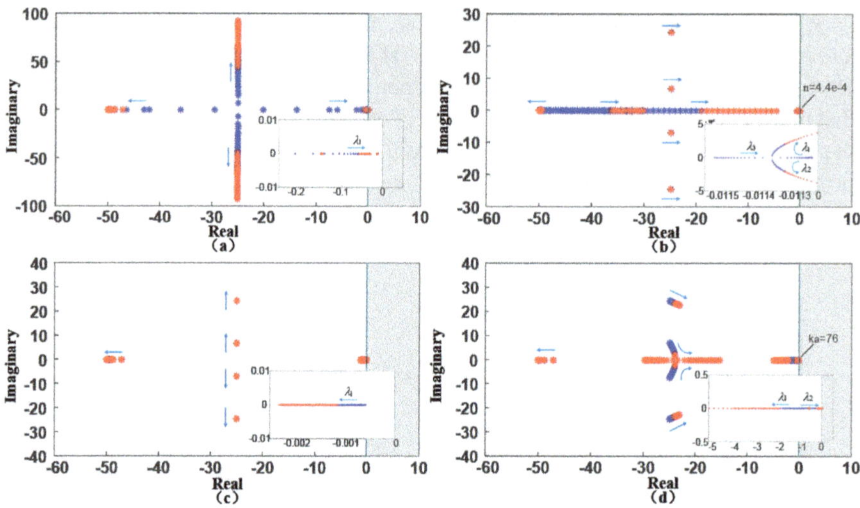

Fig. 16.5 Graphical diagram of the proposed communication network

in Fig. 16.5b. Let $m/C = 5\text{e}-5$, $k_a = 1$, and $k_c = 500$. It shows that two conjugate poles move from the left half-plane to the right half-plane with n increasing from 1e−5 to 4.4e−4, in which blue part shows $n \in$ [1e−5,2e−4) and red part shows $n \in$ [2e−4,4.4e−4]. When n is larger than 4.4e−4, the system is unstable. Thus, n should not be set too large.

The SOC regulator coefficients are significant for the stability of ESS. In Fig. 16.5c, when k_c increasing from 100 to 2000, in which blue part shows $k_c \in$ [100,1000) and red part shows $k_c \in$ [1000,2000], with $m/C = 5\text{e}-5$, $n = 2.8\text{e}-4$ and $k_a = 1$, the real axis coordinates of some dominant poles decrease. The stability and the dynamic performance of the system is improved with k_c increasing. Figure 16.5d shows that k_a should be limited in 0.1–76 when $m/C = 5\text{e} - 5$, $n = 2.8\text{e}-4$ and $k_c = 500$. The increasing of k_a, shorten the stabilization time. However, the frequency of each ESU should be limited within 49–51 Hz. The reasonable range of k_a is 0.1–1.5.

A simulation is designed based on MATLAB/Simulink. The simulation results of the small-signal model and the physical simulation are compared to verify the accuracy of the model. The system control parameters of the physical simulation are given by the eigenvalue analysis, which are presented in Table 16.1.

The system becomes stable before 500 s. The comparison of load mutation ($R = 0.65\ \Omega$ load step) at 500 s is shown in Fig. 16.6. After the load mutation, the output active powers change from 885 W to 845 W in Fig. 16.6a, and the output powers change from 885 W to 852 W in Fig. 16.6b. Compare to the physical simulation, the error of the active power is 0.8%. The error is mainly introduced by the linearization process. In addition to the steady-state error in the allowable range, the small-signal

Table 16.1 Parameters of simulations

Items	Value
Control coefficient	$m = 1.05\text{e} - 4, n = 2.8\text{e} - 4, k_a = 1, k_c = 500$
Line impedance (Ω)	$Z_1 = 0.02 + j0.628, Z_2 = 0.01 + j0.314, Z_3 = 0.03 + j0.942$
DC voltage (V)	$V_{DC} = 200$ V
Voltage reference (V)	$V^* = 150$
Frequency reference (Hz)	$f^* = 50$
Load characters (Ω)	$Z = 10 + j5$

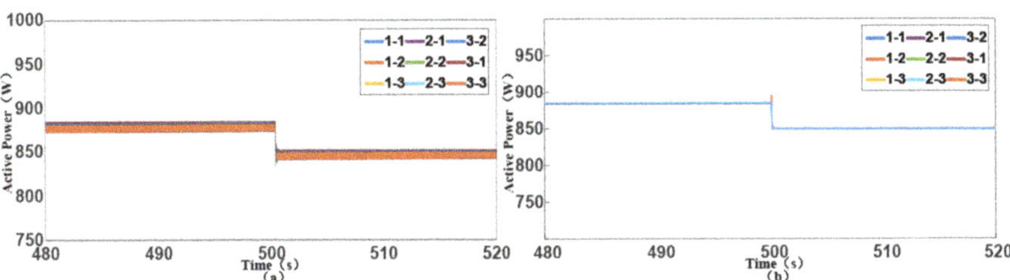

Fig. 16.6 Graphical diagram of the proposed communication network

model also shows a good dynamic performance. Thus, the small-signal model shows a high accuracy in describing the ESS.

16.4 Simulation Results

To verify the control method, a hybrid series–parallel ESS model contains 3 × 3 ESUs is developed in the MATLAB/Simulink environment. The system control parameters are listed in Table 16.1.

16.4.1 Case 1: Performance of Discharging Mode

In this case, the capacity of each ESU is the same and set $C_1 = C_2 = C_3 = 2.1$ Ah. The initial SoC values of all ESU are different and set $SoC_{11} = 0.9, SoC_{12} = 0.87, SoC_{13} = 0.84, SoC_{21} = 0.89, SoC_{22} = 0.86, SoC_{23} = 0.83, SoC_{31} = 0.88, SoC_{32} = 0.85, SoC_{33} = 0.82$. The voltage and the current waveforms are shown in Fig. 16.7. The waveforms are similar in every case. The voltage and current are analyzed only in this case. From Fig. 16.7, in the dynamic process the SOC balancing operation did not change the voltage of each string. The currents amplitude is positive related to the estimated average SOC of strings. This control method ensures the dynamic voltage quality. The other simulation curves are shown in Fig. 16.8. Let $\Delta SOC = SOC_{\max} - SOC_{\min}$ and $\Delta P = P_{\max} - P_{\min}$, where

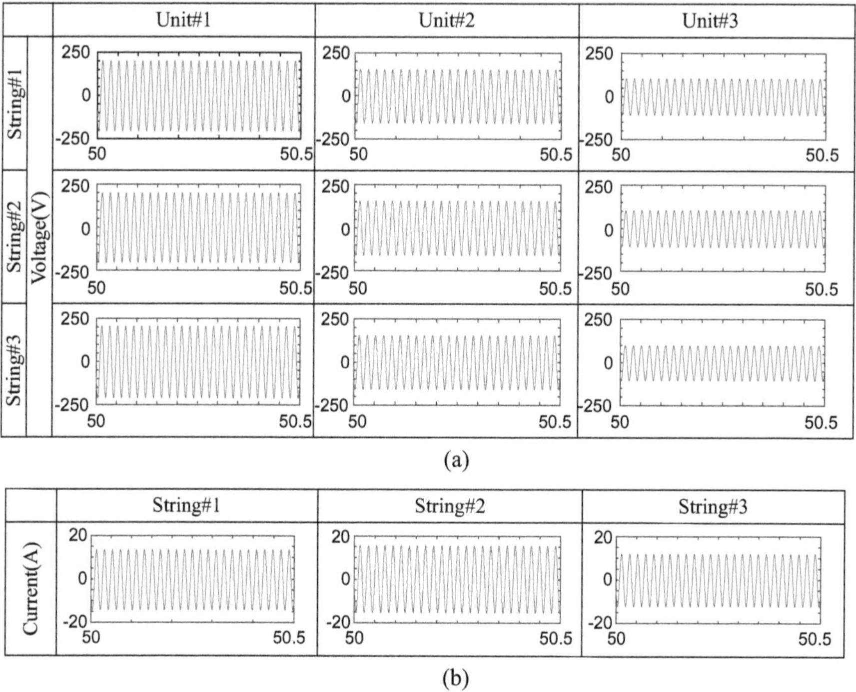

Fig. 16.7 Waveforms of voltage and current: (**a**) Voltage (**b**) Current

$SOC_{\max}$ and $P_{\max}$ are the max SOC and active power value among all ESUs in steady state. $SOC_{\min}$ and $P_{\min}$ are the minimum. Figure 16.8a shows that SOC values are convergent in the discharging process and, Fig. 16.8b shows that active power is shared accurately under the RL load. Figure 16.8c shows the reactive power allocation process, in which reactive power sharing can only be achieved in the same string. There has close relation between line impedance and reactive power sharing in the ESS. In Fig. 16.8d, all frequencies synchronize, though there is a little deviation to the rated frequency. The ΔSOC and ΔP gradually decrease and finally equal to zero. The targets of the designed control method are presented to be achieved.

16.4.2 Case 2: Performance of Plug-and-Play

In this case, the series string #3 is not connected to the ESS before 100 s and unplugged at 400 s. Figure 16.9 shows the simulation results. The capacity of ESUs and the initial SOC values of connected strings are set the same as case1. The initial SOC of ESU string #3 are set $SoC_{31} = 0.82$, $SoC_{32} = 0.80$, and

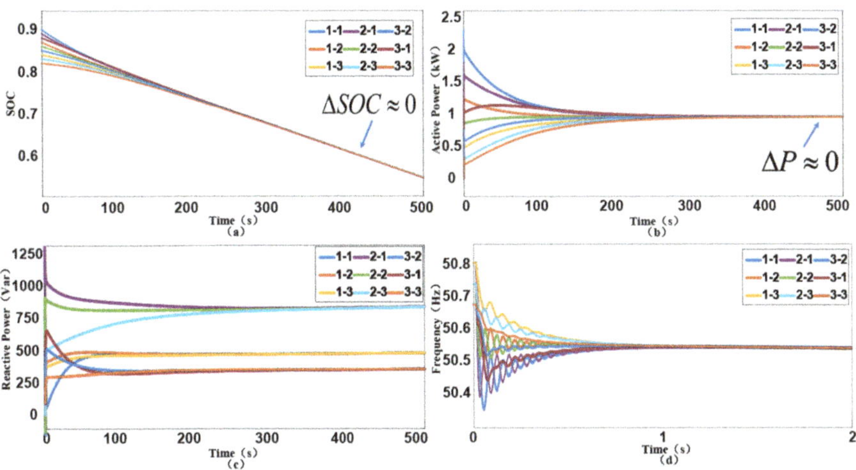

Fig. 16.8 Performance of discharging mode: (**a**) SOC (**b**) Active Power (**c**) Reactive Power (**d**) Frequency

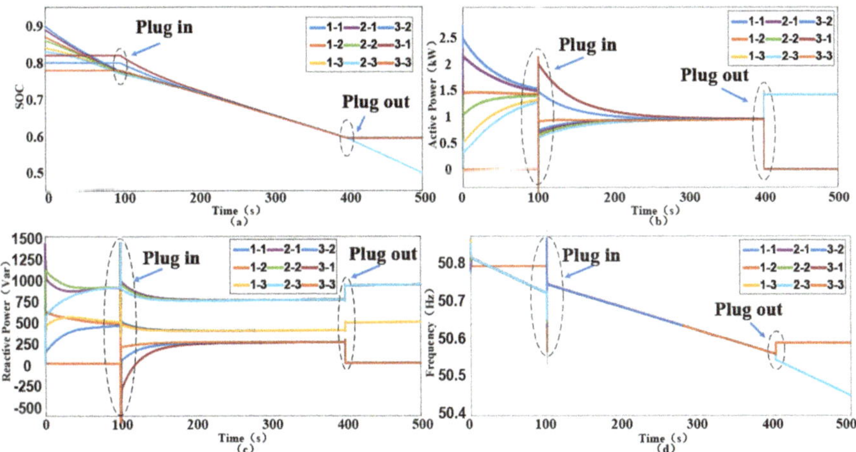

Fig. 16.9 Performance of Plug-and-Play: (**a**) SOC (**b**) Active Power (**c**) Reactive Power (**d**) Frequency

$SoC_{33} = 0.78$. When ESU string#3 is plugged, the system gradually reaches a steady state. Figure 16.9a, c show that the SOC values converged and the in-string reactive power sharing are achieved. It is shown in Fig. 16.9b that the active power of ESUs are converged to the same value of case 1 before 400 s. After the unplugging, the output power of ESUs in ESU string #3 decrease to zero, while power sharing is still maintained among the remaining ESUs. When string#3 is plugged in or plugged out, the frequencies are quickly synchronized in Fig. 16.9d. The simulation results prove the plug-and-play capability of ESU strings.

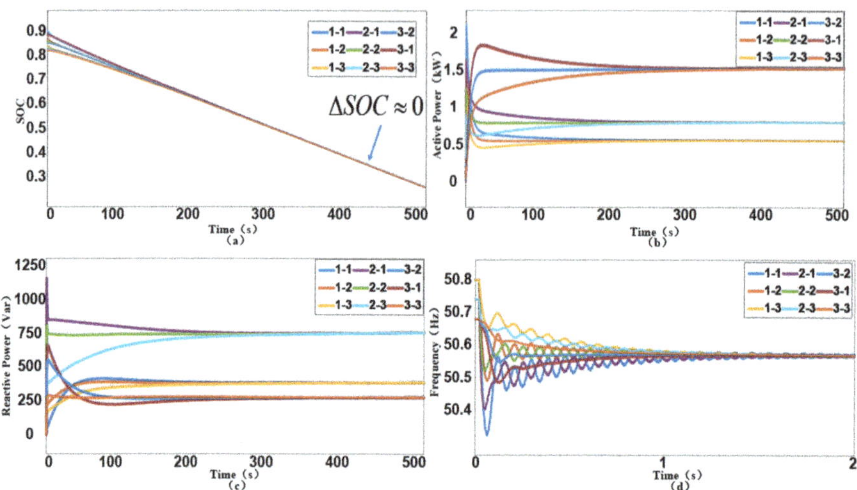

Fig. 16.10 Performance of different capacity: (**a**) SOC (**b**) Active Power (**c**) Reactive Power (**d**) Frequency

16.4.3 Case 3: Performance of ESU with Different Capacity

Figure 16.10 shows the SOC and active power waveforms when the capacity of ESUs are different. In this case, the capacity of ESUs is set $C_1 = 1.4$ Ah, $C_2 = 2.1$ Ah, $C_3 = 4.2$ Ah and the initial SoC values are set the same as case 1. It can be seen from Fig. 16.10a, c, and d that the SOC balancing, the in-string reactive power sharing, and the frequency synchronization are achieved with the scheme adopted. The active power is shared proportionally to the capacity of ESU. The objective mentioned above of the control method is verified to be reached by the simulation results.

16.4.4 Case 4: Performance of Charging Mode

In charging mode, the capacity of ESU is set the same as case 1. And initial SoC values SoC_{11}, SoC_{12}, SoC_{13}, SoC_{21}, SoC_{22}, SoC_{23}, SoC_{31}, SoC_{32}, SoC_{33} is fixed to be 0.2, 0.23, 0.26, 0.21, 0.24, 0.27, 0.22, 0.25, 0.28 respectively. The simulation results are represented in Fig. 16.11. It is shown in Fig. 16.11b that the active power is converged rapidly and in steady state is shared accurately. In Fig. 16.11d, frequencies are synchronized as authors expected. From Fig. 16.11c, the in-string reactive power is shared. Also, Fig. 16.11a shows that the SOC value of each ESU will finally be balanced. The ΔSOC and ΔP equaling to 0 in steady state reveals the effectiveness of the control method. This case proves that this method can unify both charging and discharging mode.

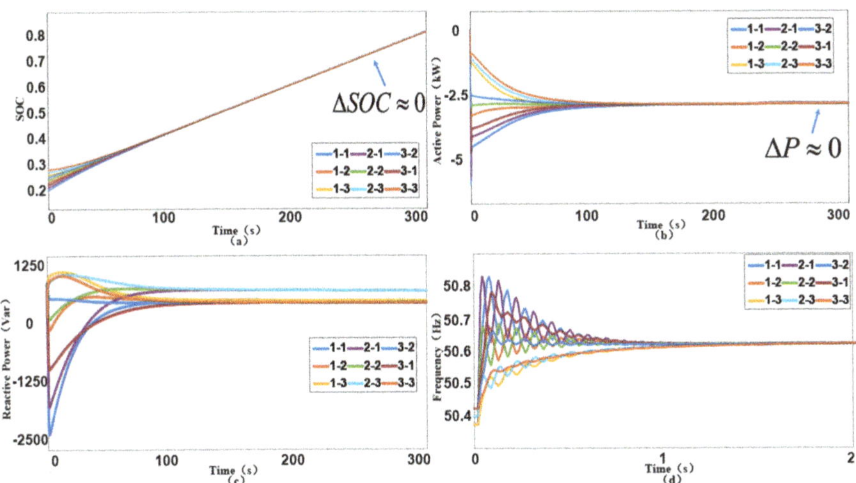

Fig. 16.11 Performance of charging mode: (**a**) SOC (**b**) Active Power (**c**) Reactive Power (**d**) Frequency

16.4.5 Case 5: Performance of Communication Link Failure

Distributed control is usually used to reduce communication dependence. Based on the case I, the communication link between ESU #1 $-$ 1 and ESU #1 $-$ 2 is destroyed. With the other parameters keeping the same as case 1, the waveforms of the SOC and active power are shown in Fig. 16.12. From Fig. 16.12, with the same simulation time, the SOC balancing and active power sharing can be roughly realized. However, the convergence speed is slower than that in case 1. ΔSOC and ΔP also cannot decrease to zero, which means the accuracy is decreased. But the ESS can still work stably, with $\Delta SOC = 0.002$ is limited within the acceptable range.

16.4.6 Case 6: Comparison of Traditional Method

The existing SOC balancing strategy for hybrid series–parallel ESS is centralized. Figure 16.13 shows the comparison of distributed and centralized control method when communication link between ESU #1 $-$ 1 and ESU #1 $-$ 2 is destroyed at 250 s. It can be found in Fig. 16.13a that the convergence tendency maintains whenever the failure occurs. However, the communication link destruction not only makes the ESU crashed but also affects the convergency of the SOC. In this case, the resiliency to single point communication failure of the ESS is verified.

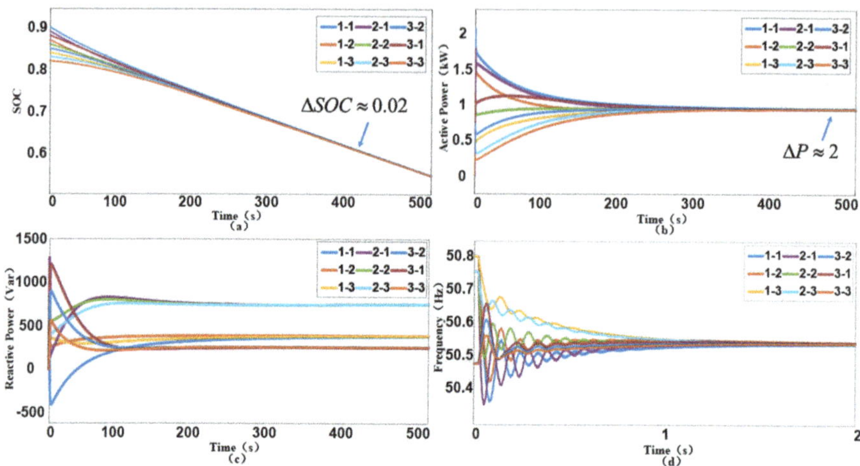

Fig. 16.12 Performance of communication failure: (**a**) SOC (**b**) Active Power (**c**) Reactive Power (**d**) Frequency

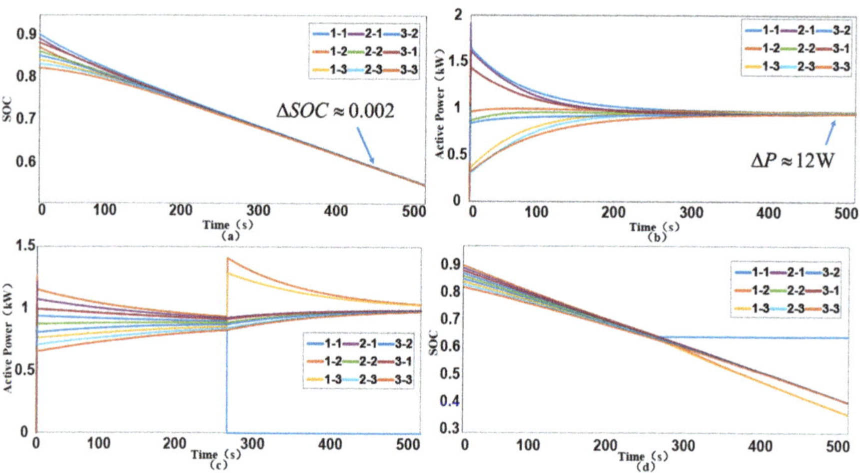

Fig. 16.13 Comparison of single communication link destroyed: Distributed (**a**) SoC (**b**) Active Power; Centralized (**c**) SOC (**d**) Active Power

16.5 Conclusion

In this chapter, a local-distributed and global-decentralized SOC balancing method is presented for hybrid series–parallel ESS. This scheme achieves global SOC balancing and power sharing of energy storage units. Because only a low-bandwidth distributed communication network used in the local ESU string, the method has the advantages of higher reliability and lower communication cost. When one single communication link is destroyed, SOC balancing and power sharing can also be achieved within acceptable ranges. And, a variant droop control implemented within ESU strings improves the capability of the ESS. The stability of the system is proved, and the proposed method's effectiveness is verified by simulation results.

Appendix

Parameter matrixes $\mathbf{T}_{pv}$, $\mathbf{T}_{p\theta}$, $\mathbf{T}_{qv}$, and $\mathbf{T}_{q\theta}$ in (16.33) are given as follows:

$$\begin{cases} \mathbf{T}_{pv} = \left[\mathbf{T}_{pv}^{11} \cdots \mathbf{T}_{pv}^{1M} \; \mathbf{T}_{pv}^{21} \cdots \mathbf{T}_{pv}^{2M} \cdots \mathbf{T}_{pv}^{N1} \cdots \mathbf{T}_{pv}^{NM} \right]^T; \\ \mathbf{T}_{p\theta} = \left[\mathbf{T}_{p\theta}^{11} \cdots \mathbf{T}_{p\theta}^{1M} \; \mathbf{T}_{p\theta}^{21} \cdots \mathbf{T}_{p\theta}^{2M} \cdots \mathbf{T}_{p\theta}^{N1} \cdots \mathbf{T}_{p\theta}^{NM} \right]^T; \\ \mathbf{T}_{qv} = \left[\mathbf{T}_{qv}^{11} \cdots \mathbf{T}_{qv}^{1M} \; \mathbf{T}_{qv}^{21} \cdots \mathbf{T}_{qv}^{2M} \cdots \mathbf{T}_{qv}^{N1} \cdots \mathbf{T}_{qv}^{NM} \right]^T; \\ \mathbf{T}_{q\theta} = \left[\mathbf{T}_{q\theta}^{11} \cdots \mathbf{T}_{q\theta}^{1M} \; \mathbf{T}_{q\theta}^{21} \cdots \mathbf{T}_{q\theta}^{2M} \cdots \mathbf{T}_{q\theta}^{N1} \cdots \mathbf{T}_{q\theta}^{NM} \right]^T; \end{cases} \tag{16.34}$$

Where

$$\begin{cases} \mathbf{T}_{pv}^{ij} = \left[\frac{\partial p_{ij}}{\partial V_{11}} \cdots \frac{\partial p_{ij}}{\partial V_{1M}} \; \frac{\partial p_{ij}}{\partial V_{21}} \cdots \frac{\partial p_{ij}}{\partial V_{2M}} \cdots \frac{\partial p_{ij}}{\partial V_{N1}} \cdots \frac{\partial p_{ij}}{\partial V_{NM}} \right]; \\ \mathbf{T}_{p\theta}^{ij} = \left[\frac{\partial p_{ij}}{\partial \theta_{11}} \cdots \frac{\partial p_{ij}}{\partial \theta_{1M}} \; \frac{\partial p_{ij}}{\partial \theta_{21}} \cdots \frac{\partial p_{ij}}{\partial \theta_{2M}} \cdots \frac{\partial p_{ij}}{\partial \theta_{N1}} \cdots \frac{\partial p_{ij}}{\partial \theta_{NM}} \right]; \\ \mathbf{T}_{qv}^{ij} = \left[\frac{\partial q_{ij}}{\partial V_{11}} \cdots \frac{\partial q_{ij}}{\partial V_{1M}} \; \frac{\partial q_{ij}}{\partial V_{21}} \cdots \frac{\partial q_{ij}}{\partial V_{2M}} \cdots \frac{\partial q_{ij}}{\partial V_{N1}} \cdots \frac{\partial q_{ij}}{\partial V_{NM}} \right]; \\ \mathbf{T}_{q\theta}^{ij} = \left[\frac{\partial q_{ij}}{\partial \theta_{11}} \cdots \frac{\partial q_{ij}}{\partial \theta_{1M}} \; \frac{\partial q_{ij}}{\partial \theta_{21}} \cdots \frac{\partial q_{ij}}{\partial \theta_{2M}} \cdots \frac{\partial q_{ij}}{\partial \theta_{N1}} \cdots \frac{\partial q_{ij}}{\partial \theta_{NM}} \right]; \end{cases} \tag{16.35}$$

Parameter matrixes **M**, **N**, $\mathbf{K}_a$, $\mathbf{K}_c$, and **H** in (16.33) are given in (16.36). And parameter matrixes **L** and **C** in (16.33) are given in (16.37)

$$\begin{aligned} &\mathbf{M} = diag\left\{ diag\left\{m/C_1\right\} \cdots diag\left\{m/C_N\right\} \right\}; \\ &\mathbf{N} = diag\left\{n\right\}; \\ &\mathbf{K}_a = diag\left\{k_a\right\}; \\ &\mathbf{K}_c = diag\left\{k_c\right\}; \\ &\mathbf{H} = diag\left\{ \mathbf{H}_1 \cdots \mathbf{H}_N \right\}; \end{aligned} \tag{16.36}$$

where $\mathbf{H}_i = s(s\mathbf{I}_M + \mathbf{L}_i)^{-1}$

$$\begin{cases} \mathbf{L} = diag\left\{ diag\left\{\mathbf{L}_1\right\} \cdots diag\left\{\mathbf{L}_N\right\} \right\} \\ \mathbf{C} = diag\left\{ diag\left\{ \frac{1}{V_{DC}C_1} \right\} \cdots diag\left\{ -\frac{1}{V_{DC}C_N} \right\} \right\} \end{cases} \tag{16.37}$$

References

1. Z. Miao et al., An SOC-based battery management system for microgrids. IEEE Trans. Smart Grid **5**(2), 966–973 (2013)
2. J.M. Carrasco, L.G. Franquelo, J.T. Bialasiewicz, E. Galvan, R.C.P. Guisado, M. Angeles, M. Prats, J.I. Leon, N. Moreno-alfonso, Power-electronic systems for the grid integration of renewable energy sources : a survey. IEEE Trans. Ind. Electron. **53**(4), 1002–1016 (2006)
3. S. Boudoudouh, M. Ouassaid, M. Maâroufi, Multi agent system in a distributed energy management of a multi sources system with a hybrid storage, in *2015 3rd International Renewable and Sustainable Energy Conference (IRSEC), Marrakech* (2015), pp. 1–6
4. W. Huang, J.A. Abu Qahouq, Energy sharing control scheme for state-of-charge balancing of distributed battery energy storage system. IEEE Trans. Ind. Electron. **62**(5), 2764–2776 (2015). https://doi.org/10.1109/TIE.2014.2363817
5. A. Hongke, W. Junyong, T. Mingjie, et al., Research on two-stage SOC self-balancing control strategy in hybrid cascade energy storage system[J]. Power Syst. Prot. Control **22**, 75–80 (2014)
6. Q. Wu, R. Guan, X. Sun, Y. Wang, X. Li, SoC balancing strategy for multiple energy storage units with different capacities in islanded microgrids based on droop control. IEEE J. Emerg. Sel. Top. Power Electron. **6**(4), 1932–1941 (2018)
7. L. Maharjan et al., State-of-charge (SOC)-balancing control of a battery energy storage system based on a cascade PWM converter. IEEE Trans. Power Electron. **24**(6), 1628–1636 (2009)
8. H. Wang, Z. Wu, G. Shi, Z. Liu, SOC balancing method for hybrid energy storage system in microgrid, in *2019 IEEE 3rd International Conference on Green Energy and Applications (ICGEA), Taiyuan* (2019), pp. 141–145
9. N. Ghanbari, S. Bhattacharya, SOC balancing of different energy storage systems in DC microgrids using modified droop control, in *Proc IECON 2018 - 44th Annu Conf IEEE Ind Electron SOC*, vol. 1 (2018), pp. 6094–6099
10. H. Han et al., A local-distributed and global-decentralized SoC balancing method for hybrid series-parallel energy storage system. IEEE Syst. J. **PP.99**, 1–11 (2021)

Chapter 17
Leader-Distributed Follower-Decentralized Control Strategy for Economic Dispatch

17.1 Control Objectives

Compared with the parallel-type and series-type microgrids, the economic dispatch problems (EDP) of SPMGs are the most challenging due to its complex structure and transmission characteristics. To the best of the authors' knowledge, the EDP has not received any considerations in SPMGs now. The main contributions of this work are summarized as follows.

1. LDFD control based ED scheme is presented for large-scale SPMGs under different load types. The global optimal solution of EDP and frequency synchronization are obtained iteratively with low-bandwidth communication links between neighboring leaders, which can significantly reduce the infrastructure costs and improve system reliability.
2. Voltage quality is ensured, which is realized by constructing an output power estimator and a unified voltage regulator.
3. Small-signal stability, dynamic, and steady-state performances of the hybrid system are analyzed in theory and simulation, which has important guiding significance for control of microgrids with complex structure.

17.2 Economic Dispatch Control Strategy

The targets of the control strategy include: (1) the communication cost is low enough; (2) all DGs are dispatched to ensure the minimum generation cost; (3) voltage quality of the microgrid must be ensured.

Y. Sun et al., *Series-Parallel Converter-Based Microgrids*, Power Systems,
https://doi.org/10.1007/978-3-030-91511-7_17

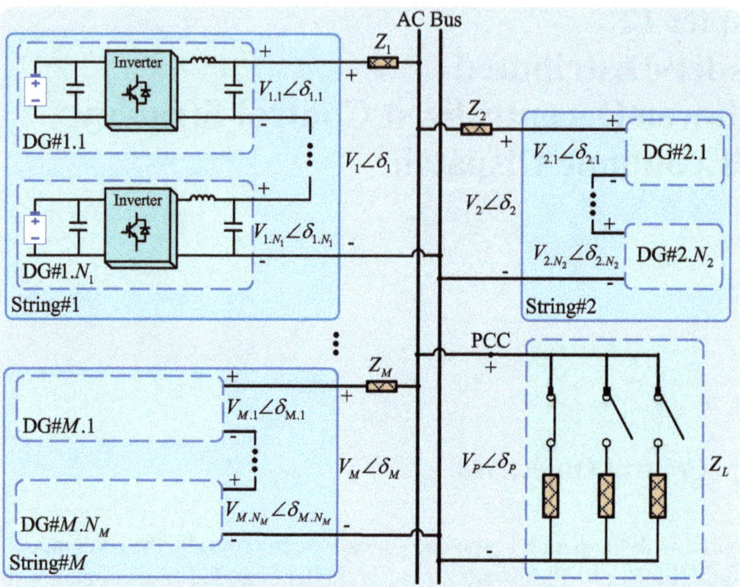

Fig. 17.1 Structure for series–parallel hybrid microgrid

17.2.1 Economic Dispatch Problem in Series–Parallel Microgrid

The configuration of the SPMG is shown in Fig. 17.1. DG#$i.j$ ($i = 1, 2, , M$; $j = 1, 2, , N_i$) denotes the j-th DG in String#i, where N_i denotes the number of DGs in String#i. Different from the cascaded H-bridge multilevel converters [1], the DG in the hybrid microgrid is interfaced with an output LC filter for ease of independent control. A number of DGs connect in series to form a medium/high voltage string, then several strings supply power to the point of common coupling (PCC) in parallel. Therefore, low voltage power sources can be integrated to form a high voltage network without back stage converters, which promotes flexible and efficient applications of microgrids.

Assume that all the DGs in this study are dispatchable. The DGs include diesel generators, micro-turbines, and energy storage units. And their cost functions could be expressed as the following quadratic function [2, 3].

$$C_{i.j}(P_{i.j}) = \frac{1}{2}\alpha_{i.j}P_{i.j}^2 + \beta_{i.j}P_{i.j} + \gamma_{i.j} \quad (\alpha_{i.j} > 0) \tag{17.1}$$

where $\alpha_{i.j}$, $\beta_{i.j}$, and $\gamma_{i.j}$ are constant coefficients of cost function of DG#$i.j$. $P_{i.j}$ is the output active power of DG#$i.j$.

Since the goal of EDP is to minimize the total generation cost while meeting the load demand, the EDP can be formulized as

$$\min \sum_{i=1}^{M} \sum_{j=1}^{N_i} C_{i.j}(P_{i.j})$$
$$\text{subject to} \quad \sum_{i=1}^{M} \sum_{j=1}^{N_i} P_{i.j} = P_L \tag{17.2}$$

where P_L is the equivalent load demand. Considering all cost functions are smooth and convex, that is, $\partial^2 C_{i.j} \big/ \partial P_{i.j}^2 > 0$ [4]. Then the solution to the EDP described by (17.2) is achieved when the following equation is satisfied [4].

$$\lambda_{1.1} = \lambda_{2.1} = \cdots = \lambda_{M.1} = \lambda_{1.2} = \cdots = \lambda_{M.N_M} \tag{17.3}$$

where $\lambda_{i.j} = \partial C_{i.j} / \partial P_{i.j}$ is the incremental cost rate (ICR) of DG#$i.j$.

17.2.2 Communication Graph

Due to the complex transmission characteristics of the hybrid microgrid, communication is required to ensure the coordinated operation of the system. From control-flow perspective, microgrid can be regarded as a multi-agent cyber-physical system. At the cyber-layer, considering the sparsity and reliability, a low-bandwidth communication network that facilitates a distributed control scheme shown in Fig. 17.2. is designed to achieve the first target. The communication graph is composed of a set of nodes $V_G = v_{1.1}, v_{2.1}, \ldots, v_{M.1}$ connected via a set of edges, which represent communication links that are bidirectional to form an undirected graph.

In the network, only DG#$i.1$ ($i = 1, 2, \ldots, M$), set as the leader of the i-th string, exchanges information with its own neighboring leaders that are represented by $N_{i.1}$. Besides, minimum redundancy is configured in the network so that when any link fails, the remaining network still contains a spinning tree, where there exists a path including all the DGs.

Note that the leaders can be selected for different purposes. In this study, the DGs with the shortest distance to its neighbors are chosen as the leaders, which decreases the communication costs. In addition, for ease of analysis, assume that the number of DGs of each string is the same and the leader of each string is numbered as #1, that is, the leader of String#1 is DG#$i.1$.

A communication graph with N nodes is represented by an associated adjacency matrix $A_G = [a_{ij}] \in \Re^{N \times N}$, where a_{ij}=1 if an edge is connected from node $v_{j.1}$ to node $v_{i.1}$, and a_{ij}=0, otherwise. For an undirected graph, the information change

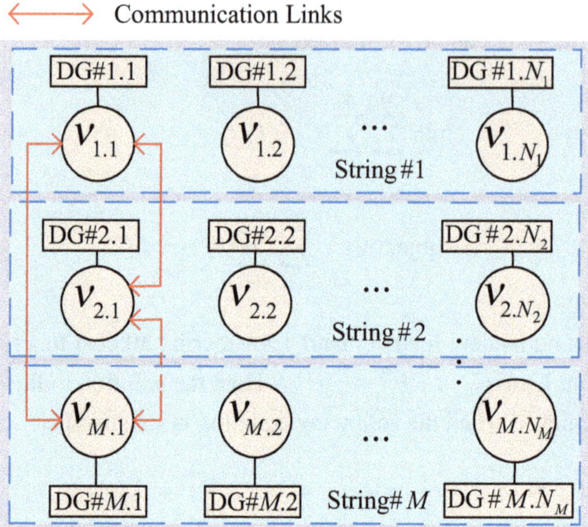

Fig. 17.2 Designed communication network between strings

is reciprocal, that is, if $v_{j.1} \in N_{i.1}$, then $v_{i.1} \in N_{j.1}$ must be true. The diagonal in-degree matrix is $D_G^{\text{in}} = \text{diag}\{\, d_i^{\text{in}}\}$, where $d_i^{\text{in}} = \sum_{j \in N_i} a_{ij}$ denotes the in-degree of the node $v_{i.1}$. The Laplacian matrix is then constructed as $\mathbb{L} = D_{\text{G}}^{\text{in}} - A_{\text{G}}$, of which eigenvalues determine the global dynamics [5].

17.2.3 LDFD Control Strategy

To reduce the communication links, a leader-based network is designed as shown in Fig. 17.2. In this network, only the leader in each string exchanges its information with its neighboring leaders, while other DGs as followers operate in a decentralized way based on local measurements [6].

Based on the designed network, to achieve the control targets, an LDFD control strategy as shown in Fig. 17.3 is presented, which is expressed as (17.4) and (17.5).

$$\omega_{i.j} = \begin{cases} \omega^* + \text{sgn}(Q_{i.j}) m \bar{\lambda}_{i.1} + \mu \sum\limits_{k \in \mathcal{N}_i} \left(\lambda_{k.1} - \lambda_{i.j} \right) \quad (j = 1) \\ \\ \omega^* + \text{sgn}(Q_{i.j}) m \lambda_{i.j} \quad (j \neq 1, j \leqslant N_i) \end{cases} \tag{17.4}$$

$$V_{i.j} = \frac{P_{i.j}}{\hat{P}_{i.j}} V^* \tag{17.5}$$

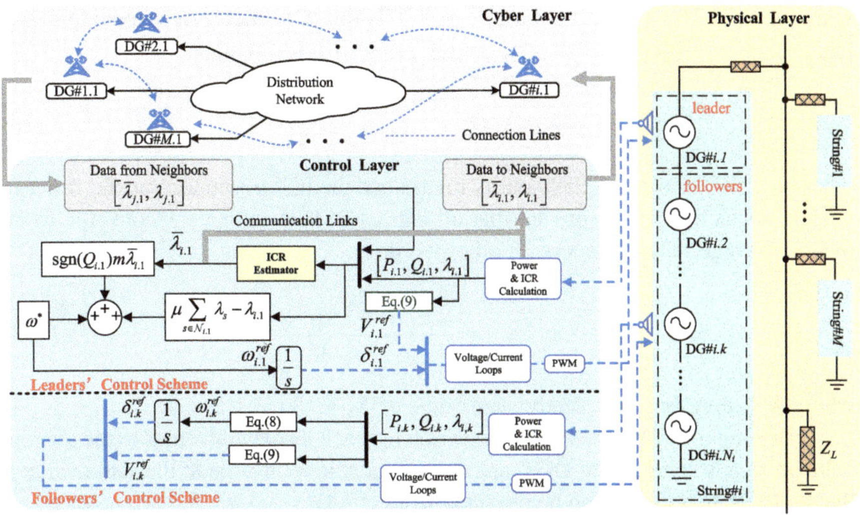

Fig. 17.3 Block diagram of the leader-distributed follower-decentralized control scheme

where ω^* is the rated angular frequency. m and μ are constant control parameters. sgn() is the signum function, and sgn($Q_{i.j}$) is employed to unify the control modes under different load types. $\bar{\lambda}_{i.1}$ represents the estimate of the average ICR among leaders. V^* is the rated voltage amplitude of common bus. $\hat{P}_{i.j}$ represents the estimate of total active power of DGs in String#i. The LDFD control strategy could realize the goals of both ED and voltage regulation, which are elaborated below.

(1) Economic Dispatch

From (17.2) and (17.3), the EDP can be transformed into a control problem about how to make the ICR of each DG equal. To achieve the ED, DG#$i.j$ adjusts its angular frequency $\omega_{i.j}$ according to (17.4). The leader ($j = 1$) regulates the angular frequency in a distributed way by obtaining its neighboring leader's information, and followers ($j \neq 1$) operate in a decentralized manner by dealing their own information through an improved droop control scheme.

$\bar{\lambda}_{i.1}$ is observed by an ICR estimator constructed in DG#i.1, which updates by processing its neighboring leaders' estimates $\bar{\lambda}_{j.1}(j \in N_{i.1})$ received from the communication links and its local measurements

$$\bar{\lambda}_{i.1}(t) = \lambda_{i.1}(t) + \int_0^t \sum_{j \in \mathcal{N}_i} a_{ij} \left(\bar{\lambda}_{j.1}(\tau) - \bar{\lambda}_{i.1}(\tau) \right) d\tau \tag{17.6}$$

Then, the global observer dynamic can be formulated as

$$\dot{\bar{\lambda}} = \dot{\lambda} - \left(D_G^{in} - A_G \right) \bar{\lambda} = \dot{\lambda} - \mathbb{L}\bar{\lambda} \tag{17.7}$$

where $\lambda = [\lambda_{1.1},\ \lambda_{2.1},\ \ldots,\ \lambda_{M.1}]^T \in \mathbb{R}^M$ is the ICR vector of leaders and $\bar{\lambda} = \left[\bar{\lambda}_{1.1}\ \bar{\lambda}_{2.1}\ \ldots\ \bar{\lambda}_{M.1}\right]^T \in \mathbb{R}^M$. Therefore, (17.8) can be obtained.

$$\bar{\lambda} = s(sI_l + \mathbb{L})^{-1}\lambda = H\lambda \tag{17.8}$$

where $I_l \in \mathbb{R}^M$ and H are the identity matrix and the ICR estimator transfer function matrix, respectively. It is proved that all entries of ICR estimate $\bar{\lambda}$ converge to the overall average ICR of leaders. In other word,

$$\bar{\lambda}_{1.1}^{ss} = \bar{\lambda}_{2.1}^{ss} = \cdots = \bar{\lambda}_{M.1}^{ss} = \frac{\bar{\lambda}_{1.1}^{ss} + \bar{\lambda}_{2.1}^{ss} + \cdots + \bar{\lambda}_{M.1}^{ss}}{M} \tag{17.9}$$

where X^{ss} represents the steady-state value of X.

According to (17.9), the ICR observation for each leader converges to the average value in steady state. Since DGs' angular frequencies converge to the same value in the steady state, (17.10) can be obtained from (17.4).

$$\begin{cases} \lambda_{1.1}^{ss} = \lambda_{2.1}^{ss} = \cdots = \lambda_{M.1}^{ss} \\ \lambda_{i.1}^{ss} = \lambda_{i.2}^{ss} = \cdots = \lambda_{i.N_i}^{ss}\ \ (i = 1,\ 2,\ \ldots,\ M) \end{cases} \tag{17.10}$$

Simplification leads to

$$\lambda_{1.1}^{ss} = \cdots = \lambda_{M.N_M}^{ss} \tag{17.11}$$

which shows that all the DGs in the system reach the optimal solution in steady state.

(2) Voltage Regulation

The voltage regulation algorithm is described as (17.5) in a decentralized way, in which the voltage amplitude of each DG is set as the ratio of its output active power to the estimate of total active power of each string.

$\hat{P}_{i.j}$ is obtained by an output power estimator, which is constructed as

$$\begin{cases} \hat{P}_{i.j} = \sum\limits_{k=1}^{N_i} \dfrac{\lambda_{i.j} - \beta_{i.k}}{\alpha_{i.k}} = \sigma_{i.j}^1 P_{i.j} + \sigma_{i.j}^2 \\ \sigma_{i.j}^1 = \sum\limits_{k=1}^{N_i} \dfrac{\alpha_{i.j}}{\alpha_{i.k}} \\ \sigma_{i.j}^2 = \sum\limits_{k=1}^{N_i} \dfrac{\beta_{i.j} - \beta_{i.k}}{\beta_{i.k}} \end{cases} \tag{17.12}$$

where $\sigma_{i.j}^1$ and $\sigma_{i.j}^2$ are constants related to the coefficients in cost function of DGs in String#i.

From (17.6), the ICR of each DG equal in steady state. Hence, (17.12) can be rewritten as

$$\hat{P}_{i.j}^{ss} = \sum_{k=1}^{N_i} \frac{\lambda_{i.j}^{ss} - \beta_{i.k}}{\alpha_{i.k}} = \sum_{k=1}^{N_i} \frac{\lambda_{i.k}^{ss} - \beta_{i.k}}{\alpha_{i.k}} = \sum_{k=1}^{N_i} P_{i.k}^{ss} \tag{17.13}$$

Substitution of (17.13) into (17.5) yields

$$V_{i.j}^{ss} = \frac{P_{i.j}^{ss}}{\sum_{k=1}^{N_i} P_{i.k}^{ss}} V^* \tag{17.14}$$

which means that the voltage amplitude of each DG in a string is proportional to their output active power. Then the following equation can then be derived.

$$V_{i.j}^{ss} : V_{i.k}^{ss} = P_{i.j}^{ss} : P_{i.k}^{ss} \tag{17.15}$$

Since the inverter of each DG in the same string shares the same current, their output voltage phase angle and power factor will be consistent in steady state. Thus, the voltage quality of each string is guaranteed. Then the output voltage of String#i in the steady state would be

$$V_i^{ss} = \sum_{k=1}^{N_i} V_{i.k}^{ss} = V^* \tag{17.16}$$

It can be seen that when the system is operating at the equilibrium point, the output voltage of each string is equal to the rated voltage.

17.3 Small-Signal Modeling and Stability Analysis

17.3.1 Small-Signal Modeling

(1) Network Modeling

In the SPMG system, the power generation of the DG#$i.j$ is presented as

$$p_{i.j} + jq_{i.j} = \frac{V_{i.j} e^{j\delta_{i.j}}}{2} \left(\left(\sum_{b=1}^{N_i} V_{ib} e^{j\delta_{ib}} - V_P e^{j\delta_P} \right) |Y_i| \, e^{j\varphi_i} \right)^* \tag{17.17}$$

where $|Y_i|$ and φ_i represent the amplitude and phase angle of the equivalent line admittance in String#i. According to Kirchhoff's laws, the voltage of AC bus is calculated by

$$V_p e^{j\delta_P} = \sum_{a=1}^{M} \sum_{b=1}^{N_i} \frac{Y_a V_{a.b} e^{j\delta_{a.b}}}{Y_L + \sum\limits_{c=1}^{M} Y_c} \tag{17.18}$$

Simplification leads to

$$V_p e^{j\delta_P} = \sum_{a=1}^{M} \sum_{b=1}^{N_i} \left|Y'_a\right| V_{a.b} e^{j(\delta_{a.b}+\varphi'_a)} \tag{17.19}$$

where

$$Y'_a = \frac{Y_a}{Y_L + \sum\limits_{c=1}^{M} Y_c} \tag{17.20}$$

Defining $\dot{\tilde{\delta}}_i = \dot{\delta}_i - \dot{\delta}_s = \omega_i - \omega_s$, where ω_s represents the steady-state angular frequency of the system. Combining Eqs. (17.17)–(17.20), the instantaneous power supplied by DG#$i.j$ can be obtained as shown below.

$$\begin{cases} p_{i.j} = \frac{1}{2} \sum\limits_{a=1}^{M} \sum\limits_{b=1}^{N_a} V_{i.j} V_{a.b} \left|Y'_a\right| |Y_i| \sin\left(\tilde{\delta}_{i.j} - \tilde{\delta}_{a.b} - \varphi'_a\right) \\ \quad -\frac{1}{2} \sum\limits_{c=1}^{N_i} V_{i.j} V_{i.c} |Y_i| \sin\left(\tilde{\delta}_{i.j} - \tilde{\delta}_{i.c}\right) \\ q_{i.j} = \frac{1}{2} \sum\limits_{c=1}^{N_i} V_{i.j} V_{i.c} |Y_i| \cos\left(\tilde{\delta}_{i.j} - \tilde{\delta}_{i.c}\right) \\ \quad -\frac{1}{2} \sum\limits_{a=1}^{M} \sum\limits_{b=1}^{N_M} V_{i.j} V_{a.b} \left|Y'_a\right| |Y_i| \cos\left(\tilde{\delta}_{i.j} - \tilde{\delta}_{a.b} - \varphi'_a\right) \end{cases} \tag{17.21}$$

The average active and reactive power is obtained through a low pass filter with cutoff frequency ω_c, which can be expressed as

$$\begin{cases} P_{i.j} = \dfrac{\omega_c}{s+\omega_c} p_{i.j} \\ Q_{i.j} = \dfrac{\omega_c}{s+\omega_c} q_{i.j} \end{cases} \tag{17.22}$$

Substituting (17.21) into (17.22) and according to the small-signal analysis theory, the linearized equation of the average power can be represented as follows

$$\begin{cases} \Delta \dot{P} = -\omega_c \Delta P + \omega_c K_{P\tilde{\delta}} \Delta \tilde{\delta} + \omega_c K_{PV} \Delta V \\ \Delta \dot{Q} = -\omega_c \Delta Q + \omega_c K_{Q\tilde{\delta}} \Delta \tilde{\delta} + \omega_c K_{QV} \Delta V \end{cases} \tag{17.23}$$

Where parameter matrixes $K_{P\tilde{\delta}}$, K_{PV}, $K_{Q\tilde{\delta}}$, and K_{QV} are given in Appendix. Δ denotes small perturbations around the equilibrium point and

$$\begin{cases} \Delta P = [\,\Delta P_{1.1}\ \Delta P_{2.1} \cdots \Delta P_{M.N_M}\,]^T \\ \Delta Q = [\,\Delta Q_{1.1}\ \Delta Q_{2.1} \cdots \Delta Q_{M.N_M}\,]^T \\ \Delta \tilde{\delta} = [\,\Delta \tilde{\delta}_{1.1}\ \Delta \tilde{\delta}_{2.1} \cdots \Delta \tilde{\delta}_{M.N_M}\,]^T \\ \Delta V = [\,\Delta V_{1.1}\ \Delta V_{2.1} \cdots \Delta V_{M.N_M}\,]^T \end{cases} \tag{17.24}$$

(2) Controller Modeling

Combining (17.1), (17.3), and (17.8), then the presented control scheme is linearized at the equilibrium point, which can be described by a state space equation as follows

$$\begin{cases} \Delta \dot{\tilde{\delta}}_l = D_l m H K_{al} \Delta P_l - \mu \mathbb{L} K_{al} \Delta P_l \\ \Delta \dot{\tilde{\delta}}_f = D_f m K_{af} \Delta P_f \\ \Delta \dot{V} = K_{V\tilde{\delta}} \Delta \tilde{\delta} + K_{VV} \Delta V \end{cases} \tag{17.25}$$

where parameter matrixes D_l, D_f, K_{al}, K_{af}, and K_{VV} are given in Appendix. ΔP_l and ΔP_f are the vector of state variables of leaders and followers, respectively as given in (17.26).

$$\begin{cases} \Delta P_l = [\,\Delta P_{1.1}\ \Delta P_{2.1} \cdots \Delta P_{M.1}\,]^T \\ \Delta P_f = [\,\Delta P_{1.2} \cdots \Delta P_{M.2}\ \Delta P_{1.3} \cdots \Delta P_{M.N_M}\,]^T \end{cases} \tag{17.26}$$

(3) Dynamic Model of the Whole Microgrid System

Δ_P, Δ_Q, $\Delta\tilde{\delta}$, and Δ_V are selected as state variables, and their dimensions are $N = M * NM$. Equation (17.25) shows that Δ_P and $\Delta\tilde{\delta}$ are decomposed into two parts, representing the relative states of the leader and the follower, respectively. Combining (17.23) and (17.25), the small-signal dynamic model of the entire system is formed as

$$\dot{x} = Ax \tag{17.27}$$

$$\begin{cases} x = [\Delta P \ \Delta Q \ \Delta\tilde{\delta} \ \Delta V] \\ A = \left[\begin{array}{cc|c|c|c} \multicolumn{2}{c|}{diag\{-\omega_c\}} & 0_{N\times N} & \omega_c K_{P\tilde{\delta}} & \omega_c K_{PV} \\ \hline \multicolumn{2}{c|}{0_{N\times N}} & diag\{-\omega_c\} & \omega_c K_{Q\tilde{\delta}} & \omega_c K_{QV} \\ \hline (D_l m H - \mu \mathbb{L}) K_{al} & 0_{M\times(N-M)} & \multirow{2}{*}{$0_{N\times N}$} & \multirow{2}{*}{$0_{N\times N}$} & \multirow{2}{*}{$0_{N\times N}$} \\ \cline{1-2} 0_{(N-M)\times M} & D_f m K_{af} & & & \\ \hline \multicolumn{2}{c|}{0_{N\times N}} & 0_{N\times N} & K_{V\tilde{\delta}} & K_{VV} \end{array}\right] \end{cases} \tag{17.28}$$

17.3.2 Eigenvalue Analysis

Small-signal stability is analyzed by calculating the eigenvalues of the system matrix A in (17.27). For simplicity, assume that all DGs' control parameters take the same value. Since the influence of control parameters on the dynamic response of the system is similar under different load characteristics, only resistance-inductance load situation is illustrated here. For a rotationally symmetric system, there will be an eigenvalue of zero. Therefore, only the nonzero eigenvalues are used to analyze the system stability.

(1) Eigenvalue Trajectory with Respect to m

As m increases from 1e−7 to 1e−4 with $\mu = 5e - 5$, the eigenvalue trajectory of matrix A is shown in Fig. 17.4. When m is small, all eigenvalues have a negative real part, which means that the system is stable. When m is greater than 9.6e−5, poles λ_2 and λ_4 lie on the right half-plane, indicating that the system is unstable.

(2) Eigenvalue Trajectory with Respect to μ

Figure 17.5 shows the eigenvalue trajectory as μ increases from 1e−7 to 5e−5 with $m = 5e - 6$. It is can be seen that poles λ_2 and λ_4 lie on the right half-plane when μ is small. As μ increases, λ_2 and λ_4 move toward the left half-plane. The system becomes unstable until μ is greater than 2.3e−6.

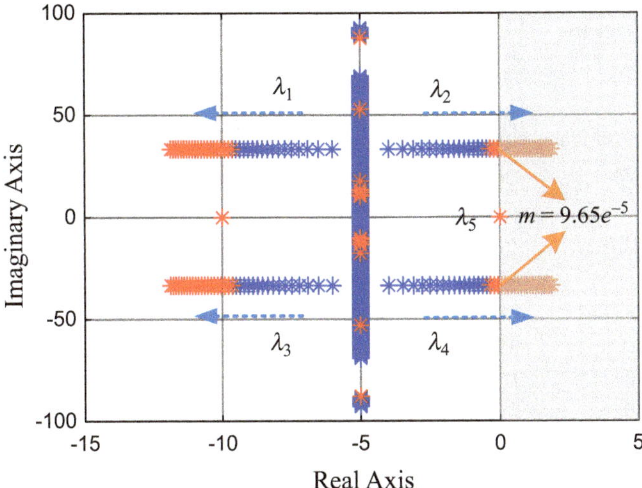

Fig. 17.4 Eigenvalue trajectory for different m

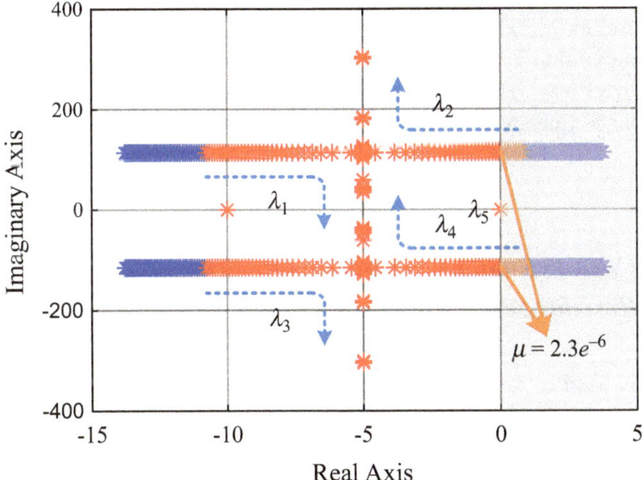

Fig. 17.5 Eigenvalue trajectory for different μ

It can be seen from the above analysis that the choice of parameters m and μ has a significant influence on stability. Considering a certain stability margin, reasonable stability ranges of those parameters under resistance-inductance and resistance-capacitance loads are listed in Table 17.1.

Table 17.1 Stability region of control parameters

Parameter	RL load	RC load
m	$[1e^{-7},\ 9.6e^{-5}]$	$[1e^{-7},\ 1.5e^{-4}]$
μ	$[2.3e^{-6},\ 5e^{-5}]$	$[3e^{-6},\ 5e^{-5}]$

Table 17.2 Cost function parameters of DGS

Item	DG#1.1	DG#1.2	DG#1.3	DG#2.1	DG#2.2	DG#2.3
a	0.1	0.1	0.3	0.1	0.25	1
b	0.01	10	1	0.01	0.05	0.1
c	1	5	1.5	1	1	5
Item	DG#3.1	DG#3.2	DG#3.3	DG#4.1	DG#4.2	DG#4.3
a	0.1	0.3	0.25	0.12	0.12	1
b	10	1	0.05	0.01	0.01	0.1
c	10	2	1	1	2	5

17.4 Simulation Results

To verify the effectiveness of the control strategy, a 4×3 DGs microgrid is built in Simulink/MATLAB. Leaders communicate with their neighboring leaders through the communication links and the other two DGs in each string operate in a decentralized way. The voltage reference of each string is set as 311 V and the rated angular frequency is 50 Hz. The cost function parameters of DGs are given in Table 17.2.

17.4.1 Case 1: Switching Between Resistance-Inductance and Resistance-Capacitance Load

This case is used to verify the effectiveness of the unified control strategy when different types of load are switched. There is a resistance-inductance load at the beginning, and switches to a resistance-capacitance load at $t = 6$ s, and then switches to a resistance-inductance load at $t = 12$ s. The output reactive power and voltage frequency of DGs are shown in Fig. 17.6a and b, respectively. It can be seen from the simulation results that the system can switch back and forth between resistive and capacitive load, and the frequency synchronization is guaranteed. Therefore, the proposed strategy can achieve stable operation under different load characteristics.

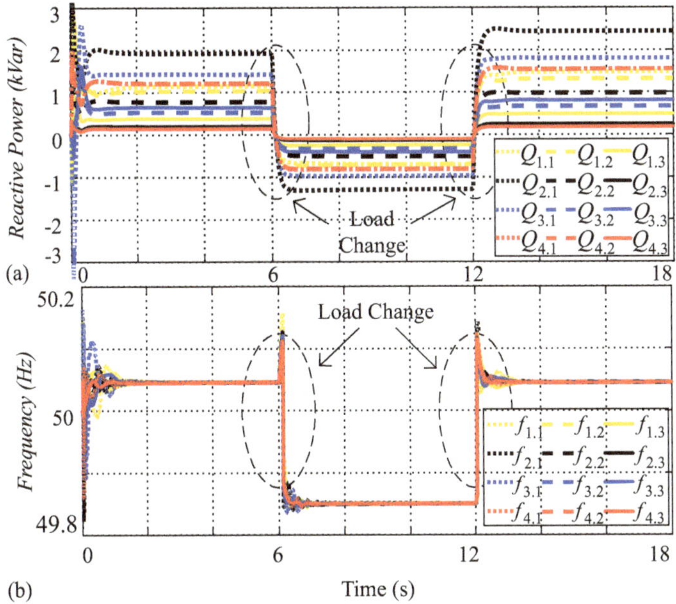

Fig. 17.6 Simulation results of case 1. (**a**) Reactive power. (**b**) Voltage frequency

17.4.2 Case 2: Performance under Resistance-Inductance Load

Figure 17.7 shows simulation results of the series–parallel system with resistance-inductance loads. At $t = 0$ s, load impedance $Z_L = 2+j1.6$, then Z_L changes to $1+j0.8$ at $t = 6$ s and Z_L recovers to $2 + j1.6$ when $t = 12$ s. As shown in Fig. 17.7a, the output voltage frequency of DGs are greater than 50 Hz due to the inverse droop control caused by inductive load characteristic. Figure 17.7b shows that all DGs achieve ED in the steady state as the ICRs converge to $\lambda^* = 189$ as $Z_L = 2 + j1.6$, and $\lambda^* = 347.3$ when $Z_L = 1 + j0.8$, and the active power dispatch among DGs is shown in Fig. 17.7c. The output voltage of DGs in String#2 in time interval [8.9 s, 9.1 s] is depicted in Fig. 17.7d, which shows that their output voltage phase angle are consistent in steady state. Moreover, the output voltage amplitude of each string equal to the rated value in steady state, as shown in Fig. 17.7e. Clearly, the system responds quickly to load disturbance and exhibits good steady-state performances under resistance-inductance load.

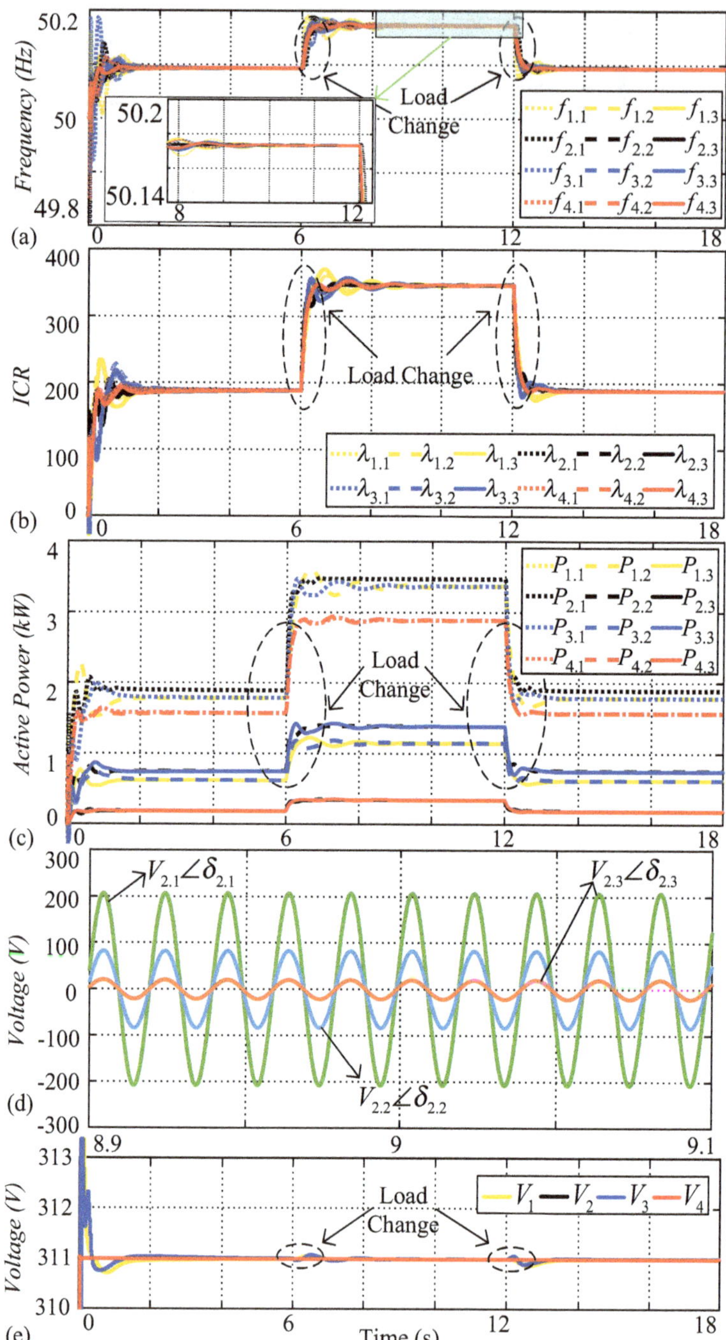

Fig. 17.7 Simulation results of case 2. (**a**) Voltage frequency. (**b**) ICR of DGs. (**c**) Active power. (**d**) Output voltage of DGs in String#2. (**e**) Voltage amplitude of strings

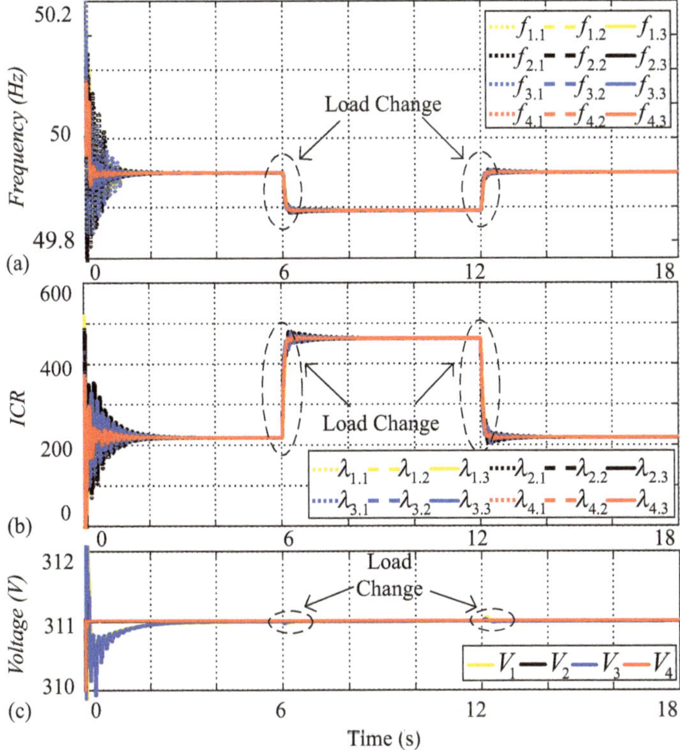

Fig. 17.8 Simulation results of case 3. (**a**) Voltage frequency. (**b**) ICR of DGs. (**c**) Voltage amplitude of strings

17.4.3 Case 3: Performance under Resistance-Capacitance Load

To verify the performance of the presented scheme under resistance-capacitance load, simulation test is carried out with $Z_L = 2 - j1.6$. The load changes to $1 - j0.8$ at $t = 6$ s and recovers at $t = 12$ s. The voltage frequencies shown in Fig. 17.8a are less than 50 Hz, because the droop control performed under capacitive load characteristic. As shown in Fig. 17.8b, the ICRs converge to $\lambda^* = 219.2$ as $Z_L = 2 - j1.6$ and $\lambda^* = 464.5$ when $Z_L = 1 - j0.8$, which means the system operate with minimum generation costs. Figure 17.8c depicted the waveform of the voltage amplitude of strings, which converges to the rated value in steady state. As seen, the proposed strategy can also achieve ED and voltage regulation under resistance-capacitance load.

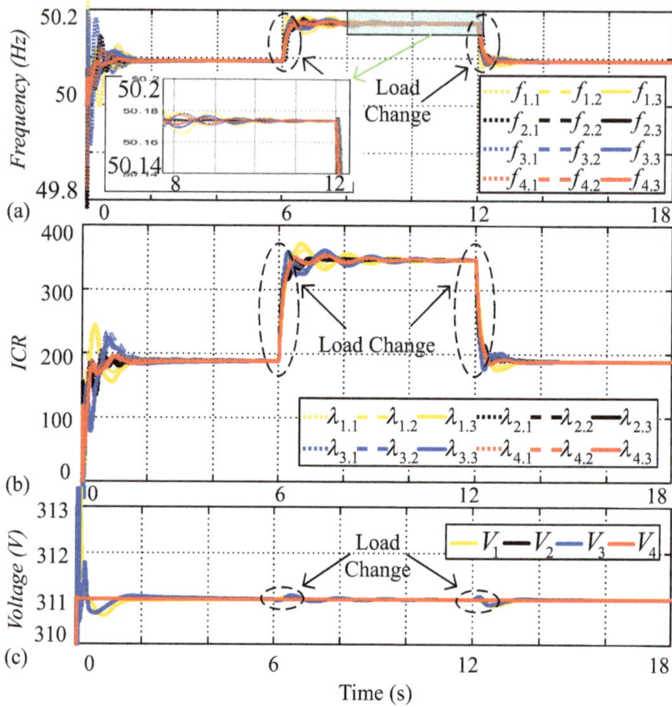

Fig. 17.9 Simulation results of case 4. (**a**) Voltage frequency. (**b**) ICR of DGs. (**c**) Voltage amplitude of strings

17.4.4 Case 4: Communication Link Failure

In this case, resiliency to the communication link failure is studied. For analysis, communication link between the leader of String#4 and String#1 is broken based on case 1 to make the system with the minimum connectivity. Since there remains a path containing all the DGs through the communication links, there should have no impact on the steady-state performance. The simulation results are shown in Fig. 17.9a–c. The waveform of interval [8 s, 12 s] in Fig. 17.9a shows that the link failure will slow down the system dynamics to some extent. However, it can be seen from Fig. 17.9b, c that single link failure will not affect the steady-state performance.

17.5 Conclusion

This chapter presents an LDFD control strategy for the economic dispatch in SPMGs under different load types. To reduce the communication costs, a low-bandwidth communication network with minimum redundancy is designed, where each leader exchanges information with their neighbors, while other DGs only depend on their local information. With the help of the improved droop control, the ICR of DGs in one string can converge to the same value. Thus, the followers of each string can realize self-synchronization and load sharing without communication. Simulation results show that the optimal economic dispatch is achieved, and the quality of voltage is guaranteed even in the presence of a single communication link failure.

Appendix

Parameter matrixes $K_{P\tilde{\delta}}$, K_{PV}, $K_{Q\tilde{\delta}}$, and K_{QV} in (17.23) are given in (17.29)–(17.31).

$$\left\{ K_{P\tilde{\delta}} = \begin{bmatrix} k_{p\tilde{\delta}1.1}^{1.1} & k_{p\tilde{\delta}2.1}^{1.1} & \cdots & k_{p\tilde{\delta}M.N_M}^{1.1} \\ k_{p\tilde{\delta}1.1}^{2.1} & k_{p\tilde{\delta}2.1}^{2.1} & \cdots & k_{p\tilde{\delta}M.N_M}^{2.1} \\ \vdots & \vdots & \ddots & \vdots \\ k_{q\tilde{\delta}1.1}^{M.N_M} & k_{q\tilde{\delta}2.1}^{M.N_M} & \cdots & k_{q\tilde{\delta}M.N_M}^{M.N_M} \end{bmatrix}, K_{PV} = \begin{bmatrix} k_{pv1.1}^{1.1} & k_{pv2.1}^{1.1} & \cdots & k_{pvMN_M}^{1.1} \\ k_{pv1.1}^{2.1} & k_{pv2.1}^{2.1} & \cdots & k_{pvM.N_M}^{2.1} \\ \vdots & \vdots & \ddots & \vdots \\ k_{pv1.1}^{M.N_M} & k_{pv2.1}^{M.N_M} & \cdots & k_{pvM.N_M}^{M.N_M} \end{bmatrix} \right. \tag{17.29}$$

$$\left\{ K_{Q\tilde{\delta}} = \begin{bmatrix} k_{q\tilde{\delta}1.1}^{1.1} & k_{q\tilde{\delta}2.1}^{1.1} & \cdots & k_{q\tilde{\delta}M.N_M}^{1.1} \\ k_{q\tilde{\delta}1.1}^{2.1} & k_{q\tilde{\delta}2.1}^{2.1} & \cdots & k_{q\tilde{\delta}M.N_M}^{2.1} \\ \vdots & \vdots & \ddots & \vdots \\ k_{q\tilde{\delta}1.1}^{M.N_M} & k_{q\tilde{\delta}2.1}^{M.N_M} & \cdots & k_{q\tilde{\delta}M.N_M}^{M.N_M} \end{bmatrix}, K_{QV} = \begin{bmatrix} k_{qv1.1}^{1.1} & k_{qv2.1}^{1.1} & \cdots & k_{qvM.N_M}^{1.1} \\ k_{qv1.1}^{2.1} & k_{qv2.1}^{2.1} & \cdots & k_{qvM.N_M}^{2.1} \\ \vdots & \vdots & \ddots & \vdots \\ k_{qv1.1}^{M.N_M} & k_{qv2.1}^{M.N_M} & \cdots & k_{qvM.N_M}^{M.N_M} \end{bmatrix} \right. \tag{17.30}$$

Where

$$\left\{ \begin{array}{l} k_{p\tilde{\delta}a.b}^{i.j} = \dfrac{\partial p_{i.j}}{\partial \tilde{\delta}_{a.b}}; k_{pva.b}^{i.j} = \dfrac{\partial p_{i.j}}{\partial V_{a.b}} \\ k_{q\tilde{\delta}a.b}^{i.j} = \dfrac{\partial p_{i.j}}{\partial \tilde{\delta}_{a.b}}; k_{qva.b}^{i.j} = \dfrac{\partial q_{i.j}}{\partial V_{a.b}} \end{array} \right. \begin{pmatrix} i = 1, 2, \ldots M; j = 1, 2, \ldots, N_i \\ a = 1, 2, \ldots M; b = 1, 2, \ldots, N_a \end{pmatrix} \tag{17.31}$$

Parameter matrixes D_l, D_f, K_{al}, K_{af}, $K_{V\tilde{\delta}}$, and K_{VV} in (17.25) are given in (17.32)–(17.33)

$$\begin{cases} D_l = diag\{\text{sgn}(Q_{i.1})\},\ D_f = diag\{\text{sgn}(Q_{i.j})\}\quad (j \neq i) \\ K_{al} = diag\{a_{i.1}\},\ K_{af} = diag\{a_{i.j}\}\quad (j \neq i) \\ K_{V\tilde{\delta}} = \begin{bmatrix} k_{v\tilde{\delta}1.1}^{1.1} & k_{v\tilde{\delta}2.1}^{1.1} & \cdots & k_{v\tilde{\delta}M.N_M}^{1.1} \\ k_{v\tilde{\delta}1.1}^{2.1} & k_{v\tilde{\delta}2.1}^{2.1} & \cdots & k_{v\tilde{\delta}M.N_M}^{2.1} \\ \vdots & \vdots & \ddots & \vdots \\ k_{v\tilde{\delta}1.1}^{M.N_M} & k_{v\tilde{\delta}2.1}^{M.N_M} & \cdots & k_{v\tilde{\delta}M.N_M}^{M.N_M} \end{bmatrix}, \\ K_{VV} = \begin{bmatrix} k_{vv1.1}^{1.1} & k_{vv2.1}^{1.1} & \cdots & k_{vvM.N_M}^{1.1} \\ k_{v\tilde{\delta}1.1}^{2.1} & k_{v\tilde{\delta}2.1}^{2.1} & \cdots & k_{vvM.N_M}^{2.1} \\ \vdots & \vdots & \ddots & \vdots \\ k_{vv1.1}^{M.N_M} & k_{vv2.1}^{M.N_M} & \cdots & k_{vvM.N_M}^{M.N_M} \end{bmatrix} \end{cases} \tag{17.32}$$

Where

$$\begin{cases} k_{v\tilde{\delta}a.b}^{i.j} = \dfrac{\omega_c\left(V^* - \sigma_{1i.j}V_{i.j}\right)}{\sigma_{2i.j}} k_{p\tilde{\delta}a.b}^{i.j} \\ k_{v\tilde{\delta}a.b}^{i.j} = \begin{cases} \dfrac{\omega_c\left(V^* - \sigma_{1i.j}V_{i.j}\right)}{\sigma_{2i.j}} k_{pva.b}^{i.j};\ (i.j \neq a.b) \\ \dfrac{\omega_c V^*}{\sigma_{2i.j}} k_{pvi.j}^{i.j} - \dfrac{\omega_c\sigma_{1i.j}}{\sigma_{2i.j}}\left(k_{pvi.j}^{i.j}V_{i.j} + p_{i.j}\right) - \omega_c;\ (i.j = a.b) \end{cases} \\ (i = 1, 2, \ldots M;\ j = 1, 2, \ldots, N_i;\ a = 1, 2, \ldots M;\ b = 1, 2, \ldots, N_a) \end{cases} \tag{17.33}$$

References

1. X. Ge, H. Han, W. Xiong, et al., Locally-distributed and globally-decentralized control for hybrid series-parallel microgrids. Int. J. Electr. Power Energy Syst. **116**, 105537 (2020)
2. W.T. Elsayed, E.F. El-Saadany, A fully decentralized approach for solving the economic dispatch problem. IEEE Trans. Power Syst. **30**(4), 2179–2189 (2015)
3. W. Zhang et al., Online optimal generation control based on constrained distributed gradient algorithm. IEEE Trans. Power Syst. **30**(1), 35–45 (2015)
4. L. Li, Y. Sun, H. Han, et al., Communication-free optimal economical dispatch scheme for cascaded-type microgrids with capacity constraints. IET Power Electron. **13**(13), 2866–2873 (2020)
5. V. Nasirian, A. Davoudi, F.L. Lewis et al., Distributed adaptive droop control for DC distribution systems. IEEE Trans. Energy Convers. **29**(4), 944–956 (2014)
6. H. Han et al., Leader-distributed follower-decentralized control strategy for economic dispatch in cascaded-parallel microgrids. Int. Trans. Electr. Energy Syst. **31**(9), e12964 (2021)

Index

Y. Sun et al., *Series-Parallel Converter-Based Microgrids*, Power Systems,
https://doi.org/10.1007/978-3-030-91511-7

www.ingramcontent.com/pod-product-compliance
Ingram Content Group UK Ltd.
Pitfield, Milton Keynes, MK11 3LW, UK
UKHW021833270726
14058UKWH00001B/119

* 9 7 8 3 0 3 0 9 1 5 1 3 1 *